Bioluminescence Methods and Protocols

METHODS IN MOLECULAR BIOLOGY™

John M. Walker, SERIES EDITOR

METHODS IN MOLECULAR BIOLOGY™

Bioluminescence Methods and Protocols

Edited by

Robert A. LaRossa

DuPont Co., Wilmington, DE

Humana Press ✳ Totowa, New Jersey

Library of Congress Cataloging in Publication Data

Main entry under title:

Methods in molecular biology™.

Bioluminescence Methods and Protocols/edited by Robert A. LaRossa
        p.   cm.—(Methods in molecular biology™; vol. 102)
    Includes bibliographical references and index.
    ISBN 0-89603-520-4
    1.Bioluminescence assay. I. LaRossa, Robert A. II. Series: Methods in molecular biology
(Totowa, NJ); 102.
QP519.9B55B536 1998
572'.36--DC21
DNLM/DLC
for Library of Congress   97-44152
                                                                                                       CIP

Preface

The biological emission of light has fascinated humankind for millennia; who is not enchanted by the flashing of fireflies during summer's evenings? Their collection by children in the area surrounding Baltimore, Maryland in the middle of this century for the purpose of biochemical isolation and characterization of light producing proteins and cofactors may constitute the initiation of serious, quantitative studies of this phenomenon. The biochemistry of light production is certainly well-understood today. In bacterial systems it is coupled to the reducing and energy states of the cell, whereas in eukaryotic systems it is dependent upon the availability of ATP. These differences are reflected in the unrelated substrate requirements and structures of eukaryotic and prokaryotic luciferases. These biochemical studies have been aided by the cloning of luciferase-encoding structural genes, the structural genes for cofactor generation, and genes encoding proteins that interact with these light-producing enzymes. These genes also provide starting materials for a wide range of applications that are the subject of this volume.

Many of these advances have earlier been catalogued in two volumes of the *Methods in Enzymology* series. It is thus necessary to provide a rationale for the publication of *Bioluminescence Methods and Protocols*. The format of the *Methods in Molecular Biology* series differs significantly from that of the above-mentioned, classic compilation. It is my belief that the Notes sections of each chapter in this volume will make the methods far easier to exploit in other experimentalists' laboratories by alerting researchers to potential pitfalls and by detailing how they may be adapted for use in related studies.

The choice of contributed material represents the current status of luminescent assays; they are used in fundamental biochemical and microbiological research, analytical biochemistry, basic and molecular toxicology, medicine, and environmental biotechnology. It is my belief that such broad applications provide both an intellectual challenge and great opportunity. By sampling representative applications from such a smorgasbord, the researcher will be encouraged to investigate unexplored territory or to adapt established technologies to their purpose. In either case I hope that this collection will prove a practical, user-friendly reference, one that will lower the "activation energy," and thus foster the use of bioluminescent methods and protocols by a broader segment of the scientific community.

Bioluminescence research integrates genetic, biochemical, molecular, biological, and metabolic concepts; the training provided by my mentors, Dieter Söll, Dale Kaiser, Brooks Low, and Phil Hartman, was invaluable in my entry into this arena. My studies have benefited from the collaboration and support of several people. They include Dana Smulski, Tina Van Dyk, Will Majarian, Roz Young, Dave Elsemore, Tim Reed, Mary Jane Reeve, Tricia Watson, Jim Romesser, Prasad Dhurjati, Sameer Rupani, Man Bock Gu, Konstanin Konstantinov, Amy Vollmer, and Shimshom Belkin. I am also grateful to the support of the bioluminescence community; among others, Pete Greenberg, Rheinhardt Rosson, Anne Summers, Moni Ulitzer, Anthony Bulich, Ted Meighen, and Ken Nealson have made me feel most welcome. The contributors are to be thanked for providing manuscripts that are complete, interesting, and conforming to the standards of the *Methods in Molecular Biology* series. I am most grateful to John Walker for providing me with the opportunity to edit this volume. Both he and the staff of Humana Press have given sound guidance, forthright support, and the needed flexibility to bring this project to fruition. I am most appreciative of their help.

For 17 years I have been privileged to work within Central Research and Development at DuPont. The multidisciplinary atmosphere of this section, which defines the crossroads of a large corporation with only indirect accountability to its profit centers, has been most stimulating. I am thankful to DuPont, my colleagues, and family for allowing me to follow my dreams.

Robert A. LaRossa

Contents

Contributors

DAVID C. ALEXANDER • *Department of Microbiology and Immunology, McGill University, Montreal, Canada*

TAMAR BARKAY • *Department of Molecular Microbiology and Biotechnology, Tel Aviv University, Ramat Aviv, Israel*

SHIMSHON BELKIN • *Environmental Sciences, The Freddy & Nadine Herrmann Graduate School of Applied Science, The Hebrew University of Jerusalem, Israel*

ROBERT S. BURLAGE • *Environmental Sciences Division, Oak Ridge National Laboratory, Oak Ridge, TN*

MICHAEL A. COSTANZO • *Department of Microbiology and Immunology, McGill University, Montreal, Canada*

IAN A. CREE • *Department of Pathology, Institute of Ophthamology, University College, London, UK*

MICHAEL S. DuBow • *Department of Microbiology and Immunology, McGill University, Montreal, Canada*

DAVID A. ELSEMORE • *Small Molecule Therapeutics, Monmouth Junction, NJ*

YVAN FISCHER • *Institute of Physiology, Medical Faculty, Aachen, Germany*

SHARON R. FORD • *Department of Biochemistry and Molecular Biology, Oklahoma State University, Stillwater, OK*

VANESSA GURTU • *Cell Biology and Vectorology Group, CLONTECH Laboratories, Palo Alto, CA*

JULIE GUZZO • *Department of Microbiology and Immunology, McGill University, Montreal, Canada*

JANET K. JANSSON • *Arrhenius Laboratories for Natural Sciences, Department of Biochemistry, Stockholm University, Sweden*

EBERHARD JÜNGLING • *Institute of Physiology, Medical Faculty, Aachen, Germany*

STEVEN R. KAIN • *Cell Biology and Vectorology Group, CLONTECH Laboratories, Palo Alto, CA*

HELMUT KAMMERMEIER • *Institute of Physiology, Medical Faculty, Aachen, Germany*

MATTI KARP • *Department of Biotechnology, University of Turku, Finland*

CAROL A. KELLY • *Department of Microbiology, University of Manitoba, Winnipeg, Canada*

PAUL A. KITTS • *Cell Biology and Vectorology Group, CLONTECH Laboratories, Palo Alto, CA*
MATTI KORPELA • *Department of Biotechnology, University of Turku, Finland*
Franklin R. Leach • *Department of Biochemistry and Molecular Biology, Oklahoma State University, Stillwater, OK*
ROBERT L. MATTS • *Department of Biochemistry and Molecular Biology, Oklahoma State University, Stillwater, OK*
ANNELIE MÖLLER • *Arrhenius Laboratories for Natural Sciences, Department of Biochemistry, Stockholm University, Sweden*
LASSE D. RASMUSSEN • *Department of General Microbiology, The University of Copenhagen, Denmark*
REINHARDT A. ROSSON • *Bio-Technical Resources, Manitowoc, WI*
JOHN W. M. RUDD • *Freshwater Institute, Winnipeg, Canada*
SISKO TAURIAINEN • *Department of Biochemistry and Pharmacy, Åbo Akademi University, Turku, Finland*
VANITHA THULASIRAMAN • *Department of Biochemistry and Molecular Biology, Oklahoma State University, Stillwater, OK*
RICCARDO TOMBOLINI • *Arrhenius Laboratories for Natural Sciences, Department of Biochemistry, Stockholm University, Sweden*
RALPH R. TURNER • *Frontier Geosciences, Seattle, WA*
TINA K. VAN DYK • *Central Research and Development, DuPont Company, Wilmington, DE*
MARKO VIRTA • *Department of Biotechnology, University of Turku, Finland*
AMY CHENG VOLLMER • *Swarthmore College, Swarthmore, PA*
L. WINONA WAGNER • *Central Research and Development, DuPont Company, Newark, DE*
GUOHONG ZHANG • *Cell Biology and Vectorology Group, CLONTECH Laboratories, Palo Alto, CA*

I

THE BASICS

Improvements in the Application of Firefly Luciferase Assays

Sharon R. Ford and Franklin R. Leach

1. Introduction

1.1. Firefly Luciferase Assay Differs from Usual Enzyme Assays

The firefly luciferase-based assay differs from most familiar enzyme-based determinations. Most enzyme assays are based either on the production of a product or the disappearance of a substrate. Usually the compound measured is stable so that its concentration can be determined after a specific time. At low adenosine 5'triphosphate (ATP) concentrations, firefly luciferase is a stoichiometric reactant rather than a catalyst. In the case of the firefly luciferase reaction, AMP, PP_i, CO_2, and oxyluciferin are typical products that accumulate, but the product that is most often and most easily determined is light. The photons of light are not accumulated in the measuring technique unless film or some electronic summation procedure is used in photon counting.

The two-step firefly luciferase reaction sequence is shown below. Step one forms an enzyme-bound luciferyl adenylate. Either MgATP or LH_2 (luciferin) can add first to the enzyme LUC.

$$LH_2 + MgATP + LUC \longleftrightarrow LUC\text{-}LH_2\text{-}AMP + MgPP_i \tag{1}$$

Step two is the oxidative decarboxylation of luciferin with the production of light on decay of the excited form of oxyluciferin.

$$LUC\text{-}LH_2\text{-}AMP + O_2 + OH^- \longrightarrow LUC\text{-}OL + CO_2 + AMP + light + H_2O \tag{2}$$

The oxyluciferin product, OL, is released slowly from the enzyme–product complex. This gives the flash kinetic pattern observed with high ATP concentrations, under which conditions firefly luciferase acts catalytically. The initial flash of light emission observed with high ATP concentration is owing to a

From: *Methods in Molecular Biology, Vol. 102: Bioluminescence Methods and Protocols*
Edited by: R. A. LaRossa © Humana Press Inc., Totowa, NJ

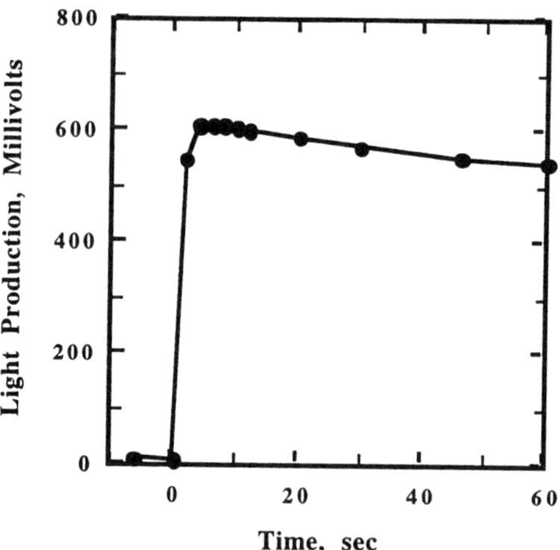

Fig. 1. Time-courses with nanomolar ATP.

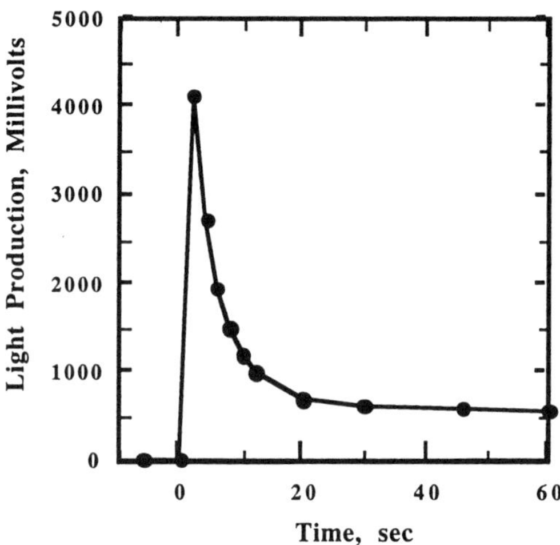

Fig. 2. Time-courses with micromolar ATP.

"first round" of enzyme activity. This flash rapidly decays to a relatively constant light emission, similar to that seen at low ATP concentrations, which is thought to be the result of the enzyme slowly turning over by releasing the oxyluciferin.

1.2. Kinetic Pattern Varies with ATP Concentration

The two kinetic patterns of light production are shown in **Figs. 1** and **2**. This property can be a source of experimental difficulties. When measuring light emission using high ATP concentrations, the delay between starting the reaction and starting the measurement of light emitted, as well as the length of time that the light emission is measured become critical. In this case, it is essential that the reaction be initiated while the sample is within the counting chamber of the luminometer, that the initiating reagent be rapidly and completely mixed with the components already in the reaction cuvet, and that the light emission always be measured over the same period of time.

1.3. Origin of the Use of Firefly Luciferase to Determine ATP

Firefly luciferase was first applied to the determination of ATP in 1947 by McElroy *(1)*. Given the status of instrumentation available for the measurement of light in the 1940s and 1950s, some procedural compromises evolved. One was the use of arsenate buffer in the reaction mixture, which reduced light emitted and changed the time-course of the reaction. In 1952 Strehler and Trotter *(2)* recommended the use of arsenate buffer to prevent precipitation that occurred when phosphate buffer and Mg were used. The application of firefly luciferase to the assay of ATP was described by Strehler and McElroy *(3)* and further amplified by Strehler *(4)*.

1.4. Modern Development

New instrumentation with fast response times is now readily available, and many ATP determinations require great sensitivity. Those two factors obviate the need to use arsenate-based assay systems and, in fact, make them undesirable. The use of arsenate-inhibited systems persists because of precedence and the fact that some commercial suppliers still provide firefly luciferase in an arsenate buffer. McElroy *(5)* cautions against using the commercially prepared luciferase with arsenate, because it lowers sensitivity, is an inhibitor, and is not required with current instrumentation.

1.5. The Response Is Determined by the Ratio of Reactants

Since the reaction occurs in a defined volume, increasing the concentration of either luciferase or luciferin increases the light production achieved with a given concentration of ATP. This concentration increase makes collisions of molecules more likely. Thus, a change in the ratio of the components changes light production, shifting the light emission vs ATP concentration standard curve either to the right (reduced sensitivity) or left (enhanced sensitivity). This is illustrated in **Table 1**. When using a reaction mixture that contains both luciferase and luciferin added together in a single volume (such as in a commercially available mix), the counts observed decrease as the square of any dilution of the reaction mix *(7)*. The reaction requires three substrates: luciferin, MgATP, and oxygen. In addition, several stabilizing compounds are added to a typical assay system. **Table 2**

Table 1
Effect of Changing of Reactant
Proportions on Light Production[a]

Firefly luciferase, nM	Luciferin, μM	KRLU
54	110	5.0
54	280	6.6
108	110	8.6
108	280	12
216	110	16
216	280	23

[a]Sigma luciferase (L 5256) and D-luciferin (L 6882) were used in a 300-μL vol in the Model 2010A Biocounter. [ATP] = 67 pM. KRLU = 1,000,000 counts. Modified from **ref. (6)**.

Table 2
Reaction Requirements for Firefly Luciferase[a]

Component omitted	Light production, light units/10s
None	52.0 ± 1.1
–MgSO$_4$, 5 mM	2.1 ± 0.2
–DTT, 0.5 mM	52.5 ± 0.7
–EDTA, 0.5 mM	54.0 ± 1.2
–Luciferin, 0.358 mM	0.002
–ATP, 321 nM	0.002

[a]Crystalline native luciferase from Sigma was used in a 300-μL vol. The effect of omission of the indicated component was determined in triplicate assays on a Model 2010A Biocounter. A light unit is 1000 counts produced. [ATP] = 321 nM. Modified from **ref. (8)**.

shows what occurs with the omission of each component. The buffer maintains the enzyme at its optimum pH of 7.8 *(9)*. —SH compounds are added to ensure that the cysteine residues of firefly luciferase are not oxidized (there are no disulfide linkages present in the protein). EDTA is added to prevent any metal ions from interfering with the reaction. The presence of metals can change the wavelength of light produced. Firefly luciferase preparations (particularly those sold in kit form) are often stabilized by the addition of bovine serum albumin, trehalose, glycerol, or other compound(s).

As shown in **Table 2**, light production by firefly luciferase is completely dependent on the presence of Mg^{2+}, ATP, and luciferin in the reaction mixture. Dithiothreitol (DTT) and ethylenediaminetetraacetic acid (EDTA) are added to the reaction mixture to prevent inhibition of the reaction.

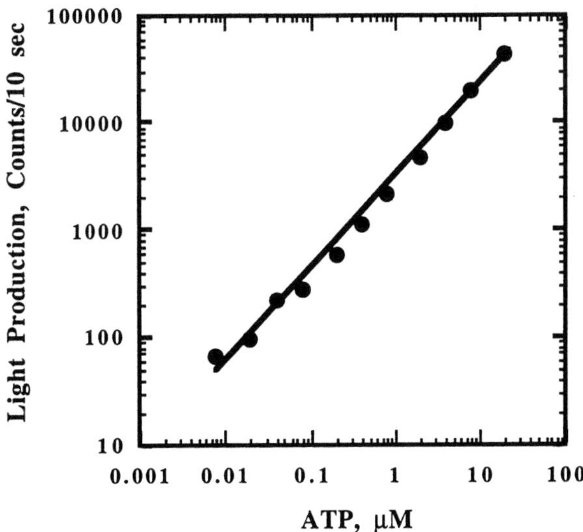

Fig. 3. Light production as a function of ATP concentration. Note that the plot has log vs log scales.

The light production response from firefly luciferase is linear over a range of four to five logs of ATP concentration (**Fig. 3**). As little as 50 fg of ATP was measured *(10)*.

1.6. Optimum Assay Conditions

1.6.1. pH

The optimum pH for the reaction is pH 7.8 *(9)*. We have shown that Tricine buffer, which has a pK_a of 8.15 and offers the greatest buffering capacity of any common buffer, works well for firefly luciferase *(11)*. **Table 3** shows the functionality of several buffers with firefly luciferase.

The necessity for pH maintenance was clearly demonstrated by the following experiment. When ATP solutions were not neutralized, we observed that 10 m*M* ATP inactivated luciferase during incubation before addition of luciferin and assay. This occurred when 6 m*M* Tris-succinate buffer was used. When ATP was prepared in a buffer, incubation of firefly luciferase with 10 m*M*-concentrations of ATP did not inactivate the enzyme.

1.6.2. Temperature

The optimum temperature for the firefly luciferase is 25°C. At temperatures >30°C, native *Photinus pyralis* luciferase is rapidly inactivated. Mutants of luciferase have been isolated with increased temperature stability, but most commercially available firefly luciferases are based on the native *P. pyralis* enzyme.

Table 3
Effect of Buffer on Light Production[a]

Buffer, 25 mM	pK$_a$ 20°C	Act. relative to HEPES
MOPS	7.20	0.65
Phosphate	7.21	0.09
TES	7.50	0 54
HEPES	7.55	1.00
HEPPS	8.00	0.68
Tricine	8.15	1.25
Glycine amide	8.20	0.80
Tris	8.30	1.00
Glycylglycine	8.40	0.72

[a]The assays were done a Model 2010A Biocounter. Values obtained with three different ATP concentrations were averaged and expressed relative to the value obtained with HEPES. All were assayed at pH 7.8. From **ref. *(6)***.

1.6.3. Effect of Products on the Reaction

PP$_i$ has little effect at low concentrations (~0.13, μM), activates when used at moderate concentrations (~1.3–13 μM), and inhibits at high concentrations (>1.3 mM) *(12)*. AMP at 1 mM inhibits firefly luciferase. At low ATP concentration (0.24 μM), light production is inhibited by about 70%. At high ATP concentration (0.24 mM), the peak of light production is inhibited by about 30%, but there is little effect on light production at times greater than 1 min.

1.6.4. Effect of Additives on the Reaction

Several substances have been found that change the flash of light production into a linear production of light that lasts for at least a minute as shown in **Figs. 4** and **5**.

1. Coenzyme A (CoA). Airth and colleagues *(13)* found that CoA addition to a reaction mixture after the flash stimulated light production; this was presumably through removal of oxyluciferin from luciferase. The observed enhancement of light production was proportional to CoA concentration *(14)*. The effect of CoA was recently reinvestigated by Wood *(15–17)*, who observed that addition of CoA prevented the rapid inhibition of light production and elicited a nearly constant production of light. He found that dethioCoA was a competitive inhibitor, suggesting that the sulfhydryl group of CoA was required. Pazzagli et al. *(18)* observed no effect of CoA on peak light intensity, but found that 0.66 mM CoA significantly modified the kinetics of light emission. They concluded that "despite the present inability to explain the role of CoA in the bioluminescent reaction of the firefly luciferase, the addition of CoA to the reaction mixture for the firefly luci-

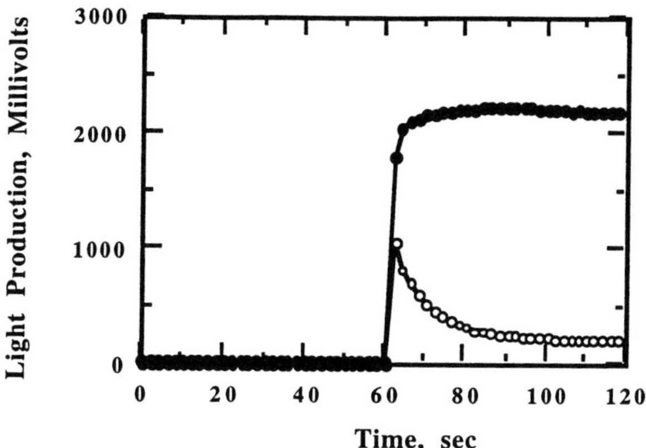

Fig. 4. Effect of CoA on light production by firefly luciferase. Light production was initiated by injection of ATP at 60 s. The time-course of light production was determined in an LKB 1251 luminometer. —o— Control, —•— 0.05 mM CoA.

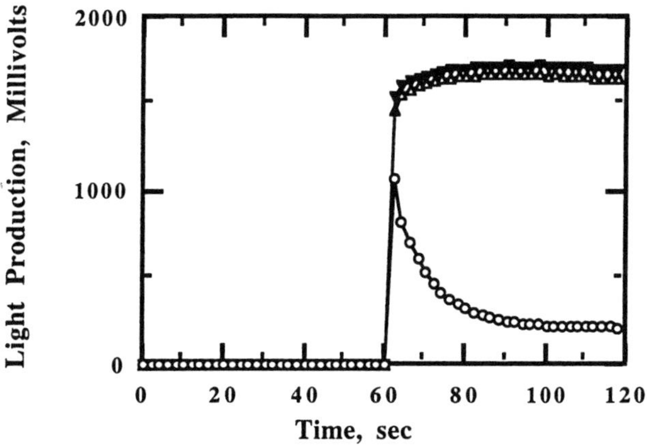

Fig. 5. Effect of PP_i and periodate-oxidized and sodium borohydride-reduced ADP on light production by firefly luciferase. Light production was initiated by injection of ATP at 60 s. The time-course of light production was determined in an LKB 1251 luminometer. —▼— 0.013 mM PP_i, —△— 1 mM orADP, —o—Control.

ferase assays has allowed assay conditions of enhanced sensitivity, excellent reproducibility, and a maintained linearity of the calibration curve to be established."

2. Nucleotide analogs: Ford et al. *(12,19)* found that cytidine triphosphate and other nucleotides enhanced firefly luciferase activity in a manner similar to that of CoA. DethioCoA inhibited the activation by both cytidine nucleotides and CoA. The enhancement of light production with CoA or nucleotides occurred only with high ATP concentrations.

3. Triton X-100: Gandelman et al. *(20)* found that 25 mM Triton X-100 increased both luciferase light production and the rate of destruction of the enzyme. It presumably allows formation of a more active, though more labile, enzyme conformation. An additive effect of CoA and Triton X-100 has been observed by Wang and Andrade *(21)*.

4. Other detergents: Simpson and Hammond *(22)* found that anionic detergents inhibited firefly luciferase, cationic detergents stimulated activity with a sharply defined concentration optimum, but they also inactivated the enzyme, and nonionic and zwitterionic detergents increased reaction rate without affecting stability until high concentrations were used. Stability of the enzyme was measured during a 20-s incubation. Kricka and DeLuca *(23)* found that a number of solvents stimulated the firefly luciferase reaction by promoting the dissociation of inhibitory products. These experiments were done in a phosphate-buffered reaction mixture (phosphate inhibits activity), and the time-course of light production was not significantly altered. There is no clear evidence that detergents can improve the routine assay of ATP.

5. PP$_i$ and L-luciferin combination: Lundin *(24)* has shown that addition of 1 μM PP$_i$ and 16 μM L-luciferin (*Note:* this is not the normal substrate) to a firefly luciferase reaction mixture containing 1 μM ATP stabilized light production for ~2 min. This reagent was available from LKB (Stockholm, Sweden), and is now available from BioOrbit Oy (Turku, Finland), and BioThema (Dalorö, Sweden).

6. Polyphosphates: Lundin *(25)* reported that 20 μM PP$_i$ gives an optimum sustained light emission over an extended period of time (up to 12 min) at 0.2 mM ATP. We (Ford et al. *[12]*) found similar results using 13 μM PP$_i$. Lower and higher PP$_i$ concentrations were less effective. We also found that tripolyphosphate, tetrapolyphosphate, and trimetaphosphate (all at 1 mM) gave a sustained enhanced light emission.

1.7. Use of Additives in Quantitation of Firefly Luciferase

When using the firefly luciferase assay to measure the amount of enzyme in a sample, maximum sensitivity is needed. Thus, the assay must be done using high ATP concentrations (~0.2 mM) and preferably with additives to increase the light production. Several methods to do this have been developed. Lundin *(25)* established an optimized assay for firefly luciferase using 20 mM PP$_i$ as an additive to enhance light production. Boehringer Mannheim (Mannheim, Germany) sells a kit (cat. no. 1669 893) containing CoA, that yields a constant rate of light production for at least 60 s, and allows the detection of 5 fg of firefly luciferase. Promega's (Madison, WI) luciferase assay system (cat. no. E1500) contains 270 μM CoA. Ford et al. *(19)* report that 0.18 mM periodate oxidized CTP increased the sensitivity of luciferase determination fourfold and were able to measure 1.5 pg of luciferase. Prolonged incubation of luciferase with periodate oxidized CTP (>5 min) inactivated the enzyme. However, Ford et al. *(12)* found that the activating activity of periodate-oxidized and then sodium borohydride-reduced ADP was retained for at least a 150-min incubation of additive with firefly luciferase.

1.8. Mechanisms of Action

Ford et al. *(12)* interpreted that the increased turnover of firefly luciferase through release of oxyluciferin is the mechanism by which the nucleotide analogs and CoA enhance firefly luciferase activity. There was an increase from 0.97 to ~5.23 photons of light produced/min/molecule of luciferase with 0.24 mM ATP. McElroy et al. *(26)* had previously ascribed the mechanism of action of pyrophosphate to the same phenomenon.

2. Materials

2.1. Water and Glassware

Water quality is of paramount importance. Minute contamination of reagents (especially bacterial contamination) will cause high background luminescence because of the sensitivity of the technique. We routinely prepare the water used in all reagents as follows: The building's reverse osmosis and UV-treated water is passed through two mixed-bed ion-exchange resins (Barnstead/ Thermolyne D 8902 Ultrapure Cartridges, Dubuque, IA, glass-distilled, pressure-filtered through a sterile 0.45-μm Millipore® (Bedford, MA) filter into sterile bottles, and then autoclaved. After opening, a bottle of water can be used for several days if handled using good sterile technique.

We recommend as a minimum standard that "Milli-Q-quality" water be additionally filtered through a sterile 0.45-μm filter and autoclaved before use. Backgrounds in the standard ATP assay containing 100 μL of Firelight® and no ATP in a 500-μL total volume should be <100 counts/10 s in a Lumac Model 2010A Biocounter. If backgrounds are high, the "Milli-Q" water should be distilled before filtering and autoclaving.

We recommend that all glassware used for reagents for these assays be washed in phosphate-free detergent, soaked in Pierce (Rockford, IL) brand RBS-pf®, rinsed in reverse-omosis-treated (RO) or deionized water, and sterilized.

2.2. Chemicals

Prepare all stocks in sterile glass- or plasticware using sterile water as described in **Subheading 2.1.**, and store frozen to reduce the chance of bacterial contamination.

1. Tricine: We find that Tricine buffer yields a system giving the greatest light production under our laboratory conditions. The optimum pH is 7.8. We use Sigma (St. Louis, MO) T 9784. Prepare stock solution of 1.0 M, and dilute as needed to make Tricine-containing reagents.

2. Bovine serum albumin (BSA): Fraction V Powder ($\geq$96%) is adequate. We use Sigma A 2153. BSA is present in many commercial preparations to stabilize firefly luciferase by reducing proteolytic degradation and adsorption to surfaces. The stock solution is 100 mg/mL in water.

3. $MgSO_4$: Use ACS-grade salts. A 50-mM stock is prepared in water.

4. DL-Dithiothreitol (Cleland's reagent, DTT). Use the highest purity available. We use Sigma D 5545 to prepare a 50-mM stock.

5. EDTA: Use the highest grade available. We use Sigma E 1644, disodium salt. When preparing the 50-mM stock solution, check pH, and titrate to neutrality with NaOH.

6. Luciferin: D-Luciferin is the natural, functional configuration. We recommend Sigma L 6882 sodium salt, because it is readily soluble in water. Alternatively, the free acid form (Sigma L 9504) is more economical, but it must be titrated with NaOH. Dissolve the free acid form at 5.0 mg/mL in 20 mM Tricine, pH 7.8, titrate with NaOH to return the pH to 7.8, and ensure that all the luciferin is in solution. Protect luciferin from light while the solutions are being prepared. Purge the atmosphere above the solution with N_2, and store frozen and protected from light (we store in brown bottles, capped with Parafilm® and wrapped in foil). For use, dilute the luciferin to 1.0 mg/mL in 20 mM Tricine, pH 7.8. Unused diluted luciferin can be purged with N_2 and stored frozen.

   L-Luciferin supports light production only under special conditions. This isomer competes with the natural form. It has been used to linearize the time-course of light production. This is one of the components used in the LKB ATP Monitoring reagent, produced now by BioOrbit Oy *(25)*.

7. ATP: Use crystalline, 99–100% pure, disodium salt (<1 ppm vanadium). We use Sigma A 5394. ATP solutions can be prepared either in 20 mM Tricine buffer, pH 7.8, or in water. Check the pH of ATP solutions and neutralize, if necessary, with NaOH.

8. Pyrophosphate: Use the highest purity available, such as Sigma P 9146 or Sigma S 9515 tetrasodium salts (decahydrate), 1 mM stock pyrophosphate solutions must be titrated to neutrality.

9. CoA: Use either the lithium or the sodium salt (Sigma C 3019 or C 3144, respectively). We have always prepared only enough of the 5-mM stock to satisfy a single day's need by dissolving in water. We have not determined the stability of CoA solutions on storage.

10. Nucleotide analogs: Periodate-oxidized CTP (Sigma C 5150, oCTP) and periodate-oxidized, sodium borohydride-reduced ADP (Sigma A 6910, orADP), among others, can be used to linearize the assay. Prepare only enough of the analogs for a single day of use by dissolving in water. These are prepared as 10-mM stocks.

11. Enzyme stabilizer: AuthentiZyme™ Enzyme Stabilizer from Innovative Chemistry (Marshfield, MA) is a proprietary product that protects enzymes from inactivation by oxidation and heavy metals. Make solutions according to the manufacturer's instructions.

2.3. Firefly Luciferase

We recommend Firelight®, catalog no. 2005 from Analytical Luminescence Laboratory (Ann Arbor, MI) for routine assays. Dissolve enzyme in 50 mM

Tricine, pH 7.8, containing 10 mM MgSO$_4$, 1 mM DTT, 1 mM EDTA, and 1 mg/mL BSA. Let enzyme "age" for ≥1 h at 0–4°C before use. Unused enzyme can be stored at 4°C overnight, with some loss of activity (*see* **Note 1**).

When purified firefly luciferase is needed, we use Sigma L 5256, crystallized and lyophilized powder. This preparation is no longer available, but is replaced by L 2533, which is prepared without arsenate. Dissolve it at 0.1 to 1 mg/mL in 50 mM Tricine, pH 7.8, containing 10 mM MgSO$_4$, 1 mM DTT, 1 mM EDTA, and 1 mg/mL BSA or in a 1:1 mixture of 250 mM Tricine, pH 7.8, containing 50 mM MgSO$_4$, 5 mM DTT, 5 mM EDTA, and Authenti-Zyme® Enzyme Stabilizer (*see* **Note 2**). This preparation is not easily soluble: To dissolve the protein, add the desired solvent and let sit on ice, with occasional gentle mixing, for at least 1 h. Visually check that the protein has all gone into solution before use. Alternatively, Sigma L 9009 and L 1759 are soluble preparations containing buffer and salts.

2.4. Luminometer

A high-quality luminometer that allows injection of reactant into the sample while the sample is in the measuring chamber is needed. We recommend the Lumac Model 2010A Biocounter (Luma, Landgraf, The Netherlands; recently purchased by Celsis, Cambridge, UK) or equivalent (*see* **Note 3**).

3. Methods

3.1. Caution

The great sensitivity (50 fg) and wide dynamic range (four decades) of the firefly luciferase determination of ATP make a robotic application of the procedure relatively easy. Numbers can be obtained, but their meaning could be misleading. It is our contention that the operator needs to know the nuances of the assay components and instrumentation to obtain maximally reliable data. The mind needs to be engaged while doing the measurements. A monograph on *Bioluminescence Analysis* has been written by Brolin and Wettermark that outlines and discusses the particularities of the technique *(27)*.

3.2. Basic Reaction Components

Depending on the parameters of the instrument to be used, we recommend a reaction volume of from 200–500 μL containing the following:

25 mM Tricine buffer, pH 7.8;
5 mM MgSO$_4$;
0.5 mM EDTA;
0.5 mM DTT;
1 mg/mL BSA;

0.05 mg/mL D-luciferin (if using purified firefly luciferase);
ATP as required;
Firefly luciferase/luciferin (Firelight®) or purified firefly luciferase as required;
Water to desired total volume.

A 10X reaction mixture containing 250 mM Tricine buffer, pH 7.8; 50 mM $MgSO_4$; 5 mM EDTA; and 5 mM DTT is convenient to use. This mixture can be prepared ahead, aliquoted in amounts to be used in a single day, and stored frozen. We recommend using Firelight instead of purified luciferase plus luciferin for routine assays because of the ease of use and consistency of results.

3.3. General Protocol

The reaction is carried out at room temperature (25°C), preferably in semidarkness.

1. Set up reaction cuvets containing for a 500-µL reaction: 50 µL of 10X reaction mixture, BSA, and water as needed to bring the final volume (after subsequent addition of ATP, luciferin, and enzyme) to 500 µL. These components can be added to all cuvets before starting the assays.
2. Just before placing the cuvet into the counting chamber, add ATP (at room temperature) and luciferin (kept on ice) if needed.
3. Mix by vortexing, place cuvet into the instrument and start the reaction by injecting the enzyme preparation (at room temperature). Alternatively, enzyme can be added to the cuvet before placing it in the sample chamber and the reaction initiated by the injection of ATP. This is more economical if using a luminometer with an automatic dispenser because of losses of reagent in the lines of the automatic dispenser.
4. Determine light emitted for desired time. For routine assays, a 10-s counting time is usually sufficient. The Lumac instrument gives the rate of counting averaged over the time period selected. Thus, a 30-s counting time will give the same value as a 10-s counting, but with improved precision (*see* **Note 4**).

3.4. ATP Determination

To measure ATP in biological samples, replace ATP in the general protocol with the biological sample for which the ATP content is to be determined. If it is necessary to keep the samples cold until just before they are assayed (when they are warmed to room temperature), the volume of sample assayed should be kept to a minimum (no more than 10% of the total reaction volume). For each biological sample assayed, run a second determination with 0.1–0.5 ng of ATP added to the biological sample to determine the extent of inhibition, if any, of the assay itself. Inhibition is calculated by comparing the difference in light emitted in the biological sample with and without added ATP to the light emitted from the same concentration of ATP in the absence of biological sample. For ATP determinations, it is usually most practical to start the reaction by injecting enzyme. An ATP standard curve must be run each day to determine the absolute amount of ATP in samples.

3.5. Firefly Luciferase Determination

To measure firefly luciferase in biological samples, replace the Firelight or purified firefly luciferase in the general protocol with the biological sample to be assayed. If the biological sample must be kept cold, keep the volume of the sample to no more than 10% of the total reaction volume. Include D-luciferin (0.05 mg/mL) in the assay mixture. Assay with a high concentration of ATP (0.5 mM). Add the biological sample to the assay tube before placing in the luminometer, and begin the reaction by injecting the ATP.

3.6. Supplementation to Linearize Light Production

When high concentrations of ATP are measured, a flash of light followed by a decay of light emitted is the normal pattern. This pattern can be converted to a linear production of light at the high rate of the flash by addition of any number of compounds as discussed in **Subheading 1**. To linearize light production, add one of the following supplements to the basic reaction mixture:

13–20 µM PP$_i$ (used by Lundin and this laboratory);
0.18 mM oCTP (used in this laboratory);
1 mM orADP (used in this laboratory);
270–500 µM CoA (used by Analytical Luminescence and Promega);
1 µM PP$_i$ and 16 µM L-luciferin (used by BioOrbit Oy).

4. Notes

1. Firefly luciferase: Three grades of firefly luciferase with different degrees of purity are commercially available. Crude lantern extracts contain sufficient pyrophosphatase, so that PP$_i$ does not accumulate *(28)*. These preparations also contain adenylate kinase, and nucleoside diphosphate kinase, which enable nucleotides other than ATP to be enzymatically converted to ATP and thus produce light in the assay system. These preparations are not recommended for sensitive determination of ATP. Purification procedures have been developed that remove the adenylate kinase, pyrophosphatase, and nucleoside diphosphate kinase. These preparations can be used for the sensitive determination of ATP. Many are supplemented with sufficient luciferin, so that no additional luciferin is required. Crystalline luciferase is purer, but is somewhat more difficult to handle. There is little difference between crystalline native and recombinant firefly luciferases. The slight differences in conformation and lability to proteolytic enzymes that exist for these two luciferases are not significant *(8)*.

   Although firefly luciferase can be fairly stable when stored properly after making a solution *(29)*, we recommend the use of a commercial preparation (such as Analytical Luminescence Laboratory's Firelight) made fresh and pooled each day. The use of a commercial preparation with its stabilizers and quality control means that the individual laboratory does not need its own reagent quality-control program. This laboratory has operated both systems and finds the use of commercial kits better for routine studies. The use of commercial kits is now much more accepted with the advent of molecular biology's cloning kits—it is more time-efficient to let the supplier provide the quality control. This means carefully selecting a supplier of reagents. This laboratory evaluated the commercially available reagents in 1986 *(6)*. Much progress has been made in commercial firefly luciferase reagent kits during the subsequent decade. Many of the suppliers listed in Table 1 of our comparison no longer supply the reagents, and there are also many new suppliers. The techniques and experiments used in the comparative evaluations are still appropriate to evaluate those products. The commercial firms whose products have survived probably have done so because of good quality. Beginning in 1993, Stanley has published lists of commercial firms providing luminescence kits based on information provided by the supplier *(30–34)*. There is no experimental comparison of the kits and reagents in Stanley's listing. Wang and Andrade *(35)* have added 100 mg/mL of trehalose to stabilize solutions of firefly luciferase particularly when preparing films.

2. Enzyme stabilizer: Firefly luciferase dissolved in a mixture of salts and AuthentiZyme™ Enzyme Stabilizer is stable frozen for several months, even with repeated thawing and freezing *(29)*.

3. Instrumentation—luminometer: Although relatively expensive and specialized, we recommend the use of an instrument designed for bioluminescent/chemiluminescent measurements. These instruments have a wide range of specific properties (such as geometry of the detector) and design criteria (temperature control and sample size). Some permit variation of the high voltage supplied to the photomultiplier, whereas others have fixed voltage; some allow temperature regulation, but others operate at room temperature. Ten commercially available instruments have been experimentally compared by Jago and associates *(36)*. The most sensitive instruments were the Lumac Model 2010A and the Turner 20 TD photometers, which had actual limits of 0.09 and 0.12 pg ATP/sample, respectively. George Turner *(37)* presents a provocative assessment of instrument development from the viewpoint of a person trained in physics and electronics trying to get the most out of the instrument/reagent system. Van Dyke *(38)* reviews the manufacturers' provided information for photometers that were available in 1985. Further review of the commercial instrumentation has been made by Phil Stanley in a continuing series of articles *(30–34,39–41)*.

    If the investigator desires to construct a photometer, Anderson et al. *(42)* give complete instructions. These instructions were updated in 1985 *(43)* with "the strong recommendation that in most cases a researcher would be better served to purchase a commercial instrument."

    For calibration of light production, please refer to the methods described by O'Kane and coworkers *(44)* and by Lee and Seliger *(45)*.

4. Protocol: We recommend that preliminary experimentation be done to establish that the reagents, instruments, and protocols are working in your laboratory, and meet the desired quality-control characteristics. What is the instrument background, and what are the reagent backgrounds? Is the response to known (standard) amounts of ATP and/or luciferase in line with published values? Is the response linear over several orders of magnitude? Is the slope of the standard curve one? Are the reagents stable over the desired assay period? What is the response when a know standard amount of either ATP or luciferase is added to an experimental reaction mixture (in other words, what is the extent of inhibition in the assay mix itself)?

Several of the commercial manufacturers have published detailed protocols or quality-control information for the use of their reagents. These include:

> Luciferase Assay Guide Book, Protocols and Information for Measuring Firefly Luciferase Expressed in Cells, Analytical Luminescence Laboratory, 1180 Ellsworth Road, Ann Arbor, MI 48108 (1-800-854-7050).

> Luminescence Analysis, Application Note 100; and The Bioluminescent Assay of ATP, Application Note 201 Bio-Orbit Oy, Box 36 SF-20521 Turku, Finland, Voice +358 21 510666; Fax +358 21510150.

> Luciferase, ATP Bioluminescence Assay Kit HS II, and Luciferase Reporter Gene Assay protocol are available from Boehringer Mannheim Biochemicals, P.O. Box 50816, Indianapolis, IN 46250 (1-800-428-5437) (Internet: http://biochem.boehringer-mannheim.com).

> Luciferase Assay System (Part# TB 101) Promega, 2800 Woods Hollow Road, Madison, WI, 53711-5399 (1-800-356-9526) (Internet http://www.promega.com) Protocols and application notes are available on-line.

> Sigma Quality Control Test Procedure for Products L1759, L5256, and L9009, available at Internet: http://www.sigma.sial.com/sigma/enzymes/lucifera.htm.

> Luciferase protocol, Tropix, Inc. (1-800-542-2369) Internet: http://www.tropix.com/luciptl.htm

> Turner Instrument Literature (http://www.turnerdesigns.com/mono_lst.htm).

Acknowledgments

This research was supported in part by the Oklahoma Agricultural Experiment Station (Project 1806) and is published with the approval of the Director. Robert Matts and E. C. Nelson read the manuscript and made useful suggestions.

References

1. McElroy, W. D. (1947) The energy source for bioluminescence in an isolated system. *Proc. Nat. Acad. Sci.* USA. **33,** 342–345.
2. Strehler, B. L. and Trotter, J. R. (1952) Firefly luminescence in the study of energy transfer mechanism. I. Substrate and enzyme determination. *Arch. Biochem. Biophys.* **40,** 28–41.

3. Strehler, B. L. and McElroy, W. D. (1957) Assay of adenosine triphosphate. *Methods Enzymol.* **3,** 871–873.
4. Strehler, B. L. (1968) Bioluminescence assay: principles and practice. *Methods Biochem. Anal.* **16,** 99–181.
5. McElroy, W. D. (1977) Comments on the history of the firefly system, in *2nd Bi-Annual ATP Methodology Symposium* (G. A. Borun, ed.), SAI Technology, San Diego, CA, pp. 405–413.
6. Leach, F. R. and Webster, J. J. (1986) Commercially available firefly luciferase reagents. *Methods Enzymol.* **133,** 51–70.
7. Webster, J. J. and Leach, F. R. (1980) Optimization of the firefly luicferase assay for ATP. *J. Appl. Biochem.* **2,** 469–479.
8. Ford, S. R., Hall, M. L., and Leach, F. R. (1992) Comparison of properties of commercially available crystalline native and recombinant firefly luciferase. *J. Biolumin. Chemilumin.* **7,** 185–193.
9. DeLuca, M. (1976) Firefly luciferase. *Adv. Enzymol.* **44,** 37–63.
10. Webster, J. J., Chang, J. C., and Leach, F. R. (1980) Sensitivity of ATP determination. *J. Appl. Biochem.* **2,** 516, 517.
11. Webster, J. J., Chang, J. C., Manley, E. R., Spivey, H. O., and Leach, F. R. (1980) Buffer effects on ATP analysis by firefly luciferase. *Anal. Biochem.* **106,** 7–11.
12. Ford, S. R., Chenault, K. H., Bunton, L. S., Hampton, G. J., McCarthy, J., Hall, M. S., Pangburn, S. J., and Leach, F. R. (1996) Use of firefly luciferase for ATP measurement: other nucleotides enhance turnover. *J. Biolumin. Chemilumin.* **11,** 149–167.
13. Airth, R. L., Rhodes, W. C., and McElroy, W. D. (1958) The function of coenzyme A in luminescence. *Biochim. Biophys. Acta.* **27,** 519–532.
14. McElroy, W. D. (1957) Chemistry and physiology of bioluminescence, in *The Harvey Lectures,* 1955–56. Academic, NY, pp. 240–266.
15. Wood, K. V. (1990) Novel assay of firefly luciferase providing greater sensitivity and ease of use. *J. Cell Biol.* **111,** 380a.
16. Wood, K. V. (1991) The origin of beetle luciferases, in *Bioluminescence and Chemiluminescence: Current Status* (Stanley, P. E. and Kricka, L. J., eds.) John Wiley, Chichester, UK, pp. 11–14.
17. Wood, K. V. (1991) Recent advances and prospects for use of beetle luciferase as genetic reporter, in *Bioluminescence and Chemiluminescence: Current Status* (Stanley, P. E. and Kricka, L. J., eds.), John Wiley, Chichester, UK, pp. 543–546.
18. Pazzagli, M., Devine, J. H., Peterson, D. O., and Baldwin, T. O. (1992) Use of bacterial and firefly luciferases as reporter genes in DEAE-dextran-mediated transfection of mammalian cells. *Anal. Biochem.* **204,** 315–323.
19. Ford, S. R., Hall, M. S., and Leach, F. R. (1992) Enhancement of firefly luciferase activity by cytidine nucleotides. *Anal. Biochem.* **204,** 283–291.
20. Gandelman, O. A., Brovko, L. Y., Bowers, K. C., Cobbold, P. H., Polenova, T. Y., and Ugarova, N. N. (1993) Kinetics of enzymic oxidation of firefly luciferin in vitro and in cytoplasm, in *Bioluminescence and Chemiluminescence: Status Report* (Szalay, A. A., Kricka, L. J., and Stanley, P. E., eds.) John Wiley, Chichester, UK, pp. 84–88.

21. Wang, C. Y. and Andrade, J. D. (1996) Surfactants and coenzyme A as cooperative enhancers of the activity of firefly luciferase. *J. Biolumin. Chemilumin.* **11,** 25.

22. Simpson, W. J. and Hammond, J. R. M. (1991) The effect of detergents on firefly luciferase reactions. *J. Biolumin. Chemilumin.* **6,** 97–108.

23. Kricka, L. J., and DeLuca, M. (1982) Effect of solvent on the catalytic activity of firefly luciferase. *Arch. Biochem. Biophys.* **217,** 674–681.

24. Lundin, A. (1982) Application of firefly luciferease, in *Luminescent Assays: Perspectives in Endocrinology and Clinical Chemistry* (Serio, M. and Pazzagli, M., eds.), Raven, New York, NY, pp. 29–45.

25. Lundin, A. (1993) Optimised assay of firefly luciferase with stable light emission, in *Bioluminescence and Chemiluminescence: Status Report* (Szalay, A. A., Kricka, L. J., and Stanley, P., eds), John Wiley, Chichester, UK, pp. 291–295.

26. McElroy, W. D., Hastings, J. W., Couloombre, J., and Sonnenfeld, V. (1953) The mechanism of action of pyrophosphate in firefly luminescence. *Arch. Biochem. Biophys.* **46,** 399–416.

27. Brolin, S. and Wettermark, G. (1991) *Bioluminescence Analysis.* VCH Weinheim, Germany, 151 pp.

28. DeLuca, M. and McElroy, W. D. (1978) Purification and properties of firefly luciferase. *Methods Enzymol.* **57,** 3–15.

29. Hall, M. S. and Leach, F. R. (1988) Stability of firefly luciferase in Tricine buffer and in a commercial enzyme stabilizer. *J. Biolumin. Chemilumin.* **2,** 41–44.

30. Stanley, P. E. (1993) A survey of some commercially available kits and reagents which include bioluminescence or chemiluminescence for their operation. *J. Biolumin. Chemilumin.* **8,** 51–63.

31. Stanley, P. E. (1993) Commercially available luminometers and imaging devices for low-light measurements and kits and reagents utilizing chemiluminescence or bioluminescence: Survey update 1. *J. Biolumin. Chemilumin.* **8,** 234–240.

32. Stanley, P. E. (1993) Commercially available luminometers and imaging devices for low-light measurements and kits and reagents utilizing chemiluminescence or bioluminescence: Survey update 2. *J. Biolumin. Chemilumin.* **9,** 51–53.

33. Stanley, P. E. (1993) Commercially available luminometers and imaging devices for low-light measurements and kits and reagents utilizing chemiluminescence or bioluminescence: Survey update 3. *J. Biolumin. Chemilumin.* **9,** 123–125.

34. Stanley, P. E. (1993) Commercially available luminometers and imaging devices for low-light measurements and kits and reagents utilizing chemiluminescence or bioluminescence: Survey update 4. *J. Biolumin. Chemilumin.* **11,** 175–191.

35. Wang, C.-Y., and Andrade, J. D. (1994) Purification and preservation of firefly luciferase, in *Bioluminescence and Chemiluminescence: Fundamental and Applied Aspects* (Campbell, A. K., Kricka, L. J., and Stanley, P. E., eds.), John Wiley, Chichester, UK, pp. 423–426.

36. Jago, P. H., Simpson, W. J., Denyer, S. P., Evans, A. W., Griffiths, M. W., Hammond, J. R. M., Ingram, T. P., Lacey, R. F., Macey, N. W., McCarthy, B. J., Salusbury, T. T., Senior, P. S., Sidorowicz, S., Smithers, R., Stanfield, G., and Stanley, P. E. (1989) An evaluation of the performance of ten commercial luminometers. *J. Biolumin. Chemilumin.* **3,** 131–145.

37. Turner, G. K. (1985) Measurement of light from chemical or biochemical reactions, in *Bioluminescence and Chemiluminescence: Instruments and Application,* vol. I (Van Dyke, K., ed.), CRC, Boca Raton, FL, pp. 43–78.
38. Van Dyke, K. (1985) Commercial instruments, in *Bioluminescence and Chemiluminescence: Instruments and Application,* vol. I (Van Dyke, K., ed.), CRC, Boca Raton, FL, pp. 83–128.
39. Stanley, P. E. (1985) Characteristics of commercial radiometers. *Methods Enzymol.* **133,** 587–603.
40. Stanley, P. E. (1992) A survey of more than 90 commercially available luminometers and imaging devices for low light measurement of chemiluminescence and bioluminescence, including instruments for manual, automatic and specialized operation for HPLC, LC, GLC and microplates. Part 1 descriptions. *J. Biolumin. Chemilumin.* **7,** 77–108.
41. Stanley, P. E. (1992) A survey of more than 90 commercially available luminometers and imaging devices for low light measurement of chemiluminescence and bioluminescence, including instruments for manual, automatic and specialized operation for HPLC, LC, GLC and microplates. Part 1 photographs. *J. Biolumin. Chemilumin.* **7,** 157–169.
42. Anderson, J. M., Faini, G. J., and Wampler, J. E. (1978) Construction of instrumentation for bioluminescence and chemiluminescence assays. *Methods Enzymol.* **57,** 529–540.
43. Wampler, J. E., and Gilbert, J. C. (1985) The design of custom radiometers, in *Bioluminescence and Chemiluminescence: Instruments and Application,* vol. I (Van Dyke, K., ed.), CRC, Boca Raton, FL, pp. 129–150.
44. O'Kane, D. J., Ahmad, M., Matheson, I. B. C., and Lee, J. (1986) Purification of bacterial luciferase by high-performance liquid chromatography. *Methods Enzymol.* **133,** 109–127.
45. Lee, J. and Seliger, H. H. (1972) Quantum yields of the luminol chemiluminescence reaction in aqueous and aprotic solvents. *Photochem. Photobiol.* **15,** 109–127.

2

Visualization of Bioluminescence

Amy Cheng Vollmer

1. Introduction

There are an increasing number of specialized instruments that may be used for the purpose of measuring bioluminescence. **Table 1** contains a representative list of different luminometers and cameras that are available. These instruments have been used to detect bioluminescence in a number of organisms using either bacterial luciferase (*lux*; *1,2*) or firefly luciferase (*luc*; *3,4*) as reporters. Sensitivity of the newer luminometers ranges from six to eight logs. Options such as temperature control and agitation of samples are usually available at an extra cost. Most of the systems can be driven by computer with commercially available or customized software. Storage, display, and analysis of data involve the same or additional software packages. Sample containers have also become more specialized. In the case of the multiplate luminometers, opaque plates are available in either white or black. Black plates are recommended for bright samples where reflection into neighboring wells results in "crosstalk." White plates are recommended for samples that are lower light emitters, since the reflective surface enhances detection. Opaque plates are also available with transparent bottoms. Samples in these microplates may be read in a spectrophotometer (such as an ELISA reader) to measure optical density of the sample, as an indicator of cell number particularly in the case of bacterial cells. In some applications, opaque microplates containing samples may be stacked in alternation with transparent microplates, if samples require a light source. This is essential for many of the studies involving photosynthetic microorganisms *(5,6)* as well as for those studying circadian rhythms for which light entrainment is needed *(5–7)*.

On the other hand, it is possible to measure and document bioluminescence without purchasing a dedicated instrument. In most laboratories, equipment

From: *Methods in Molecular Biology, Vol. 102: Bioluminescence Methods and Protocols*
Edited by: R. A. LaRossa © Humana Press Inc., Totowa, NJ

Table 1
Commercial Luminometers, Listed by Sample Format

Tube/vial	96-Well Microplate	Camera[c]	Manufacturer	
Monolight® 2010[a]	Monolight® 9600	Camlight®	Analytical Luminescence Lab.	San Diego, CA
	lucy 1		Anthos Labtec, Inc.	Frederick, MD
1250[a], 1251[b], 1253[a]	1258 (Galaxy®)		BioOrbit Oy	Tonawanda, NY
			(Man-Tech Assoc.)	
	Lumstar®		BMG Lab Technologies	Durham, NC
	7700 series		Cambridge Technology, Inc.	Watertown, MA
	WELLTECH		Denely Instruments, Inc.	Research Triangle, NC
	ML2200, ML2250,		Dynatech Laboratories, Inc.	Chantilly, VA
	ML3000			
	Luminoskan®		Labsystems	Needham Heights, MA
Optocomp® 1[a]			MGM Instruments	Hamden, CT
		TE/CCD512BIC5	Princeton Instruments	Trenton, NJ
		Hamamatsu	Photonics Co.	JAPAN
		ICL901	Tropix, Inc.	Bedford, MA
TD 20e[a]			Turner Designs	Mountain View, CA
	TopCount®		Packard Instrument Co.	Meriden, CT
Lumat™ LB 9507[a]	MicroLumat™ LB96P		EG&G Berthold/Wallac	Turku, Finland
Multi-Lumat™ LB9507[b]			LKB/Wallac	Gaithersburg, MD
McroBeta™ PLUS	MicroBeta™	NightOWL® LB 981		

[a]Single sample.
[b]Multiple sample.
[c]Not all are CCD cameras.

and supplies that can be used successfully in many applications already exist. There are certainly limitations to their sensitivity, especially since these instruments were usually designed with some other application in mind. This chapter will focus on the use of such instrumentation for visualization of bioluminescence in the following ways. A liquid scintillation counter can be used to measure bioluminescence from *Escherichia coli* strains carrying stress promoter::*lux* fusions on recombinant plasmids. (We have used a 1219 RackBeta® from LKB/Wallac, Gaithersburg, MD, driven by UTMac software.) Screening of bioluminescent bacterial colonies can be performed easily using X-ray film. Photography of bioluminescent bacterial colonies can be accomplished with prolonged exposure times using Polaroid type 57 film or Kodak T-MAX P3200 35-mm roll film with the appropriate cameras and lenses.

2. Materials

1. Fresh bioluminescent bacterial cultures, grown on appropriate media: Liquid cultures should be used for measurements in the scintillation counter; agar media should be used for photographic documentation.
2. Sterile 1.5-mL microcentrifuge tubes without caps: These are available commercially, or the caps can be cut off of standard 1.5-mL microcentrifuge tubes.
3. Glass vials (or otherwise transparent ones with tight-fitting lids) suitable for the scintillation counter used: These vials need to be washed and dried one time, since they will not come into direct contact with the bacterial sample. The vials must be large enough to accommodate a 1.5-mL microcentrifuge tube without its cap. Alternatively, one can use smaller vials and 0.5-mL microcentrifuge tubes (*see* **Note 1**).
4. X-ray film, such as Kodak XAR or DuPont Reflections®.
5. Polaroid type 57 film with appropriate film holder and photostand or Kodak T-MAX P3200 35-mm high-speed roll film and a 35-mm camera with an assortment of lenses.

3. Methods
3.1. Use of the Scintillation Counter

1. Scintillation counters have programs that can be set by the operator. LKB/Wallac calls these "parameter groups." Set one parameter group to read chemiluminescence, a standard setting for most scintillation counters. Bioluminescent samples will be read with that setting. The time interval over which the sample is to be counted can be varied between 10 s and several minutes. Set this interval to meet the needs of the reporter system that is being used and the amount of light that is emitted. Intervals that are <1 min are typical. Set one other parameter group to read some other window. Be sure the time interval for this parameter group is about 10–20 min. If The LKB/Wallac system assigns numbers to each parameter group. Each rack of samples can be identified by a code plug clipped to the leading edge of the rack.

2. Place one sterile, capless 1.5-mL microcentrifuge tube inside each glass scintillation vial (*see* **Note 1**).

3. Carefully place aliquots of bacterial samples into the tube. The volume of the sample placed into each tube can vary from 10–100 µL. (Volumes >100 µL may result in a reduced level of oxygenation of the sample. This may or may not be an important consideration; *see* **Note 2**).

4. Place and tighten lids on the scintillation vials. After tightening the lids, loosen by one-quarter turn to allow for the exchange of air (*see* **Note 3**).

5. Place bacterial samples into a sample rack that bears the correspondingly numbered identification code plug for that parameter group. If there are more samples than the number of places in the rack, place additional sample in another rack that bears no identification code plug. The counter will consider samples in this next rack as components of the first parameter group mode.

6. Place an empty scintillation vial (with a lid) into another rack. This rack should have a code plug that identifies the second parameter group. By inserting this rack after the bioluminescent samples, a time delay is introduced so that the samples will be read once every 10–20 min. This reading cycle will continue until the counter is stopped by the insertion of a rack bearing stop code plug or by interrupting the program through a keyboard command to the UTMac software on the computer.

7. Data saved on UTMac can be most easily formatted as a Simpletext table, which can be easily exported and "parsed" into spreadsheets or graphic programs for analysis. It is possible to record the actual times that the sample readings took place. It is also convenient to delete data recorded from counting the "dummy" sample.

8. After readings are completed, samples may be removed from the scintillation vials for plating or disposal (*see* **Note 4**).

3.2. Screening Bioluminescent Bacterial Cultures Using X-Ray Film

1. Plate bacteria on suitable agar medium. Place plates, agar side up, inside of a light-tight box that has a removable lid. Use transparent tape to secure the plates to the bottom of the box.

2. Alternatively, a microtiter plate containing liquid bacterial cultures in the wells may be taped to the bottom of the box. Care should be taken not to tilt the plate or the box.

3. In the darkroom, place one piece of X-ray film on top of the plates. Secure the film to the side of the box with transparent tape. Be careful not to place the rest of the unexposed film near the plates. Very bright emitters produce significant amounts of light and may expose the film if it is too close. Using scissors, cut one corner of the film to help to orient it later. Mark the corresponding corner of the box.

4. Place the lid of the box on top and place the box carefully inside a cabinet or drawer.

5. Exposure times are highly variable. Bright emitters need only a few seconds of exposure. Low light emitters require overnight exposure. Exposure time also depends on the concentration of bacteria inoculated onto the agar.
6. When removing the film from the box, be sure to remove any pieces of transparent tape that may have been securing the film to the box. Develop the film, and then orient it with the plates in the box, aligning the marked corner of the box with the cut corner of the film. Additional exposures may be done subsequently (*see* **Note 5**).

3.3. Photographing Bacteria on Agar Plates

3.3.1. Using Polaroid Film and Camera

1. Place plate with colonies or other visible bacterial growth under the camera, aligning the plate so that it is centered in the focal field (*see* **Note 6**).
2. With visible light illuminating the plate, take a photograph of the plate. Insert a piece of Polaroid type 57 film. Expose the film by pulling the protective barrier away from the film and opening the shutter. Exposure setting should be set to allow limited light ($f = 32$, $1/125$ s). Develop the film according to manufacturer's instructions.
3. Insert a piece of Polaroid type 57 film into the film holder. Darken the room.
4. Expose the film by pulling the protective barrier away from the film and opening the shutter. Settings for exposure should allow for maximum light to enter the lens ($f = 4.5$); exposure times will range from minutes to hours (*see* **Note 7**).
5. After closing the shutter to terminate exposure, develop film as usual (*see* **Note 8**).

3.3.2. Using High-Speed 35-mm Film and Camera

1. Load Kodak T-MAX P3200 35-mm film into a 35-mm camera (*see* **Note 9**).
2. Place plate with colonies or other visible bacterial growth under the camera, aligning the plate so that it is centered in the focal field (*see* **Note 6**).
3. Darken the room, and expose the film. Several different settings should be used. Adjust the *f*-stop on the camera to allow maximum light to the lens. Exposure times will vary between 1 and 10 min. Differences in lenses, distance, and brightness of colonies will affect the quality of the photograph.
4. Develop the film as per manufacturer's instructions using T-MAX Developer.

3.4. Results

Data collected by a scintillation counter are comparable to that collected by luminometers. Kinetics are revealed by plotting relative light units as a function of time. A comparison of the linear ranges of a luminometer and scintillation counter has been made following the methods of Burlage and Kuo *(8)*, the only difference being the range of linear response. **Figure 1** shows a photograph (panel A) as well as the exposed X-ray film image (panel B) of *E. coli* carrying a plasmid bearing promoter::*lux* fusions. The results on the X-ray film demonstrate a greater level of sensitivity than those on Polaroid film. Light

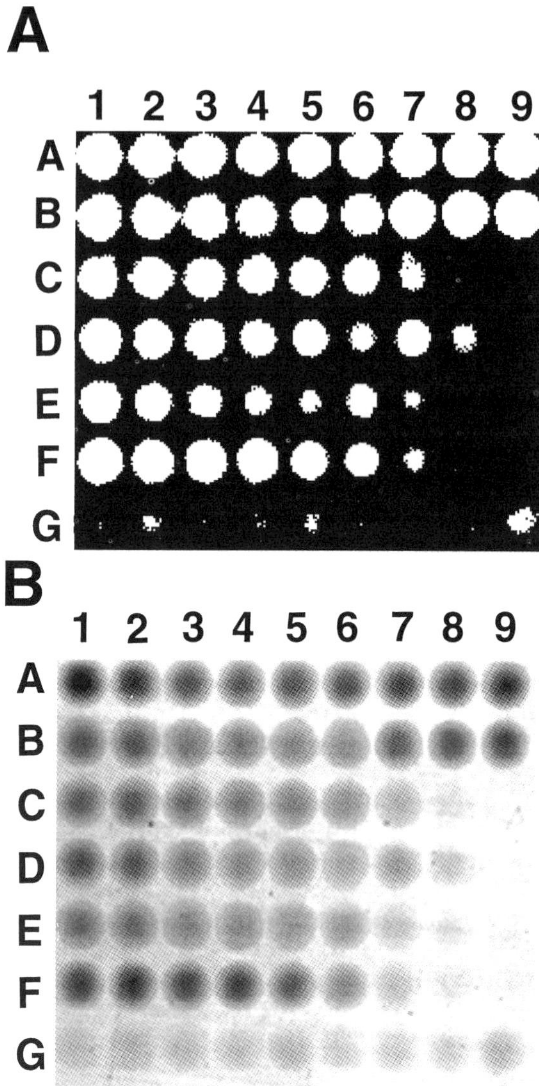

Fig. 1. All wells contained 50 µL of bacterial cultures, grown to midexponential phase (36 klett units) in LB. Rows A–F contained strain DPD 2794, *E. coli* carrying a plasmid bearing a *recA::lux* fusion. Rows A and B contained successive twofold dilutions of mitomycin C, starting with 1 µg/mL in column 1; column 9 contained no mitomycin C. Rows C and D contained successive twofold dilutions of CdCl$_2$ starting with 2 m*M* in column 1; column 9 contained no CdCl$_2$. Row E contained successive two-fold dilutions of ethidium bromide, starting with 1 mg/mL in column 1; column 9 contained no ethidium bromide. Row F contained successive twofold dilutions of

production is correlated with concentration. It is evident that the 30-s exposure of the X-ray film was too long to distinguish a dose-dependent *recA* response (**Fig. 1**, rows A, B). This is owing to the high consitutive expression of *recA* (in the absence of mitomycin C [**Fig. 1**, column 9]). **Figure 2** compares a Polaroid photograph of an agar plate with an X-ray image. The ring of light was produced by *E. coli* strain DPD2794, which carries a *recA* promoter fused to *luxCDABE* induced by mitomycin C *(9)*. A zone of growth inhibition is apparent in the photograph. The ring of light in the X-ray image emanates from cells growing just beyond the zone of inhibition. **Figure 3** compares a Polaroid photograph of a spread culture of *E. coli* DPD 2794 on an agar plate illuminated by room light with a Polaroid photograph of that plate taken in the dark. Once again, a clear zone of growth inhibition is apparent in the photograph. The circle of light is produced by cells just beyond the edges of that zone. **Figure 4** compares a Polaroid photograph of a streak culture on an agar plate of *E. coli* TV 1058 carrying a *lac::lux* plasmid *(10)* with a 35-mm photograph of that plate taken in the dark. Since O_2 is required for the production of light by bacterial luciferase, it is not surprising to *see* maximal light emitted by colonies that have less competition for O_2.

4. Notes

1. Colorless and transparent or nearly transparent vials or tubes should be used in order to allow maximum light to be detected. Use of color-tinted microcentrifuge tubes reduces sensitivity. Neutral colored microcentrifuge tubes may be purchase without attached caps. Alternatively, attached caps can be easily removed by cutting at the hinge area.
2. If exogenous aldehyde substrate needs to be introduced for *luxAB* assays, it is possible to pipet the substrate into the scintillation vial, outside of the microcentrifuge tube. If luciferin is to be added, it may be added directly into the 1.5-mL microcentrifuge sample tube.
3. It is important to bear in mind that the bacteria in the microcentrifuge tubes are not necessarily kept at constant temperature unless the chamber in which the samples are housed can be thermally regulated. Adequate mixing and agitation do occur when the sample racks are processed in the housing area.
4. If the microcentrifuge tubes are removed carefully and if no reagents have been added to the scintillation vials themselves, the vials can be immediately recycled for use.

(*Fig. 1, continued from previous page*) H_2O_2, starting with 0.0002%; column 9 contained no H_2O_2. Row G contained 50 µL of TV 1058, *E. coli* carrying a plasmid bearing a *lac::lux* fusion with no addition of any other chemicals. The Polaroid photograph (panel **A**) and DuPont Reflections film, exposed for 30 s in the dark (panel **B**) show corresponding levels of light produced. The film was developed using an automated film processor.

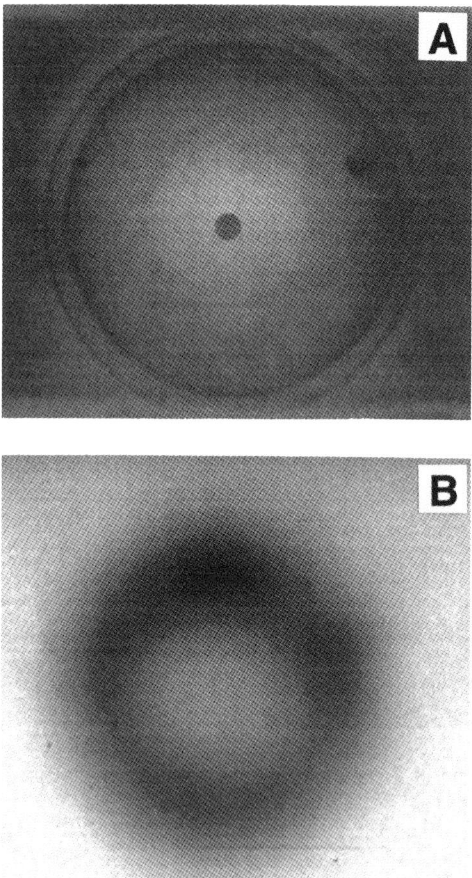

Fig. 2. An agar plate inoculated with DPD2794 *(recA::lux)*. A filter disk containing 10 μL of mitomycin C (2 mg/mL) was placed on the agar. The agar plate was incubated at 37°C overnight. Kodak XAR film was placed over the plate for 10 s in the dark and then developed. Panel **A** shows the image of a Polaroid photograph of the plate taken in room light. Panel **B** is an image of the developed XAR film.

5. It is also possible to place several pieces of film on top of the plates at once, developing each piece after intervals of exposure. In our hands, it is too easy to jar lower pieces of film or plates, if they are inadequately secured, resulting in a blurred image.

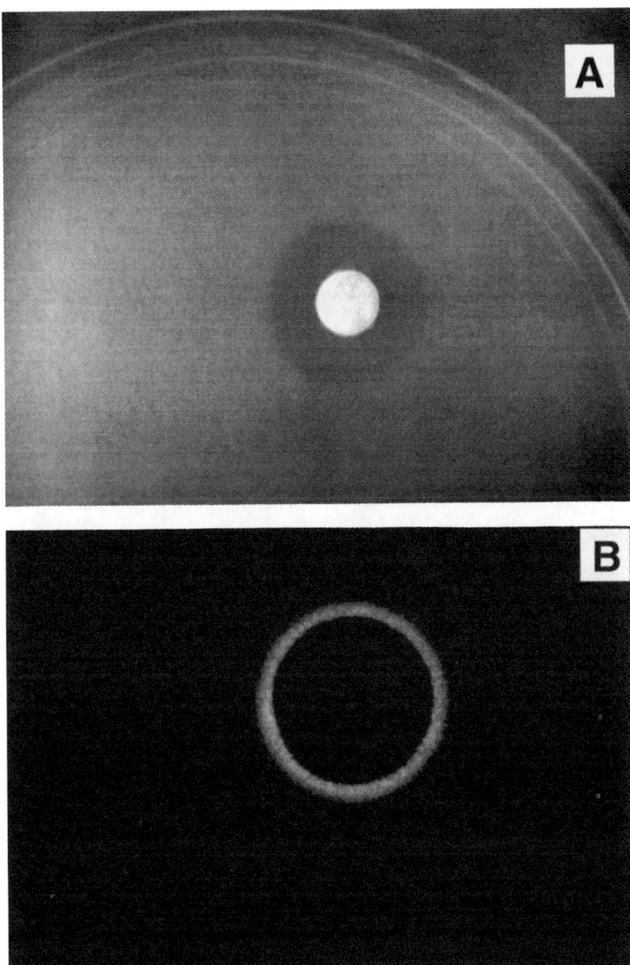

Fig. 3. DPD2794 *(recA::lux)* was inoculated on an LB agar. A filter disk containing 10 µL of mitomycin C (1 mg/mL) was placed on the agar. The agar plate was incubated at 37°C overnight. Panel **A** shows photograph of the plate taken in room light. Panel **B** is an image of Polaroid type 57 film developed following a 30-min exposure.

6. The camera should be mounted on a stand that rests on a vibration-resistant table.
7. Prolonged exposure will result in the chemical in the Polaroid packet, becoming dehydrated and ineffective. Be sure that there is no draft of air from a vent that is aimed at the camera. Humidity level in the darkroom should be moderate.

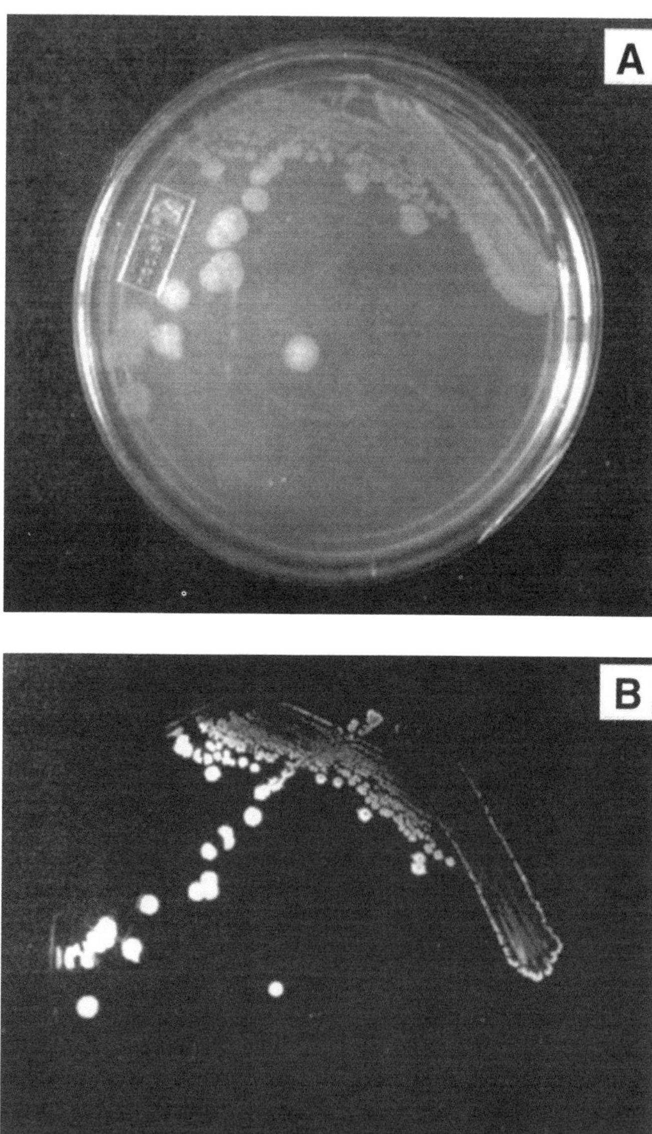

Fig. 4. TV 1058 *(lac::lux)* was inoculated by streaking on LB agar. Panel **A** shows an image of a Polaroid photograph of the plate taken in room light. Panel **B** is an image of the developed T-Max film taken during a 5-min exposure in the dark.

8. Take care to return to protective covering over the film packet and develop the film by evenly pulling with moderate speed. This ensures that the developing packet contents are distributed evenly over the surface of the film.

9. According to the manufacturer, this film is "multispeed panchromatic film with very high to ultra high speed and finer grain than other fast films."

References

1. Van Dyk, T. K., Belkin, S., Vollmer, A. C., Smulski, D. R., Reed, T. R., and LaRossa, R. A. (1994) Fusions of *Vibrio fischeri lux* genes to *Escherichia coli* stress promoters: Detection of environmental stress, in *Bioluminescence and Chemiluminescence: Fundamentals and Applied Aspects* (Campbell, A. K., Kricka, L. J., and Stanley, P. E., eds.), John, Chichester, UK, pp. 147–150.

2. Belkin, S., Vollmer, A. C., Van Dyk, T. K., Smulski, D. R., Reed, T. R., and LaRossa, R. A. (1994) Oxidative and DNA damaging agents induce luminescence in *E. coli* harboring *lux* fusions to stress promoters, in *Bioluminescence and Chemiluminescence: Fundamentals and Applied Aspects* (Campbell, A. K., Kricka, L. J., and Stanley, P. E., eds.), John, Chichester, UK, pp. 509–512.

3. Virta, M., Lampinen, J., and Karp, M. (1995) A luminescence-based mercury biosensor. *Anal. Chem.* **67,** 667–669.

4. Dunlap, P. (1993) Genetic analysis of circadian clocks. *Annu. Rev. Physiol.* **55,** 683–728.

5. Kondo, T., Straer, C. A., Kulkari, R., Taylor, W., Ishiura, M., Golden, S., and Johnson, C. (1993) Circadian rhythms in prokaryotes: luciferase as a reporter of circadian gene expression. *Proc. Natl. Acad. Sci. USA* **90,** 5672–5676.

6. Millar, A. J., Straume, M., Chory, J., Chua, N. -H., and Kay, S. (1995) The regulation of circadian period by phototransduction pathway in Arabidopsis. *Science* **267,** 1163–1166.

7. Brandes, C., Plautz, J. D., Stanewsky, R., Jamison, C. F., Straume, M., Wood, K. V., Kay, S., and Hall, J. C. (1996) Novel features of Drosophila period transcription revealed by real-time luciferase reporting. *Cell* **16,** 687–692.

8. Burlage, R. S. and Kuo, C. -T. (1994) Living biosensors for the management and manipulation of microbial consortia. *Annu. Rev. Microbiol.* **48,** 291–309.

9. Vollmer, A. C., Belkin, S., Smulski, D. R., Van Dyk, T. K., and LaRossa, R. A. (1997) Detection of DNA damage by use of *Escherichia coli* carrying *recA::lux*, *uvrA'::lux*, or *alkA'::lux* reporter plasmids. *Appl. Environ. Microbiol.* **63(7),** 2566–2571.

10. Van Dyk, T. K., Majarian, W. R., Konstantinov, K. B., Young, R. M., Dhurjati, P. S., and LaRossa, R. A. (1994) Rapid and sensitive pollutant detection by induction of heat shock gene-bioluminescence gene fusions. *Appl. Environ. Microbiol.* **60,** 1414–1420.

3

Microscopic Imagery of Mammalian Cells Expressing an Enhanced Green Fluorescent Protein Gene

Steven R. Kain, Guohong Zhang,
Vanessa Gurtu, and Paul A. Kitts

1. Introduction

The green fluorescent protein (GFP; *1–5*) from the jellyfish *Aequorea victoria* has emerged as an important reporter for monitoring gene expression, protein localization, cell transformation, and cell lineage in vivo and in real time. Unlike other bioluminescent reporters, the chromophore in GFP is intrinsic to the primary structure of the protein, and GFP does not require additional factors other than molecular oxygen (*see* **Note 2**) to fluoresce *(6, 7)*. GFP emits bright green light (λ_{max} = 510 nm) when excited with ultraviolet (UV) or blue light (λ_{max} = 395 nm, minor peak at 470 nm). Full-length GFP (238 amino acids; 27 kDa) appears to be required for fluorescence. However, the minimal chromophore responsible for light absorption consists of a Ser65-dehydroTyr66-Gly67 cyclic tripeptide, which is postulated to be buried inside the folded protein *(6)*. GFP fluorescence is stable (*see* **Note 5**), species-independent, and can be monitored noninvasively in living cells by either fluorescence microscopy, flow cytometry, or macroscopic imaging techniques. GFP has been used as a reporter in a wide range of species, including a number of different mammalian cell lines (**Table 1**). Moreover, a variety of N- and C-terminal protein fusions with GFP have been constructed, and shown to maintain both the fluorescence properties of native GFP and the biological function of the fusion partner *(5,8–12)*.

Wild-type GFP has several undesirable properties, including low fluorescent intensity when excited by blue light (*see* **Note 7**), a lag in the development of fluorescence after protein synthesis (*see* **Note 9**), and poor expression in

From: *Methods in Molecular Biology, Vol. 102: Bioluminescence Methods and Protocols*
Edited by: R. A. LaRossa © Humana Press Inc., Totowa, NJ

Table 1
Mammalian Cell Lines Successfully Used to Express GFP

Cell line	Cell type	Reference
293	Transformed primary embryonic kidney, human	*12, 29*
BHK-21	Hamster	*10*
CHO-K1	Ovary, Chinese hamster	*5, 10, 30*
COS-7	Kidney, SV40 transformed, African green monkey	*10, 25*
GH3	Pituitary tumor, rat	*31*
HeLa	Epitheloid carcinoma, cervix, human	*5, 11, 25*
JEG	Placenta, human	*32*
NIH/3T3	Embryo, contact-inhibited, NIH Swiss mouse	*5, 10, 25*
Pt K1	Kidney, kangaroo rat	*10*

several mammalian cell types *(4,7,13,14)*. To improve on these qualities, we have constructed the vector pEGFP-C1 (**Fig. 1**; CLONTECH Laboratories, Palo Alto, CA), which encodes a variant GFP protein previously described as GFPmut1 *(15)*. This variant contains two point mutations in the GFP chromophore: Ser65 to Thr and Phe64 to Leu. The GFPmut1 variant generates approx 35-fold brighter fluorescence relative to wild-type GFP when excited by blue light, has improved solubility, and more efficient protein folding characteristics *(15)*. GFPmut1 has a single major peak of excitation at 490 nm, making this variant more suitable than wild-type GFP for detection using fluorescein filter sets. The *egfp* gene contains more than 190 silent base mutations, which create an open reading frame composed almost entirely of preferred human codons *(13,14)*. These changes allow mammalian cells to more efficiently translate the *egfp* mRNA, thereby increasing expression of the protein. pEGFP-C1 contains the Kozak consensus sequence to increase translation efficiency in eukaryotic cells *(16)*, a neomycin/kanamycin resistance cassette, and the immediate early promoter of cytomegalovirus (CMV) for constitutive expression in mammalian cells. The vector also contains a multiple cloning site for fusion of heterologous proteins to the C-terminus of EGFP (*see* **Note 4**).

2. Materials

2.1. Tissue Culture and Transformations

1. Tissue-culture hood and CO_2 incubator.
2. 35-mm tissue-culture plates.
3. Glass cover slips and glass microscope slides.
4. Sterile plastic pipets.
5. Culture medium.
6. pEGFP-C1 vector (CLONTECH).
7. 12 × 75 mm Sterile tubes.

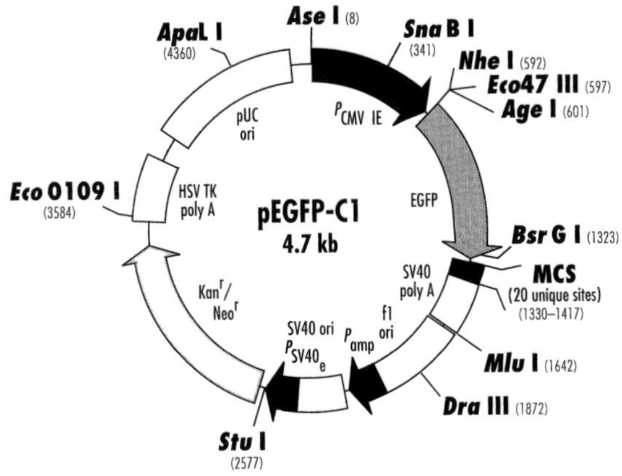

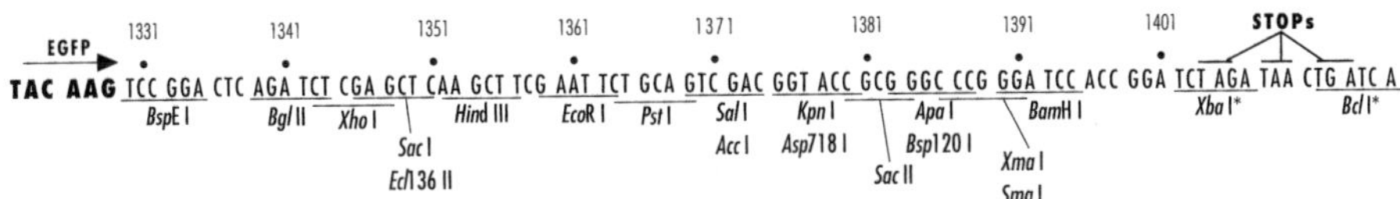

Fig. 1. Map of pEGFP-C1 and multiple cloning site (MCS). The vector pEGFP-C1 (CLONTECH Laboratories, Palo Alto, CA) contains the immediate early promoter of human CMV ($P_{CMV\ IE}$) and SV40 polyadenylation signals to drive expression of the *egfp* gene in mammalian cells. This vector contains a multiple cloning site (MCS) on the 3'-end of the *egfp* gene, and can be used to create in-frame fusions to the C-terminus of EGFP. The *egfp* gene of pEGFP-C1 encodes a variant chromophore sequence, and has been codon-optimized for maximal expression and fluorescence intensity in mammalian cells.

8. 2 *M* Calcium solution: Dissolve anhydrous $CaCl_2$ in H_2O. Store at 4°C.
9. 2X HBS: 0.05 *M* HEPES, 0.28 *M* NaCl, 1.5 m*M* Na_2HPO_4, pH 7.08 ± 0.02.

2.2. Detection of EGFP Fluorescence

1. Fluorescence microscope equipped with a fluorescein, or GFP filter set (*see* **Note 8**).
2. Cell fixative: 2% formalin, 0.05% glutaraldehyde, in 1X PBS, pH 7.4. Store at 4°C. The solution can be prepared ahead of time and used without warming.

3. Methods

3.1. Preparation of Cell Culture

BHK-21 cells (ATCC, Rockville, MD) are routinely cultured in 75-mL flasks in DMEM medium supplemented with 10% fetal bovine serum (FBS). Medium for all cultures routinely includes 100 U/mL of penicillin and 100 µg/mL of streptomycin. All media and serum, and other tissue-culture supplements can be purchased from Life Technologies (Gaithersburg, MD). Cultures were maintained at 37°C with 5% CO_2/95% air (*see* **Note 6**).

3.2. Preparation of Tissue-Culture Plate with Glass Cover Slip

1. Working in a tissue-culture hood, flame-sterilize a glass cover slip (22–25 mm^2) dipped in 95% ethanol. Be sure that the cover slip is not distorted and maintains a planar geometry.
2. Place one sterile cover slip per one 35-mm tissue-culture dish.

3.3. Transformations with pEGFP-C1

1. Working in a tissue-culture hood, plate the cells the day before the transformation experiment. The cells should be 50–80% confluent the day of transformation. We routinely plate 2–4×10^5 cells onto glass cover slips in 35-mm plates (*see* **Note 10**).
2. 0.5–3 h prior to transformation, replace culture medium on plates to be transformed with 2 mL of fresh culture medium/35-mm plate.
3. For each transformation, prepare solution A and solution B in separate sterile tubes (**Fig. 2**):
   Solution A: add components in the following order:
       2–4 µg Plasmid DNA (pEGFP-C1).
       Sterile H_2O.
       12 µL 2 *M* calcium solution.
       100 µL/total volume.
   Solution B: 100 µL 2X HBS.
   Note: To reduce variability when transforming multiple plates with the same plasmid DNA, prepare master solutions A and B sufficient for all plates.
4. Carefully and slowly vortex solution B while adding solution A dropwise. (Alternatively, blow bubbles into solution B with a 1–mL sterile pipet and an autopipeter while adding solution A dropwise.)
5. Incubate the transformation solution at room temperature for 5–20 min.
6. Briefly vortex the transformation solution, and then add solution dropwise to culture plate medium. (Add 200 µL of transformation solution per 35-mm plate.)
7. Gently move plates back and forth to evenly distribute transformation solution. Avoid circular motions with the plate, since this action may concentrate the transformation solution in the center of the plate.
8. Incubate plates at 37°C for 2–6 h in a CO_2 incubator.
9. Remove calcium phosphate-containing medium, and wash cells twice with medium, or 1X PBS.
10. Feed plate with 2 mL fresh complete growth medium and incubate at 37°C for 24–72 h.

3.4. Preparation of pEGFP-C1-Transformed Cells for Microscopy

1. Working on the lab bench (sterile conditions no longer needed), aspirate medium from dish.

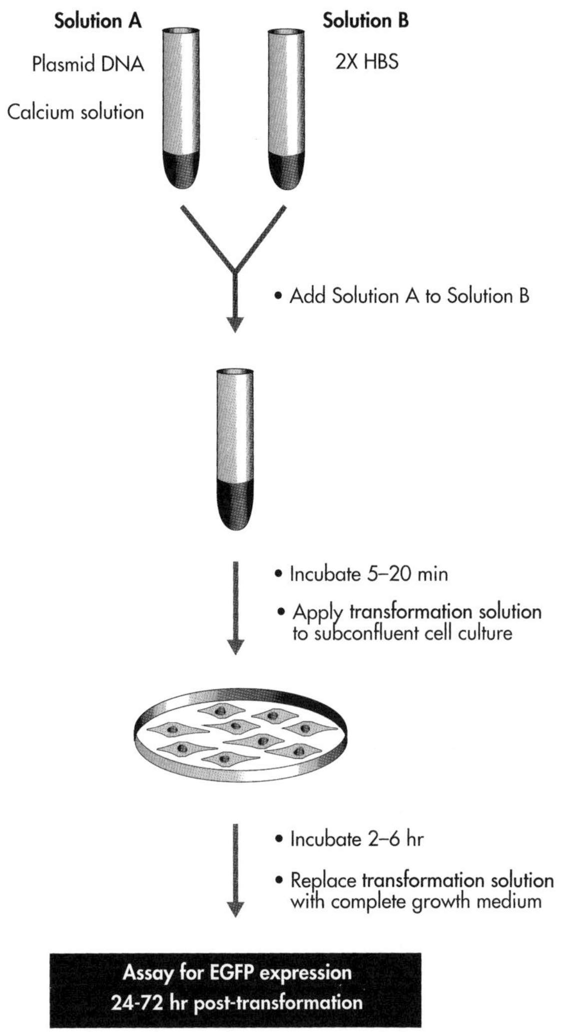

Fig. 2. Flowchart for pEGFP-C1 transformations.

2. Wash the cells twice with 2 mL of 1X PBS.
3. Fix with 2 mL cell fixative for 5 min at room temperature.
4. Wash the cells twice with 2 mL of 1X PBS.
5. Mount the cover slip cell side down in 1X PBS on glass microscope slide.
6. Blot excess PBS with a Kimwipe.
7. Seal around all four sides of cover slip with rubber cement (*see* **Note 3**).
8. Allow rubber cement to air-dry.
   Note: Air-tight-sealed slide preparations may be stored at 4°C for several weeks with no loss in GFP fluorescence.

3.5. Fluorescence Microscopy and Photography

1. View GFP-expressing cells with a Zeiss Axiolab Microscope (or equivalent) equipped with a fluorescein or GFP filter set (*see* **Note 1**). We have had good success with a GFP/FITC/PI set (Chroma Technology, cat no. CZ909).
2. Photograph GFP-expressing cells with Kodak Ektochrome Elite 400 35-mm slide film. Typical exposure times range from 4–60 s. Depending on the number and relative intensity of the fluorescing cells, use shorter exposure times for viewing fields with many, brightly fluorescing cells.

3.6. Expected Results

The CalPhos Maximizer reagent increases transformation efficiencies in a wide variety of mammalian cell types, with typical frequencies in the range of 30–70%. By following the above protocol, with the use of appropriate microscope filter sets, expression of EGFP should produce very bright green fluorescence in 24–72 h. Since EGFP is a cytoplasmic protein, the fluorescence signal should be evenly distributed throughout the cell (this includes the nuclei, since EGFP is small enough to passively transcend the nuclear pores). Background autofluorescence (*see* **Note 10**) in most cases is minimal.

4. Notes

1. Photobleaching: The fluorescence of GFP is quite stable when illuminated with 450–490 nm light. GFP is more resistant to photobleaching than is fluorescein *(8,17)*. The rate of photobleaching is less with lower-energy lamps, such as QTH or mercury lamps. High-energy xenon lamps should be avoided, since these may cause rapid photodestruction of the GFP chromophore.
2. Stability to oxidation/reduction: GFP needs to be in an oxidized state to fluoresce, since chromophore formation is dependent on an oxidation of Tyr66 *(7)*. Strong reducing agents, such as 5 mM $Na_2S_2O_4$ or 2 mM $FeSO_4$, convert GFP into a nonfluorescent form, but fluorescence is fully recovered after exposure to atmospheric oxygen *(18)*. Weaker reducing agents, such as 2% β-mercaptoethanol, 10 mM dithiothreitol (DTT), 10 mM reduced glutathione, or 10 mM L-cysteine, do not affect the fluorescence of GFP *(18)*. GFP fluorescence is not affected by moderate oxidizing agents.
3. Stability to chemical reagents: GFP fluorescence is retained in mild denaturants, such as 1% SDS or 8 M urea, and after fixation with glutaraldehyde, paraformaldehyde, or formalin, but fully denatured GFP is not fluorescent. GFP is very sensitive to some nail polishes used to seal cover slips *(1,8)*; therefore, use molten agarose or rubber cement to seal cover slips on microscope slides. GFP fluorescence is irreversibly destroyed by 1% H_2O_2 and sulfhydryl reagents, such as 1 mM 5,5'-dithio-bis (2-nitrobenzoic acid) (DTNB) *(18)*. Fluorescence is retained in the range of pH 7.0–12.0, but intensity decreases at pH 5.5–7.0 *(19)*. Many organic solvents can be used at moderate concentrations without abolishing fluorescence; however, the absorption maximum may shift *(20)*.

4. GFP dimerizes via hydrophobic interactions at protein concentrations above 5–10 mg/mL and high salt concentrations with a four fold reduction in the absorption at 470 nm *(4)*. This phenomenon is not observed with EGFP and other red-shifted GFP variants lacking a 395-nm peak of excitation *(4)*. Dimer formation is not required for fluorescence, and monomeric GFP is the form of the reporter expressed in most model systems.

5. Protein stability: GFP is exceptionally resistant to heat ($T_m = 70°C$), alkaline pH, detergents, chaotropic salts, organic solvents, and most common proteases, except pronase *(19–22)*. Fluorescence is lost if GFP is denatured by high temperature, extremes of pH, or guanidinium chloride, but can be partially recovered if the protein is allowed to renature *(19,23)*. A thiol compound may be necessary to renature the protein into the fluorescent form *(24)*.

6. Temperature sensitivity of GFP chromophore formation: Mammalian cells expressing GFP have been reported to exhibit stronger fluorescence when grown at 30–33°C compared to 37°C *(25,26)*.

7. Sensitivity: GFP, like fluorescein, has a quantum yield of about 80% *(21)*, although the extinction coefficient for GFP is much lower. Nevertheless, in fluorescence microscopy, GFP fusion proteins have been found to give greater sensitivity and resolution than staining with fluorescently labeled antibody *(8)*. GFP fusions have the advantages of being more resistant to photobleaching and of avoiding background caused by nonspecific binding of the primary and secondary antibodies to targets other than the antigen *(8)*. Although binding of multiple antibody molecules to a single target offers a potential amplification not available for GFP, this is offset because neither labeling of the antibody nor binding of the antibody to the target is 100% efficient. The EGFP chromophore variant of GFP significantly increases the sensitivity of GFP as a reporter. However, for some applications, the sensitivity of GFP may be limited by autofluorescence or limited penetration of light. Recent studies with wt GFP expressed in HeLa cells *(17)* have shown that the cytoplasmic concentration must be >~1.0 μM to discriminate signals over autofluorescence. This threshold for detection is likely to be lower with the EGFP variant, which provides enhanced fluorescent intensities.

8. Filter sets for fluorescence microscopy: Chroma Technology (Brattleboro, VT) has developed several filter sets designed for use with GFP; they claim the High Q FITC filter set (#41001) produces the best signal-to-noise ratio for visual work, and the High Q GFP set (#41014) produces the strongest absolute signal, but with some background. We have also used a Zeiss filter set (#487909) with a 450–490 nm bandpass excitation filter, 510-nm dichroic reflector, and 520–750 nm long-pass emission filter, and the Chroma filter set #31001. The best results with mammalian cells were obtained using a GFP/FITC/PI set (#CZ909). Other filter sets may give better performance, and it is necessary to match the filter set to the application.

9. The slow rate of chromophore formation and the apparent stability of GFP may preclude the use of GFP as a reporter to monitor fast changes in promoter activity *(7)*. This limitation is reduced by use of EGFP, which acquires fluorescence faster than wild-type GFP *(15)*.

10. Autofluorescence: Some samples may have a significant background auto-fluorescence, e.g., worm guts *(1,17)*. A bandpass emission filter may make the autofluorescence appear the same color as GFP; using a long-pass emission filter may allow the color of the GFP and autofluorescence to be distinguished. Use of DAPI filters may also allow autofluorescence to be distinguished *(25,27)*. Most autofluorescence in mammalian cells is owing to flavin coenzymes (FAD and FMN; *28*), which have absorption/emission = 450/515 nm. These values are very similar to those for GFP, so autofluorescence may obscure the GFP signal. The use of DAPI filters may make this autofluorescence appear blue, while the GFP signal remains green. In addition, some growth media can cause autofluorescence. When possible, perform microscopy in a clear buffer, such as PBS, or medium lacking phenol red. For mammalian cells, autofluorescence can increase with time in culture. For example, when CHO or SCI cells were removed from frozen stocks and reintroduced into culture, the observed autofluorescence (emission at 520 nm) increased with time until a plateau was reached around 48 h *(28)*. There-fore, in some cases, it may be preferable to work with freshly plated cells. For fixed cells, autofluorescence can be reduced by washing with 0.1% sodium boro-hydride in PBS for 30 min after fixation.

References

1. Chalfie, M., Tu, Y., Euskirchen, G., Ward, W. W., and Prasher, D. C. (1994) Green fluorescent protein as a marker for gene expression. *Science* **263,** 802–805.
2. Chalfie, M. (1995) Green fluorescent protein. *Photochem. Photobiol.* **62(4),** 651–656.
3. Prasher, D. C. (1995) Using GFP to see the light. *Trends Genet.* **11,** 320–323.
4. Cubitt, A. B., Heim, R., Adams, S. R., Boyd, A. E., Gross, L. A., and Tsien, R. Y. (1995) Understanding, improving and using green fluorescent proteins. *Trends Biochem.* **20,** 448–455.
5. Kain, S. R., Adams, M., Kondepudi, A., Yang, T. T., Ward, W. W., and Kitts, P. (1995) The green fluorescent protein as a reporter of gene expression and protein localization. *BioTechniques* **19,** 650–655.
6. Cody, C. W., Prasher, D. C., Westler, W. M., Prendergast, F. G., and Ward, W. W. (1993) Chemical structure of the hexapeptide chromophore of *Aequorea* green-fluorescent protein. *Biochemistry* **32,** 1212–1218.
7. Heim, R., Prasher, D. C., and Tsien, R. Y. (1994) Wavelength mutations and post-translational autoxidation of green fluorescent protein. *Proc. Natl. Acad. Sci. USA* **91,** 12,501–12,504.
8. Wang, S. and Hazelrigg, T. (1994) Implications for bcd mRNA localization from spatial distribution of exu protein in *Drosophila* oogenesis. *Nature* **369,** 400–403.
9. Flach, J., Bossie, M., Vogel, J., Corbett, A., Jinks, T., Willins, D. A., and Silver, P. A. (1994) A yeast RNA-binding protein shuttles between the nucleus and the cytoplasm. *Mol. Cell. Biol.* **14,** 8399–8407.
10. Olson, K. R., McIntosh, J. R., and Olmsted, J. B. (1995) Analysis of MAP4 func-tion in living cells using green fluorescent protein (GFP) chimeras. *J. Cell. Biol.* **130,** 639–650.

11. Kaether, C. and Gerdes, H.-H. (1995) Visualization of protein transport along the secretory pathway using green fluorescent protein. *FEBS Lett.* **369,** 267–271.

12. Marshall, J., Molloy, R., Moss, G. W. J., Howe, J. R., and Hughes, T. E. (1995) The jellyfish green fluorescent protein, a new tool for studying ion channel expression and function. *Neuron* **14,** 211–215.

13. Chiu, W., Niwa, Y., Zeng, W., Hirano, T., Kobayashi, H., and Sheen, J. (1996) *Curr. Biol.* **6,** 325–330.

14. Haas, J., Park, E.-C., and Seed, B. (1996) *Curr. Biol.* **6,** 315–324.

15. Cormack, B. P., Valdivia, R., and Falkow, S. (1996) FACS-optimized mutants of the green fluorescent protein (GFP). *Gene* **173,** 33–38.

16. Kozak, M. (1987) An analysis of 5'-noncoding sequences from 699 vertebrate messenger RNAs. *Nucleic Acids Res.* **15,** 8125–8148.

17. Niswender, K. D., Blackman, S. M., Rohde, L., Magnuson, M. A., and Piston, D. W. (1995) Quantitative imaging of green fluorescent protein in cultured cells, comparison of microscopic techniques, use in fusion proteins and detection limits. *J. Microsc.* **180(2),** 109–116.

18. Inouye, S. and Tsuji, F. I. (1994) Evidence for redox forms of the *Aequorea* green fluorescent protein. *FEBS Lett.* **351,** 211–214.

19. Bokman, S. H. and Ward, W. W. (1981) Renaturation of *Aequorea* green-fluorescent protein. *Biochem. Biophys. Res. Commun.* **101,** 1372–1380.

20. Robart, F. D. and Ward, W. W. (1990) Solvent perturbations of *Aequorea* green fluorescent protein. *Photochem. Photobiol.* **51,** 92s.

21. Ward, W. W. (1981) Properties of the Coelenterate green-fluorescent proteins, in *Bioluminescence and Chemiluminescence, Basic Chemistry and Analytical applications* (DeLuca, M. and McElroy, W. D., eds.), Academic, New York, pp. 235–242.

22. Roth, A. (1985) Purification and protease susceptibility of the green-fluorescent protein of *Aequorea victoria* with a note on Halistra *ura*. Ph.D. thesis, Rutgers University, New Brunswick, NJ.

23. Ward, W. W. and Bokman, S. H. (1982) Reversible denaturation of *Aequorea* green-fluorescent protein, physical separation and characterization of the renatured protein. *Biochemistry* **21,** 4535–4540.

24. Surpin, M. A. and Ward, W. W. (1989) Reversible denaturation of *Aequorea* green fluorescent protein—thiol requirement. *Photochem. Photobiol.* **49,** WPM-B2

25. Pines, J. (1995) GFP in mammalian cells. *Trends Genet.* **11,** 326,327.

26. Ogawa, H., Inouye, S., Tsuji, F. I., Yasuda, K., and Umesono, K. (1995) Localization, trafficking, and temperature-dependence of the *Aequorea* green fluorescent protein in cultured vertebrate cells. *Proc. Natl. Acad. Sci. USA* **92,** 11,899–11,903.

27. Brand, A. (1995) GFP in *Drosophila. Trends Genet.* **11,** 324,325.

28. Aubin, J. E., (1979) Autofluorescence of viable cultured mammalian cells. *J. Histochem. Cytochem.* **27,** 36–43.

29. Cheng, L., Fu, J., Tsukamoto, A., and Hawley, R. G. (1996) Use of green fluorescent protein (GFP) variants to monitor gene transfer and expression in mammalian cells. *Nature Biotechnol.* **14,** 606–609.

30. Yang, T. T., Kain, S. R., Kitts, P., Kondepudi, A., Yang, M. M., and Youvan, D. C. (1996) Dual color microscopic imagery of cells expressing the green fluorescent protein and a red-shifted variant. *Gene* **173,** 19–23.
31. Plautz, J. D., Day, R. N., Dailey, G., Welsh, S. B., Hall, J. C., Halpain, S., and Kay, S. A. (1996) Green fluorescent protein and its derivatives as a versatile marker for gene expression in living Drosophila, plant and mammalian cells. *Gene,* in press.
32. Yu, K. L. and Dong, K. W. (1995) Application of luciferase and green fluorescent protein as a reporter for analysis of human gonadotropin-releasing hormone gene promoters. Poster presentation at Fluorescent Proteins and Applications Meeting, Palo Alto, CA.

II

ANALYTICAL BIOCHEMISTRY

4

Luminometric Measurement of Malate and Glucose-6-Phosphate in Mammalian Tissue

Eberhard Jüngling, Helmut Kammermeier, and Yvan Fischer

1. Introduction

When using cells of mammalian origin, the amount of biological material available for analytical purposes is often limited (e.g., ~10^5 cells/sample, corresponding to few mg tissue). Since most intermediary metabolites are found in a concentration range of ~50–500 µmol/g in a wide variety of tissues (*see, e.g.,* **ref.** *1*), only nanomolar or subnanomolar quantities can be expected to be present in small samples obtained from isolated cells or cell cultures. Such quantities are at best barely detectable by conventional spectrophotometric or fluorimetric procedures.

Over the past two decades, luminometric assays for several metabolites such as, for instance, NADH *(2)*, NADP$^+$ *(3)*, pyruvate *(4)*, or malate *(5)*, have been described. However, in most cases, only measurements of aqueous standards (e.g., pure compounds in buffers) were presented, while largely ignoring analysis of metabolites from a biological matrix, the components of which could largely interfere with the assay, especially ones with an enzymatic basis *(6)*. We have therefore developed an analytical procedure allowing the measurement of subnanomolar amounts of metabolites (malate, glucose-6-phosphate) in particularly complex biological material, e.g., in extracts from isolated heart muscle cells.

Basically, measurements occur in two steps: (1) Enzymatically catalyzed reactions involving the metabolite to be measured lead to the stoichiometric production of NAD(P)H; (2) the oxidation of this NAD(P)H in a luciferase/reductase system results in light emission that is proportional to the original concentration of the metabolite. The reaction scheme is thus as follows:

From: *Methods in Molecular Biology, Vol. 102: Bioluminescence Methods and Protocols*
Edited by: R. A. LaRossa © Humana Press Inc., Totowa, NJ

1. Metabolite (malate, glucose-6-phosphate) + NAD(P)$^+$ $\longrightarrow$ X + NAD(P)H + H$^+$.
2. NAD(P)H + O$_2$ + RCOH $\longrightarrow$ NAD(P)$^+$ + RCOOH + H$_2$O + hv.

The cells used are previously subjected to an ethanolic extraction in which the cellular NAD(P)H is destroyed by acidification. Subsequent evaporation of the extracts both neutralizes and concentrates the samples. This contributes, along with other experimental maneuvers, to increasing the sensitivity of the method. With this procedure, as little as ~70 pmol of malate and ~90 pmol of glucose-6-phosphate can be detected, e.g., in cardiomyocyte samples. In principle, the procedure described can be applied to the measurement of any ethanol-extractable metabolite that can be converted in reactions involving NAD(P)$^+$.

2. Materials
2.1. Equipment

1. Sonifier (Branson, model B-12 with microtip).
2. Metal heating block with N$_2$-gassing system (for 2-mL vials).
3. Luminometric system, including compatible assay vials (Lumat 9501, Berthold, Wildbad, FRG; *see* **Note 1**).

2.2. Solutions and Buffers

Water used for all buffers and solutions should be freshly deionized with a Milli-Q Reagent Water System equipped with a bacterial filter (Millipore).

2.2.1. Cell Incubation

The following are needed for cell incubation: 6 mM KCl, 1 mM Na$_2$HPO$_4$, 0.2 mM NaH$_2$PO$_4$, 1.4 mM MgSO$_4$, 128 mM NaCl, 10 mM HEPES, 1 mM CaCl$_2$, and 2% bovine serum albumin, fatty acid free, pH 7.4, 37°C.

2.2.2. Extraction and Acidification

1. 70% Ethanol (4°C).
2. 100% Formic acid.

2.2.3. Malate Assay

1. 0.1 M phosphate buffer, pH 10.0 (*see* **Note 2**).
2. 0.1 M phosphate buffer, pH 7.6.
3. 60 mM NAD$^+$ in phosphate buffer, pH 10.0.
4. 2.5 mM acetylCoA in phosphate buffer, pH 10.0.
5. 100 U/mL citrate synthase (EC 4.1.3.7, from pig heart, Boehringer, Manheim, FRG; *see* **Note 3**) in phosphate buffer, pH 10.0.
6. 3000 U/mL L-malate dehydrogenase (EC 1.1.1.37, Boehringer, Manheim, *see* **Note 3**) in phosphate buffer, pH 10.0.

7. 12.5 mM 1,4–dithiothreitol in phosphate buffer, pH 7.6.
8. 0.02 mM flavin mononucleotide (FMN, Boehringer, Manheim) in phosphate buffer, pH 7.6 (*see* **Note 4**).
9. 42.5% Phosphoric acid.
10. 0.5% Tetradecanal (Merck, Darmstadt, FRG) emulsion: The tetradecanal emulsion is freshly prepared as follows: 100 mg α-cyclodextrin is dissolved in 5 mL water (previously degassed under vacuum; *see* **Note 7**), then gassed with nitrogen, and heated to 60°C; 25 mg tetradecanal are added, the mixture rapidly vortexed, and the emulsion obtained is kept under a stream of nitrogen (*see* **Note 8**).
11. 0.5 mg Luciferase (EC 1.14.14.3, Sigma, Munich, FRG; *see* **Note 5**) in 0.1 M phosphate buffer, pH 7.6.
12. 60 µM malate standard in phosphate buffer, pH 10.0.
13. Immediately before performing the assay, the following reaction mixture is prepared (the quantities given here are sufficient for at least 20 individual measurements):

Mixture A:

Phosphate buffer (pH 10.0)	8750 µL
NAD$^+$	375 µL
AcetylCoA	375 µL
Malate dehydrogenase	125 µL
Citrate synthase	125 µL
Dithiothreitol	125 µL
FMN	250 µL

2.2.4. Glucose-6-Phosphate Assay

1. 0.1 M phosphate buffer, pH 7.6 (*see* **Note 2**).
2. 27 mM NADP$^+$ in phosphate buffer, pH 7.6.
3. 350 U/mL glucose-6-phosphate dehydrogenase (EC 1.1.1.49, from yeast, Boehringer, Manheim) in phosphate buffer, pH 7.6.
4. 12.5 mM 1,4–dithiothreitol in phosphate buffer, pH 7.6.
5. 0.02 mM flavin mononucleotide (FMN) in phosphate buffer, pH 7.6 (*see* **Note 4**).
6. 0.5% Tetradecanal emulsion (*see* **Subheading 2.2.3.**, **item 10**).
7. 240 µM glucose-6-phosphate standard in phosphate buffer (pH 7.6).

Immediately before performing the assay, the following reaction mixture is prepared (the quantities given here are sufficient for 20 individual measurements):

Mixture B:

Phosphate buffer (pH 7.6)	8750 µL
NADP$^+$	375 µL
Glucose-6-phosphate dehydrogenase	125 µL
Dithiothreitol	125 µL
FMN	250 µL

3. Methods

3.1. Treatment of Cells

1. Cardiomyocytes *(7)* from adult Sprague-Dawley rats (180–220 g, fed ad libitum) are the source of the metabolites. The isolated cells are incubated in a shaking water bath at a density of ~2 × 10^5 cells/mL (~1.5 mg protein/mL; *see* **Note 9**) at 37°C in the buffer described under **Subheading 2.3.1.**
2. At the time-points at which the metabolite content of the cells is to be determined, aliquots of the cell suspension (containing ~1.5 × 10^5 cells; *see* **Note 10**) are rapidly spun down (45 s, 14g), washed once with 0.9% NaCl, and centrifuged again.
3. The pellet of this centrifugation is dissolved in 400 µL of cold ethanol (4°C , 70%) and homogenized for 15 s with a sonifier (Branson, 50 W; *see* **Note 11**).
4. The ethanolic samples can be kept on ice and further processed for the measurements within 2–4 h (*see* **Note 12**).
5. For calibration with internal standards, 10, 20, or 40 µL malate or glucose-6-phosphate standard (*see* **Subheadings 2.2.3.** and **2.2.4.**) are added to a parallel series of ethanolic samples from control cells and subjected to the whole extraction procedure (beginning with the sonifier treatment).

3.2. Acidic Extraction (*see* Note 13)

1. The ethanolic samples obtained as described under **Subheading 3.1.** are acidified by addition of 15 µL concentrated formic acid.
2. They are then centrifuged for 10 min at 10,000g.
3. Subsequently, the supernatants of this centrifugation are evaporated at 70°C in a heating block under a stream of nitrogen gas (*see* **Note 14**).
4. The dry material obtained can be stored, if necessary, at –20°C (stable for several months).

3.3. Malate Measurement

3.3.1. Principle

The measurement (*see* **Fig. 1**) requires conversion of malate to oxaloacetate, which also reduces NAD$^+$ to NADH. Since the reaction is not energetically favored, removal of oxaloacetate by coupling to citrate synthase is required to drive the reaction to completion. Subsequently, the protons produced in the malate dehydrogenase reaction are used to form FMNH$_2$, which is the limiting substrate for luciferase. Thus, light production reflects the amount of malate in a sample.

3.3.2. Malate Assay

1. The dry extracts (*see* **Subheading 3.2.**) with or without internal standards (*see* **Subheading 3.1.**) are dissolved in 150 µL 0.1 *M* phosphate buffer, pH 10.0 (*see* **Note 15**). Two 50-µL aliquots of each sample (i.e., for duplicate measurement) are then used in the assay. The measurement is performed as follows.

$$\text{Malate} + \text{NAD}^+ \xrightleftharpoons{\textit{Malate dehydrogenase}} \text{oxaloacetate} + \textbf{NADH} + \text{H}^+ \qquad [1]$$

$$\text{Oxaloacetate} + \text{acetylCoA} + \text{H}_2\text{O} \xrightleftharpoons{\textit{Citrate synthase}} \text{citrate} + \text{CoA-SH} \qquad [2]$$

$$\textbf{NADH} + \text{FMN} + \text{H}^+ \xrightleftharpoons{\textit{Flavinmononucleotide dehydrogenase}} \text{NAD}^+ + \text{FMNH}_2 \qquad [3]$$

$$\text{FMNH}_2 + \text{O}_2 + \text{tetradecanal} \xrightleftharpoons{\textit{Luciferase}} \text{FMN} + \text{tetradecanoic acid} + \text{H}_2\text{O} + \textbf{hv} \qquad [4]$$

Fig. 1. Enzymatic reactions linking malate consumption with light emission.

2. Pipet 405 µL mixture A (*see* **Subheading 2.2.3.** and **Note 16**) into a luminometer vial (*see* **Note 1**).
3. Add 50 µL sample (or sample with internal standard) and vortex.
4. Allow to react for 5 min at room temperature in a shaking water bath.
5. Add 1 µL phosphoric acid and vortex (*see* **Note 17**).
6. Add 10 µL tetradecanal emulsion and vortex.
7. Immediately place vial into luminometer (*see* **Note 18**), close apparatus, and start reaction by injecting 100 µL luciferase/reductase (*see* **Note 19**).
8. Measure light emission over 6 s (at room temperature; *see* **Note 20**).

3.4. Glucose-6-Phosphate Measurement

3.4.1. Principle

The reduced NADPH formed by glucose-6-phosphate dehydrogenase is used to produce FMNH_2 from FMN. The FMNH_2 is consumed in the luciferase reaction allowing light output to reflect the quantity of glucose-6-phosphate in the sample (*see* **Fig. 2**).

3.4.2 Glucose-6-Phosphate Assay

1. The dry extracts (*see* **Subheading 3.2.**) with or without internal standard (*see* **Subheading 3.1.**) are dissolved in 150 µL 0.1 *M* phosphate buffer, pH 7.6. Two 50-µL aliquots of each sample (i.e., for duplicate measurement) are then used in the assay described below. The measurement is performed as follows.
2. Pipet 385 µL mixture B (*see* **Subheading 2.3.4.**) into a luminometer vial (*see* **Note 1**).
3. Add 50 µL sample (or sample with internal standard) and vortex.
4. Allow to react for 5 min at room temperature.
5. Add 10 µL tetradecanal emulsion and vortex.
6. Immediately place vial into luminometer (*see* **Note 18**), close apparatus, and start reaction by injecting 100 µL luciferase/reductase (*see* **Note 19**).
7. Measure light emission over 6 s (at room temperature; *see* **Note 20**).

$$\text{Glucose-6-phosphate} + NADP^+ \xleftrightarrow{\quad\text{\textit{Glucose-6-phosphate dehydrogenase}}\quad} \text{6-phosphogluconolactone} + \mathbf{NADPH} + H^+$$

[5]

$$\mathbf{NADPH} + FMN + H^+ \xleftrightarrow{\quad\text{\textit{Flavinmononucleotide dehydrogenase}}\quad} NADP^+ + FMNH_2$$

[6]

$$FMNH_2 + O_2 + \text{tetradecanal} \xleftrightarrow{\quad\text{\textit{Luciferase}}\quad} FMN + \text{tetradecanoic acid} + H_2O + \mathbf{hv}$$

[4]

Fig. 2. Enzymatic reactions linking glucose-6-phosphate consumption with light emisison.

3.5. Calculations

The amount of metabolites present in cell samples is calculated from the luminescence signals as follows:

$$A = [RLU_{\text{sample}} - RLU_{\text{background}}] / m \tag{1}$$

where A is the metabolite content of the sample, RLU_{sample}, the luminescence signal (*see* **Note 21**) obtained from cell extracts in the absence of added standards, $RLU_{\text{background}}$, the background value measured in the absence of cell extract (i.e., in parallel ethanolic samples subjected to the same extraction procedure as the cell material), and *m,* the slope of the calibration curve of the internal standards (*see* **Note 22**).

4. Notes

1. The material of which the luminometer vials are made may be critical with respect to their optical properties (e.g., autoluminescence, light dispersion, and so forth); therefore, it maybe preferable to use original vials provided by the luminometer's manufacturer (e.g., Berthold).
2. Prepare this, as well as all the following solutions (i.e., 2–11) fresh, as required. Although some of them may be stable for hours to days (at least when kept at 4°C), we find it preferable for them to be freshly made to avoid, for instance, bacterial contamination.
3. Citrate synthase, as well as malate dehydrogenase (from Boehringer) are provided as suspensions in ammonium sulfate. This salt was (partly) removed by spinning down the required amount of enzyme suspension and dissolving the pellet in phosphate buffer. However, preliminary experiments have shown that the luminometric measurement of malate is apparently not affected by small amounts of ammonium sulfate (up to 20 m*M*).
4. The FMN solution is to be kept at 4°C, protected from light, and under a nitrogen atmosphere until use.
5. The luciferase preparation (from the bacterium *Photobacterium [Vibrio] fischeri*) sold by Sigma also contains FMN reductases, which are essential for both the

malate and the glucose-6-phosphate assays (reactions [3] of **Fig. 1** and [6] of **Fig. 2**); we have not tested a mixture of highly purified luciferase and FMN-reductase preparations.

6. The glucose-6-phosphate dehydrogenase (from Boehringer) we have used is provided as suspension in ammonium sulfate. The fresh solution was prepared by spinning down the required amount and dissolving the pellet in phosphate buffer.

7. Tetradecanal is readily oxidizable.

8. It is important to note that the temperature of this emulsion must be kept at a minimum of 60°C for the entire duration of the measurements; otherwise, tetradecanal tends to precipitate, which leads to a considerable loss of activity. It is also worth mentioning that we have tested alternative, previously published methods of solubilizing tetradecanal, for instance by binding it to albumin *(6)*; however, the light signals obtained were about 50% weaker than with the method described under **Subheading 2.2.3., item 10.**

9. The conditions for cell treatment (prior to the actual metabolite measurements) may, of course, vary depending on the cell type used and the experimental condition to be tested.

10. The amount of cells required to detect malate may depend on the cell type used. The quantity given here applies for the measurement of cardiomyocyte extracts.

11. These conditions specifically apply to the homogenization of isolated cardiomyocytes. Alternative homogenization methods may also be suitable for other cell or tissue types.

12. Alternatively, they can be kept at −20°C (under these conditions, the samples are stable for at least 2 w).

13. Since the measurement of malate and glucose-6-phosphate occurs via the stoichiometric production of NAD(P)H (*see* **Subheadings 3.3.1.** and **3.4.1.**), the native NAD(P)H, i.e., that originating from the cells, has to be degraded before the analytical assay is performed. Therefore, the cells are subjected to an acidic ethanolic extraction by a modification of a procedure described by Shryock et al. *(8)*. The use of a volatile solvent has the additional advantage of avoiding an unnecessary dilution of the samples, and even allowing concentration of the metabolite to be measured.

14. Evaporation must be complete, since traces of formic acid could affect the pH in the assay mixture on redissolution of the residue (*see also* **Note 22**).

15. At this pH, the formation of oxaloacetate is favored (vs the reverse reaction).

16. This mixture may be kept on ice until the beginning of the assay; however, it must be brought to room temperature before the sample is added.

17. The following reactions ([3] and [4] of **Fig. 1**) have a pH optimum in the neutral range (in contrast to the malate dehydrogenase reaction [1] of **Fig. 1**); therefore, the extracts must be neutralized at this stage.

18. The rapid mixing is important to avoid the precipitation of tetradecanal.

19. Since the light-producing reactions are rapid under these conditions, a luminometer model allowing the automated injection of the luciferase/reductase mixture into the measuring chamber may be required for accurate measurements.

20. Although longer reaction times result in larger light signals, the overall sensitivity of the method is not improved by prolonging the measuring time over 6 s (the background also increases with time).
21. Since this value is not the actual number of emitted photons, but a correlate thereof, the absolute RLU value may vary depending on the luminometer type used. With the Lumat 9501 (Berthold), background values were ~20,000 RLU in the malate assay and 10,000 RLU in the glucose-6-phosphate assay.
22. Using the protocols presented here, we could detect amounts as low as 15–20 pmol malate or glucose-6-phosphate in aqueous samples (i.e., in the absence of biological matrix). This sensitivity limit ensues from the rule that the smallest valid signal (measured in cell extracts) must be greater than or equal to the sum of the background plus 3 SEM. Note that the difference between duplicates within a particular experiment was very low <10%). In the presence of a biological matrix, the sensitivity is likely to be lower than with aqueous standards (which makes the use of internal standards necessary). Thus, in the case of measurements in extracts from cardiomyocytes (which represent a complex, protein-rich matrix), the sensitivity limit was ~70 pmol malate and ~90 pmol glucose-6-phosphate (and allowed the accurate measurement of cellular metabolite concentrations corresponding to values found by others in whole hearts; *see* refs. *9,10*). With other cell or tissue types, the detection limit may be lower or higher than with myocytes. Finally, it is also important to know that luminometric reactions (such as reactions [3], [4], or [6] of **Figs. 1** and **2**) are particularly sensitive to changes in ionic strength and pH; *see* **ref. 2**).

References

1. Williamson, D. H., and Brosnan, J. T. (1974) in *Methoden der Enzymatischen Analyse*, 3rd ed., vol. II (Bergmeyer, H. U., ed.), Verlag Chemie, Weinheim, pp. 2331–2353.
2. Stanley, P. E. (1971) Determination of subpicomole levels of NADH and FMN using bacterial luciferase and the liquid scintillation spectrometer. *Anal. Biochem.* **39,** 441–453.
3. Brolin, S. E., Agren, A., Wersall, J. P., and Hjerten, S. (1978) in *International Symposium on Analytical Applications of Bioluminiescence and Chemiluminescence* (Schram, E. and Stanley, P., eds.), State Printing and Publishing, Westlake Village, CA. pp. 109–121.
4. Agren, A., Berne, C., and Brolin, S. E. (1977) Photokinetic assay of pyruvate in islets of Langerhans using bacterial luciferase. *Anal. Biochem.* **78,** 229–234.
5. Stanley, P. E. (1978) in *Methods in Enzymology*, vol. LVII (Deluca, M. A., Colowick, S. P., and Kaplan, N. O., eds.), Academic, London, pp. 181–188.
6. Hutton, J. C., Sener, A., and Malaisse, W. J. (1978) in *International Symposium on Analytical Applications of Bioluminiescence and Chemiluminescence* (Schram, E. and Stanley, P., eds.), State Printing and Publishing, Westlake Village, CA, pp. 166–181.
7. Fischer, Y., Rose, H., and Kammermeier, H. (1991) Highly insulin-responsive isolated rat heart muscle cells yielded by a modified isolation method. *Life Sci.* **49,** 1679–1688.

8. Shryock, J. C., Rubio, R., and Berne, R. M. (1986) Extraction of adenine nucleotides from cultured endothelial cells. *Anal. Biochem.* **159,** 73–81.
9. Jüngling, E., Timmerman, M., Aretz, A., Ionescu, I., Mertens, M., Löken, C., Kammermeier, H., and Fischer, Y. (1996) Luminometric measurement of subnanomol amounts of key metabolites in extracts from isolated heart muscle cells. *Anal. Biochem.* **239,** 41–46.
10. Fischer, Y., Böttcher, U., Eblenkamp, M., Thomas, J., Jüngling, E., Rösen, P., and Kammermeier, H. (1997) Glucose transport and glucose transporter GLUT4 are regulated by product(s) of intermediary metabolism in cardiomyocytes. *Biochem. J.* **32,** 629–638.

5

Bioluminescent Assay of the Guanylates

Sharon R. Ford and Franklin R. Leach

1. Introduction

The guanine nucleotides (ribo and deoxyribo) serve many important roles in biosynthesis and biological control. They are involved in the synthesis of enzymes *(1)*, and are allosteric regulators for such enzymes as glutamate dehydrogenase *(2)* and acetyl-CoA carboxylase *(3)*. They regulate IMP dehydrogenase gene expression *(4)*. The guanine nucleotides are incorporated into DNA and RNA *(5)*, are substrates in the Krebs tricarboxylic acid cycle *(6)*, and function in microtubule assembly *(7)*. Guanine nucleotides regulate the activation of adenylate cyclase via G proteins in receptor-mediated signaling processes *(8–11)*. The G-proteins are involved in ion channel function *(12)*. Guanylates also serve as coenzymes functioning in such physiological responses as vision *(13)*, bacterial sporulation *(14)*, and the stringent response to amino acid deprivation *(15)*.

Several methods have been developed for determination of the guanylates. The most sensitive procedures involve coupled enzymatic reactions *(16)*, enzymatic cycling or amplification procedures *(17)*, a radiochemical end-point assay for GTP and GDP *(18)*, and linkage of analysis to light production by firefly luciferase *(19–23)*. To apply selective guanine determinations to environmental samples or other samples where the concentration of nucleoside triphosphate is small requires either enzyme cycling procedures or the sensitivity of firefly luciferase; either procedure must be coupled with selective destruction of ATP (and other interfering nucleotides). The procedure published here is based on an improvement *(23)* of the bioluminescence-based determination described by Karl *(19,20)* and allows the measurement of 0.1 pmol GMP.

From: *Methods in Molecular Biology, Vol. 102: Bioluminescence Methods and Protocols*
Edited by: R. A. LaRossa © Humana Press Inc., Totowa, NJ

2. Materials

2.1. Water and Glassware

Water quality is of paramount importance. Owing to the sensitivity of the technique, minute contamination of reagents (especially bacterial contamination) will cause high background luminescence. We routinely prepare the water used in all reagents as follows: The building's reverse osmosis- and UV-treated water is passed through two mixed-bed, ion-exchange resins (Barnstead/ Thermolyne D 8902 Ultrapure cartridges, Dubuque, IA), glass-distilled, pressure-filtered through a sterile 0.45-μm Millipore® (Bedford, MA) filter into sterile bottles, and then autoclaved. After opening, a bottle of water can be used for several days if handled using good sterile technique.

We recommend as a minimum standard that "Milli-Q-quality" water be additionally filtered through a sterile 0.45-μm filter and autoclaved before use. Backgrounds in the standard ATP assay containing 100 μL of Firelight® luciferase/luciferin and no ATP in a 500-μL total volume should be <100 counts/10 s in a Lumac Model 2010A Biocounter. If background counts are high, the "Milli-Q" water should be distilled before filtering and autoclaving.

We recommend that all glassware used for reagents for these assays be washed in phosphate-free detergent, soaked in Pierce brand RBS-pf® (Rockford, IL), rinsed in reverse osmosis-treated (RO) water, and sterilized (*see* **Note 1**).

2.2. Buffers

Several buffer solutions are needed for the extraction and guanylate determinations. Make buffers and all other reagents in sterile, filtered "Milli-Q-quality" water.

1. Cell extraction buffer: 20 mM Tricine, pH 7.5. Aliquot in volumes to be used in a single day and store frozen.
2. Step 1 buffer: 750 mM Tricine, pH 7.5; 50 mM MgCl$_2$; 0.125 mM KCl. Aliquot in volumes to be used in a single day and store frozen.
3. Step 2 buffer: 750 mM Tricine, pH 7.5; 50 mM MgCl$_2$; 0.125 mM KCl; and 5 mM glucose. Aliquot in volumes to be used in a single day and store frozen.
4. Step 3 buffer: Prepare a buffer containing 250 mM Tricine, pH 7.8; 50 mM MgSO$_4$; 5 mM EDTA. Aliquot in volumes to be used in a single day and store frozen.

2.3. Enzymes

1. Pyruvate kinase: Use a preparation substantially free of other kinase activities, especially adenylate kinase. We used P 1506 type II rabbit muscle enzyme from Sigma (St. Louis, MO), which contains <0.01% adenylate kinase. If using an ammonium sulfate suspension, such as Sigma P 1506, centrifuge for 5 min in a

microcentrifuge (10,000g) at 4°C, pour off the supernatant solution, and dissolve the pellet in a volume of 20 mM Tricine, pH 7.5, with 0.1% bovine serum albumin, equal to the original volume. Prepare the desired amount of enzyme fresh daily.

2. Guanylate kinase: Use a preparation as free as possible of adenylate kinase activity (*see* **Note 2**). We used Sigma G 7510, lyophilized powder from bovine brain or G 9385, 50% glycerol solution from porcine brain. Lyophilized powder can be reconstituted in water or in 20 mM Tricine, pH 7.5, with 0.1% bovine serum albumin added for improved stability. Unused enzyme can be stored frozen (–20°C) for several weeks.

3. Hexokinase and glucose-6-phosphate dehydrogenase: A convenient preparation to use for the enzymatic destruction of ATP contains both hexokinase and glucose-6-phosphate dehydrogenase. We used Sigma H 8629. Dissolve in 20 mM Tricine, pH 7.5, containing 0.1% bovine serum albumin. Unused enzyme can be stored for 2–3 d at 4°C.

4. Uridine-5'-diphosphoglucose pyrophosphorylase: Use a preparation suitable for the quantitative determination of UTP. We used Sigma U 8501 Type X from baker's yeast. Dissolve lyophilized preparation, which contains buffer salts, in 0.1% bovine serum albumin. The enzyme-containing solution can be stored frozen (–20°C) for several weeks.

5. Nucleoside-5'-diphosphate kinase: A preparation with the lowest possible adenylate kinase activity is recommended, such as Sigma N 2635 from bovine liver with <0.1% adenylate kinase. We found, however, that Sigma N 0379 from baker's yeast was adequate and is more economical. Dissolve lyophilized preparation, which contains buffer salts, in 0.1% bovine serum albumin. The enzyme solution can be stored frozen (–20°C) for several weeks.

6. Firefly luciferase: A convenient preparation to use is Firelight® from Analytical Luminescence Laboratory (Ann Arbor, MI), which contains both firefly luciferase and luciferin. Other firefly luciferase preparations can be used, but if they do not contain luciferin, it must be added separately to a concentration in the final reaction mixture of ≥0.05 mg/mL. We found the Analytical Luminescence Laboratory reagent to be stable and to behave consistently among different lots. Dissolve luciferase in 50 mM Tricine, pH 7.8; 10 mM $MgSO_4$; 1 mM dithiothreitol; and 1 mM EDTA, and age on ice ≥1 h before use. The enzyme should be warmed to room temperature before use. Unused enzyme can be stored at 4°C overnight with some loss of activity.

2.4. Other Reagents for Guanylate Determination

1. Nucleotides: The highest purity nucleotides available should be used. Make stock solutions (1–10 mM) in 20 mM Tricine, pH 7.5–7.8. Store frozen (*see* **Note 3**).

2. Substrates: Use a high-purity (≥98%) preparation of glucose (Sigma G 5000, now replaced by G 5767), glucose-1–phosphate (Sigma G 6875), and phosphoenolpyruvate (Sigma P 7002) to make stock solutions of 50, 5, and 5 mM, respectively, in sterile, "Milli-Q-quality" water. Store frozen.

3. Salts: Use ACS-grade reagents. Stocks of 500 mM $MgSO_4$, 500 mM $MgCl_2$, 1.25 mM KCl, 50 mM di- or tetra-sodium EDTA, and 50 mM dithiothreitol can be made in sterile, "Milli-Q-quality" water and stored frozen.

2.5. Luminometer

Any good-quality luminometer can be used for these assays. For best results, the reactions should be initiated within the counting chamber of the luminometer. We routinely used a Lumac Model 2010A Biocounter (Celsis, Cambridge, UK) with automatic injection of the Firelight® reagent to start the reaction (*see* **Note 4**).

3. Methods

3.1. Preparation of Cell Extracts

The cell lysates should be prepared in a manner that inactivates all enzymes as quickly as possible without destroying nucleotides. We find that, for bacteria, extracting the cells in boiling buffer works well.

1. Grow the cells under the desired conditions.
2. To extract the cells, preheat 4.5 mL of cell extraction buffer in an 18×150 mm test tube in a boiling water bath for 5 min.
3. Inject 0.50 mL of cells, using a 1-mL micropipeter.
4. Incubate in a boiling water bath for 3 min.
5. Transfer to an ice bath.
6. Cool for at least 10 min.
7. Assay immediately, or store frozen (−20°C) for later assay.

3.2. Overview of the Steps of Guanylate Determination

Three separate incubations are needed to determine the amount of guanylates in a sample; one for GTP only, a second for GTP plus GDP, and a third for GTP plus GDP plus GMP. Amounts of GDP and GMP are determined by the difference in guanylate amounts in the three incubations. The determination of guanylates for each incubation requires four separate steps. In **step 1**, GDP or GDP and GMP are converted to GTP. **Step 2** destroys any ATP and UTP in the samples that would falsely elevate the apparent concentration of guanylates in the sample. **Step 3** converts GTP into ATP, and in **step 4** ATP levels are determined in a bioluminescent assay (*see* **Fig. 1**).

3.3. Step 1: Conversion of Guanine Nucleotides to GTP

3.3.1. GTP (Reaction A)

No conversion of guanine nucleotide takes place in this incubation, but it is necessary to treat all three samples the same, so the GTP assay tube is incubated in buffer under conditions used in reactions B and C to convert GMP and GDP to GTP (*see* **Note 5**).

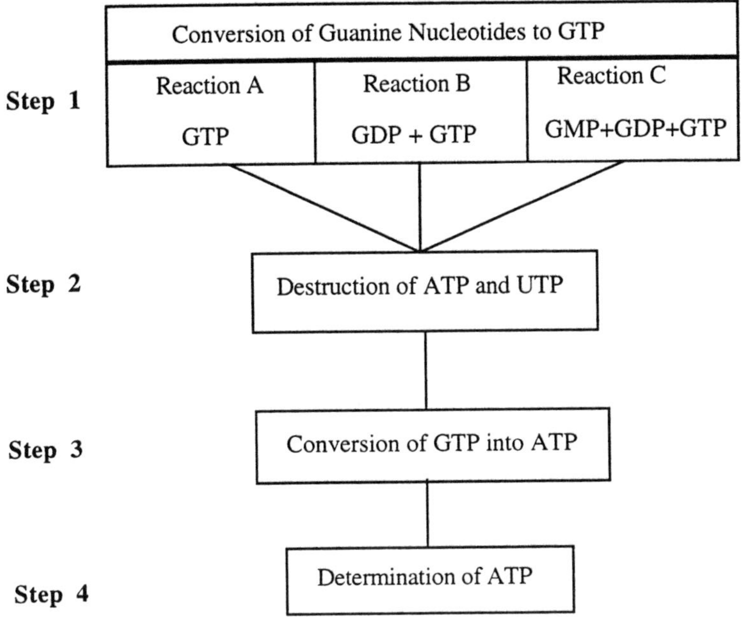

Fig. 1. The steps of guanylate determination (modified from **ref. *23***).

1. For the GTP tubes of **step 1**, make reaction mixture A containing per mL total volume: 0.10 mL of the **step 1** buffer and 0.90 mL of sterile, "Milli-Q-quality" water.
2. In 13 × 100 mm test tubes mix 50 µL of reaction mixture A and 200 µL of sample.
3. Incubate the tubes at 30°C for 90 min.
4. Stop the reactions by placing the tubes in a boiling water bath for 3 min (*see* **Note 5**).
5. Cool on ice for at least 10 min.

3.3.2. GDP (Reaction B)

To measure GDP in a sample, the total amount of GDP + GTP is determined and GDP calculated by subtracting the amount of GTP determined in the GTP-only tube. In this incubation, GDP is converted to GTP by the action of pyruvate kinase in the reaction:

$$\text{Phosphoenolpyruvate} + \text{GDP} \longleftrightarrow \text{pyruvate} + \text{GTP} \tag{1}$$

(*see* **Note 6**).

1. For the GTP + GDP tubes, make reaction mixture B containing per mL total volume: 0.10 mL of **step 1** buffer, 0.1 mL of 5 m*M* phosphoenolpyruvate, and 80 µg pyruvate kinase.
2. In 13 × 100 mm test tubes, mix 50 µL of reaction mixture B and 200 µL of sample.

3. Incubate the tubes at 30°C for 90 min.
4. Stop the reactions by placing the tubes in a boiling water bath for 3 min.
5. Cool on ice for at least 10 min.

3.3.3. GMP (Reaction C)

The GMP in a sample is measured by determining the total amount of GMP + GDP + GTP and subtracting the amount of GDP + GTP. In this incubation, GMP is converted to GDP by the action of guanylate kinase:

$$ATP + GMP \longleftrightarrow ADP + GDP \tag{2}$$

and the GDP is subsequently converted to GTP by the action of pyruvate kinase (*see* **Eq. 1** in **Subheading 3.3.2.**).

1. For the tubes measuring GTP + GDP + GMP, make a reaction mixture C containing per mL total volume: 0.10 mL of **step 1** buffer, 0.1 mL of 5 mM phosphoenolpyruvate, 80 µg of pyruvate kinase, 0.20 mL of 10 µM ATP, and 0.4 U of guanylate kinase.
2. In 13 × 100 mm test tubes mix 50 µL of reaction mixture C and 200 µL of sample.
3. Incubate the tubes at 30°C for 90 min.
4. Stop the reactions by placing the tubes in a boiling water bath for 3 min.
5. Cool on ice for at least 10 min.

3.4. Step 2: Destruction of Other Nucleoside Triphosphates

In order to measure only the guanylates, other nucleoside triphosphates must be removed from the samples. ATP, which is added in **step 1** to convert GMP to GDP, is, of course, the biggest problem. ATP is consumed in the hexokinase reaction coupled to the glucose-6-phosphate dehydrogenase reaction to ensure the complete removal of ATP:

$$Glucose + ATP \longrightarrow glucose\text{-}6\text{-}phosphate + ADP \tag{3}$$

$$Glucose\text{-}6\text{-}phosphate + NADP^+ \longrightarrow glucono\text{-}\delta\text{-}lactone\text{-}6\text{-}phosphate$$
$$+ NADPH + H^+ \tag{4}$$

UTP is also destroyed in this step by the action of uridine diphosphoglucose pyrophosphorylase:

$$UTP + glucose\text{-}1\text{-}phosphate \longrightarrow UDP\text{-}glucose + PP_i \tag{5}$$

There is currently no commercially available enzyme that can be used for the removal of CTP (*see* **Note 7**). The three samples (reaction A for GTP, reaction B for GTP + GDP, and reaction C for GTP + GDP + GMP) are treated exactly alike in this and subsequent steps.

1. Prepare a reaction mixture containing per mL: 0.10 mL of **step 2** buffer, 0.10 mL of 5 mM glucose-1-phosphate, 0.10 mL of 5 mM NADP$^+$, 0.10 mL of 150 mM MgCl$_2$, 10 U of hexokinase, 4 U of glucose-6-phosphate dehydrogenase, and 5 U

MgCl$_2$, 10 U of hexokinase, 4 U of glucose-6-phosphate dehydrogenase, and 5 U of uridine-5'-diphosphoglucose pyrophosphorylase.

2. To each sample from **step 1** add 50 μL of this reaction mixture.
3. Incubate the samples for 30 min at 30°C.
4. Stop the reaction by placing the tubes in a boiling water bath for 3 min (*see* **Note 5**).
5. Cool on ice for 10 min.

3.5. Step 3: Conversion of GTP into ATP

In this step, the enzyme nucleoside diphosphate kinase converts GTP to an equivalent amount of ATP:

$$GTP + ADP \longleftrightarrow GDP + ATP \tag{6}$$

The reaction is driven to the right by excess ADP and by the subsequent utilization of ATP in **step 4** (*see* **Note 8**). This step is carried out in the cuvet that will be placed in the luminometer for measurement of light produced in **step 4** of the reaction.

1. For each of the three reactions, add 50 μL of sample from **step 2** to a cuvet containing 50 μL of **step 3** buffer, and 300 μL of a mixture containing 0.1 U of nucleoside diphosphate kinase and 15 pmol of ADP.
2. Incubate the samples for ≥5 min at room temperature, and then determine the ATP immediately in **step 4**.
3. For analysis of very low amounts of guanine nucleotides, up to 250 μL of sample from **step 2** can be assayed, and the nucleoside diphosphate kinase and ADP added in a volume of 50 μL.

3.6. Step 4: Light Production from ATP

The final step of the reaction quantitatively determines the amount of ATP in each sample using the light produced in the firefly luciferase reaction:

$$ATP + luciferin + O_2 \longrightarrow AMP + PP_i + oxyluciferin + hv \tag{7}$$

The amount of ATP is directly proportional to the light emitted.

1. Place each sample from **step 3** into the luminometer.
2. Start the light reaction by injecting 100 μL of Firelight®.
3. Measure the light produced in a standard period of time (~ 10 s).

3.7. Calibration of ATP Assay

For calibration of the method, it is useful to measure the light production of standard amounts of ATP. The conditions of the standard ATP assay should be as close as possible to the conditions of the guanylate assay procedure for the GTP tube. It is necessary to leave the enzymes out of **step 2** of the procedure when measuring the ATP. Results of a typical ATP calibration are shown in **Fig. 2** (*see* **Note 9**).

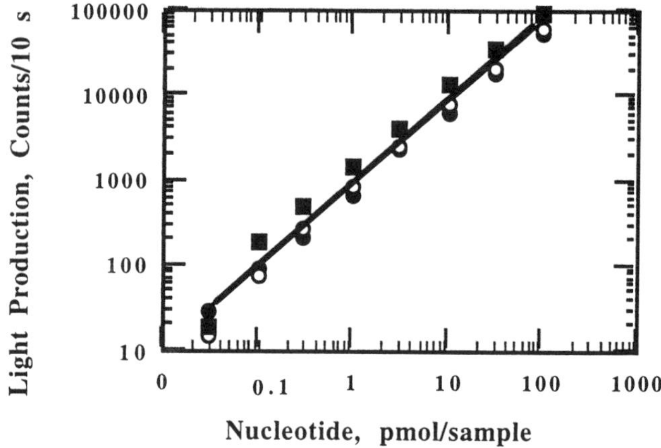

Fig. 2. Standard curve for conditions that permit measurement of 0.1 pmol of guanylate. The procedure described in the text was applied to standards containing varying concentrations of ATP, GTP, and GMP. The averages of duplicate determinations were plotted. Modified from **ref. *23*.** ■ ATP, O GTP, and ● GMP.

3.8. Guanylate Standard Curve

To determine the amounts of guanylates, a standard curve must be generated to correlate the light production with amount of nucleotide. Known concentrations of guanylates (GMP, GDP, and GTP in separate reactions) must be carried through the procedure, and the light emitted correlated with the nucleotide concentration. It is useful to compare the light production from the guanylates with light production from an equivalent amount of GTP that has not been subjected to the four-step reaction to establish the extent of metabolite loss. **Figure 2** shows the results of a typical standardization (*see* **Note 10**).

3.9. Determination of Guanylate Energy Charge (GEC)

The GEC can be calculated in the same manner as adenylate energy charge *(24)*, substituting the guanine nucleotides for the adenine nucleotides.

$$GEC = \frac{[GTP] + 0.5\,[GDP]}{[GTP] + [GDP] + [GMP]} \tag{8}$$

Standard samples of known energy charge should be assayed to calibrate the procedure and assure that there is good agreement between experimental values for standards and theoretical values. **Table 1** shows typical results for standards having theoretical GECs from 0.25 to 1.00. The experimental GEC values are calculated as:

Table 1
Determination of Guanylate Energy Charge for Standards[a]

Ratio GTP-GDP-GMP	A counts/ 10 s	B counts/ 10 s	C counts/ 10 s	Theoretical GEC	Experimental GEC
1–0–0	6973	7183	7347	1.0	0.96
0–1–0	95	6545	6845	0.5	0.49
0–0–1	0	59	5778	0	0.005
1–1–0	6153	12,427	13,641	0.75	0.68
1–0–1	6612	6694	12,855	0.50	0.52
0–1–1	57	6427	12,299	0.25	0.26
1–1–1	6244	12,175	18,387	0.50	0.50

[a]Standard amounts of nucleotides (10 pmol) were used as indicated in the first column. The counts were obtained from triplicate samples treated through **steps 1–4** as described in the text. Columns A, B, and C present: in A, GTP; in B, GTP + GDP; and in C, GTP + GDP + GMP. GEC was calculated as indicated in the text. From **ref. 23**.

$$\frac{(A \text{ counts} + B \text{ counts})}{2\ (C \text{ counts})} \tag{9}$$

3.10. Application of Bioluminescent Determination of Guanylates

Some applications of bioluminescent determination of guanylates are in the literature. These applications use the older Karl procedure *(19,20)*, but are illustrative of how this improved procedure can be used. Schaeffer and Anderson *(25)* modified the Karl procedure *(19,20)* for the determination of GTP. They used a 15-min incubation with purified nucleoside diphosphate kinase for the conversion of the GTP into ATP and used purified firefly luciferase for ATP determination. The ATP determination used arsenate buffer and was done in a liquid scintillation counter (which was not best for the purpose; *see* Chapter 1). The determination was linear from 1–200 nmol of GTP. The unmodified Karl procedure has been applied by Ostroy et al. *(26)* in a study of the GTP in toad rod photoreceptors.

3.11. Alternative Guanylate Determination Procedures

3.11.1. Enzymatic Determinations

Keppler *(29)* used both HPLC analysis and enzymatic assays to determine the nonadenylate NTPs (UTP, CTP, and GTP). The NTPs were converted to ATP using nucleoside diphosphate kinase, and the ATP was determined via hexokinase and glucose-6-phosphate dehydrogenase (measuring NADPH production). The determination was linear for concentrations between 2 and 100 nmol of NTP/cuvet (710 µL final volume). Keppler and Kaiser *(16,29)* determined GDP and GTP using guanylate kinase, nucleoside diphosphate kinase, hexokinase, and glucose-6-phosphate dehydrogenase. GMP was determined using guanylate kinase, pyruvate kinase, and lactic dehydrogenase. Adenylate kinase interfered in all determinations. The detection limits were 1.2, 0.6, and 0.3 nmol/200 µL sample for GTP, GDP, and GMP, respectively.

Enzymatic cycling can amplify signals from substrates to increase sensitivity. Cha and Cha *(17)* described a procedure for the microdetermination of the guanine ribonucleotides using enzymatic amplification techniques. The sensitivity obtained is 0.1 pmol without any separating procedure. Creatine kinase, guanylate kinase, nucleoside diphosphatase, pyruvate kinase, succinate thiokinase, and lactic dehydrogenase were used in the two-step reaction sequence. Pogson et al. *(30)* determined GTP and GDP after polyethylenimine-cellulose chromatography by using firefly luciferase. Sensitivity was 5 pmol. An enzymatic cycling procedure (with a 3×10^3-fold amplification) has been used for the measurement of GTP and GDP by De Azeredo et al. *(31)*. Succinyl CoA synthetase, pyruvate kinase, and lactic dehydrogenase were used to achieve picomolar sensitivity (down to 1 µg of freeze-dried mouse brain).

3.11.2. Chemical Determination

A combination of thin-layer chromatography and bioluminescence procedure allowed Goodrich and Burrell *(32)* to determine 0.1-pmol amounts of NTPs. Cheung and Marcus *(33)* used incorporation of [^{32}P] from phosphoenolpyruvate into GTP using pyruvate kinase as the final step in the determination of the guanylates. Cerpovicz and Ochs *(18)* have developed a radiochemical end-point assay for determining GTP and GDP. They use GTP as the limiting component for the enzymatic conversion of (U-[^{14}C]) aspartate to radioactive phosphoenolpyruvate (catalyzed by aspartate aminotransferase and phosphoenolpyruvate carboxykinase). The radioactive phosphoenolpyruvate was separated using anion-exchange chromatography. The GDP assay required prior chromatographic separation of GDP and GTP before the reactions used for GTP determination were run in the reverse direction. Radioactive aspartate was determined. About 10 pmol of guanine nucleotide can be determined, but unfortunately there is no commercial source of the key enzyme phosphoenolpyruvate carboxykinase.

4. Notes

1. It has been our experience that whenever we have trouble with high backgrounds, the most likely problem is that the water is contaminated. Investigators who have had trouble repeating our results have usually not been as careful about the preparation and storage of water and other reagents as we think is necessary. We find that careful attention paid to preparation of reagents pays off in the quality and reproducibility of results.

2. Adenylate kinase (myokinase) catalyzes the reaction:

$$2\ ADP \longleftrightarrow AMP + ATP \tag{10}$$

If any adenylate kinase is present in **Subheading 3.5.** of the guanylate determination, in which a large amount of ADP is added, falsely high ATP values will result. A major problem in the determination of GMP results if the guanylate kinase added in **Subheading 3.3.3.**, **step 1** (reaction C) to convert GMP to GDP is contaminated with adenylate kinase. We have found that many preparations of guanylate kinase are contaminated with significant levels of adenylate kinase, and the adenylate kinase activity is not inactivated by heating at 100°C. Thus, any adenylate kinase introduced in **Subheading 3.3.3.**, **step 1** of the procedure will cause the conversion of ADP added in **Subheading 3.5.**, **step 3** to ATP, which will give high backgrounds for measurement of GMP samples. The addition of inhibitors of adenylate kinase did not achieve sufficient reduction of the adenylate kinase activity of guanylate kinase preparations without also inhibiting the luciferase reaction *(23)*.

Preparations of porcine brain guanylate kinase from Boehringer Mannheim (Mannheim, Germany) (cat. no. 106 321) were superior to guanylate kinase preparations made in the US, but the US Department of Agriculture no longer permits importation of that preparation, since it has not been sterilized. At one time Boehringer considered production in the US, but has concluded that it does not have the facilities available under the current situation (email from Glenn Martin, 11/96). The preparation is available for non-US scientists under cat. no. 106 321, and we recommend that this preparation be used. The background reactions were much less with this preparation than with any other guanylate kinase tested. Use of this product greatly simplifies the determination of GMP.

For American scientists, the best solution under the circumstances is a careful selection of the lot of guanylate kinase used. Taking the quality-control analysis of the adenylate kinase activity is not sufficient; the analysis should be done in the laboratory that is using the guanylate kinase in the guanylate determination. We found no correlation between the quality-control values provided and the suitability of various preparations. Test various lots of guanylate kinase to determine which has adenylate kinase contamination that is low enough to allow easy and sensitive determination of GMP. Note that the guanylate kinase used by Karl *(19)* was that from Boehringer Mannheim, which is no longer available in the US.

The mouse and human guanylate kinases have been cloned and sequenced *(34)*. The mammalian guanylate kinases from human, mouse, bovine, and porcine are an average of 92.2% identical, whereas the yeast and *Escherichia coli* enzymes are between 40 and 50% identical with the mammalian enzymes or each other. It is not known if these enzymes will be commercially available and superior to those already available.

3. When preparing stock solutions of nucleotides remember that the nucleotides can be quite acidic. Check the pH after preparing the solutions, and titrate to neutrality if necessary. We have found that concentrated stock solutions of ATP can overwhelm the buffering capacity of the 20 mM Tricine in which they are made.

4. The geometry of the phototubes and counting chamber of each luminometer differs and greatly affects the sensitivity of the instrument. The conditions outlined in this chapter are for use with the Lumac Model 2010A Biocounter. Another luminometer might require different sample volumes. The relative concentrations of reagents should remain the same for different luminometers. For information on the relative sensitivities of currently available luminometers, please *see* Chapter 1, **Note 3**.

5. The heating steps required to inactivate the enzymes at the end of the incubation result in some chemical breakdown of GTP. A comparison of standard solutions of GTP that have been subjected to the conditions of **steps 1** and **2** (*see* **Subheadings 3.3.** and **3.4.**) with standards that have not been so treated allows estimation of the destruction of GTP during the procedure. We have found that the reaction conditions required for guanylate determinations cause the loss of approx 50% of the GTP *(23)*. It is imperative that a GTP standard curve be prepared by carrying known amounts of GTP through the four-step procedure each time guanylates of cell extracts are measured.

6. Pyruvate kinase requires K^+. This is provided by the buffer in which the reactions are run.

7. Although ATP and UTP can be efficiently removed from the samples under the conditions of **step 2** (*see* **Subheading 3.4.**), there is no commercially available enzyme that will remove CTP from the samples. Thus, CTP in environmental samples will increase the apparent GTP measured. The amount of CTP in typical environmental samples is, at most, 20–30% that of GTP *(19)*. The amount of CTP in samples can be determined by destroying GTP using 3-phosphoglycerate kinase *(35)* or succinyl thiokinase *(22)*. Alternatively, **steps 3** and **4** (*see* **Subheadings 3.5.** and **3.6.**) of the reaction can be combined by mixing the nucleoside diphosphate kinase with the luciferase. This reduces the measurement of CTP because it is converted more slowly to ATP by the nucleoside diphosphate kinase than is GTP *(19*, and unpublished results from this laboratory).

8. It is critically important that the ADP used in **step 3** (*see* **Subheading 3.5.**) of the reaction not be contaminated with ATP. Many commercially available preparations of ADP have a small percentage of ATP (0.1–1%) as a contaminant. Although this level of contamination may be tolerated for other uses of ADP, it causes extremely high backgrounds in this procedure. Therefore, check with the supplier of ADP for a preparation (cat. and lot no.) devoid of ATP contamination. We found Sigma A 4386 di(monocyclohexylammonium) salt from bacteria to have the lowest ATP contamination, but recommend checking with the vendor about the purity of the particular lot available.

9. The incubation conditions of the guanylate determination procedure do not result in nearly as much destruction of ATP as of GTP. We find approx 90% of the ATP survives the incubation as compared to 50% of GTP (*see* **Note 5**).

10. The sensitivity of the assay and the range of the standard curve will depend on the luminometer used. Each laboratory will have to determine the sensitivity achievable with the instrumentation at hand.

Acknowledgments

This research was supported in part by the Oklahoma Agricultural Experiment Station (Project 1806) and is published with the approval of the director. The manuscript was reviewed by E. C. Nelson and Vanitha Thulasiraman, who made useful suggestions.

References

1. Kaziro, Y. (1978) The role of guanosine 5'-triphosphate in polypeptide chain elongation. *Biochim. Biophys. Acta* **505,** 95–127.
2. Frieden, C. (1962) The unusual inhibition of glutamate dehydrogenase by guanosine di- and triphosphate. *Biochim. Biophys. Acta* **59,** 484–486.
3. Witters, L. A., Friedman, S. A., Tipper, S. J., and Bacon, G. W. (1981) Regulation of acetyl-CoA carboxylase by guanine nucleotides. *J. Biol. Chem.* **256,** 8573–8578.
4. Glesne, D. A., Collart, F. R., and Huberman, E. (1991) Regulation of IMP dehydrogenase gene expression by its end products, guanine nucleotides. *Mol. Cell. Biol.* **11,** 5417–5425.
5. Kornberg, A. and Baker, T. A. (1992) Biosynthesis of DNA precursors, in *DNA Replication,* 2nd ed. W. H. Freeman, New York, pp. 53,54.
6. Stryer, L. (1988) Citric acid cycle, in *Biochemistry,* 3rd ed. W. H. Freeman, New York, p. 377.
7. Kirschner, M. and Mitchison, T. (1986) Beyond self-assembly: from microtubules to morphogenesis. *Cell* **45,** 329–342.
8. Gilman, A. G. (1987) G proteins: transducers of receptor-generated signals. *Ann. Rev. Biochem.* **56,** 615–649.
9. Taylor, C. W. (1990) The role of G proteins in transmembrane signalling. *Biochem. J.* **272,** 1–13.
10. Casey, P. J. and Gilman, A. G. (1988) G protein involvement in receptor-effector coupling. *J. Biol. Chem.* **263,** 2577–2580.
11. Dohlman, H. G., Caron, M. G., and Lefkowitz, R. J. (1987) A family of receptors coupled to guanine nucleotide regulating proteins. *Biochemistry* **26,** 2657–2664.
12. Breitwieser, G. E. (1991) G-protein mediated ion channel activation. *Hypertension* **17,** 684–692.
13. Fung, B.-K., Hurley, J. B., and Stryer, L. (1982) Flow of information in the light-triggered cyclic nucleotide cascade of vision. *Proc. Natl. Acad. Sci. USA* **78,** 152–156.
14. Lopez, J. M., Dromerick, A., and Freese, E. (1981) Response of guanosine 5'-triphosphate concentration to nutritional changes and its significance for *Bacillus subtilis* sporulation. *J. Bacteriol.* **146,** 605–613.
15. Stephens, J. C., Artz, S. W., and Ames, B. N. (1975) Guanosine 5'-diphosphate 3'-diphosphate (ppGpp): positive effector for histidine operon transcription and general signal for amino acid deficiency. *Proc. Natl. Acad. Sci. USA* **72,** 4389–4393.
16. Keppler, D. and Kaiser, W. (1978) Enzymatic analysis of guanine nucleotides in tissues and cells. *Anal. Biochem.* **86,** 147–153.

17. Cha, S. and Cha, C.-J. M. (1970) Microdetermination of guanine ribonucleotides by an enzymic amplification technique. *Anal. Biochem.* **33,** 174–192.
18. Cerpovicz, P. F. and Ochs, R. S. (1991) A radiochemical enzymatic endpoint assay for GTP and GDP. *Anal. Biochem.* **192,** 197–202.
19. Karl, D. M. (1978) A rapid sensitive method for the measurement of guanine ribonucleotides in bacterial and environmental extracts. *Anal. Biochem.* **89,** 581–595.
20. Karl, D. M. (1978) Determination of GTP, GDP, and GMP in cell and tissue extracts. *Methods Enzymol.* **57,** 85–94.
21. Goodrich, G. A. and Burrell, H. R. (1982) Micromeasurement of nucleoside 5'-triphosphates using coupled bioluminescence. *Anal. Biochem.* **127,** 395–401.
22. Moyer, J. D. and Henderson, J. F. (1983) Ultrasensitive assay for ribonucleoside triphosphates in 50–1000 cells. *Biochem. Pharmacol.* **32,** 3831–3834.
23. Ford, S. R., Vaden, V. R., Booth, J. L., Hall, M. S., Webster, J. J., and Leach, F. R. (1994) Bioluminescent determination of 0.1 picomole amounts of guanine nucleotides. *J. Biolumin. Chemilumin.* **9,** 251–265.
24. Ball, W. J. and Atkinson, D. E. (1975) Adenylate energy charge in *Saccharomyces cerevisiae* during starvation. *J. Bacteriol.* **121,** 975–982.
25. Schaeffer, J. M. and Anderson, S. M. (1985) Dopamine stimulated increase of GTP levels in the rat retina. *J. Biol. Chem.* **260,** 4555–4557.
26. Ostroy, S. E., Svoboda, R. A., and Wilson, M. J. (1990) A stage in glycolysis controls the metabolic adjustments of vertebrate rod photoreceptors upon illumination. *Biochem. Biophys. Res. Commun.* **168,** 155–160.
27. Scheele, J. S., Rhee, J. M., and Boss, G. R. (1995) Determination of absolute amounts of GDP and GTP bound to Ras in mammalian cells: Comparison of parental and Ras-overproducing NIH 3T3 fibroblasts. *Proc. Natl. Acad. Sci. USA* **92,** 1097–1100.
28. Detimary, P., Van den Berg, G., and Henquin, J.-C. (1996) Concentration dependence and time course of the effects of glucose on adenine and guanine nucleotides in mouse pancreatic islets. *J. Biol. Chem.* **271,** 20,559–20,565
29. Keppler, D. (1985) Guanosine 5'-triphosphate, guanosine 5'-diphosphate, and guanosine 5'-mononphosphate. *Methods Enzymol. Anal.* **7,** 409–419.
30. Pogson, C. I., Gurnah, S. V., and Smith, S. A. (1979) A sensitive and specific assay for GTP and GDP in tissue extracts. *Int. J. Biochem.* **10,** 995–1000.
31. De Azeredo, F. A. M., Feussner, G. K., Lust, W. D., and Passonneau, J. V. (1979) An enzymatic method for the measurement of GTP and GDP in tissue extracts. *Anal. Biochem.* **95,** 512–519.
32. Goodrich, G. A. and Burrell, H. R. (1982) Micromeasurement of nucleoside 5'-triphosphates using coupled bioluminescence. *Anal. Biochem.* **127,** 395–401.
33. Cheung, C. P. and Marcus, A. (1976) Guanine nucleotide determination in extracts of wheat embryo. *FEBS Lett.* **70,** 141–144.
34. Brady, W. A., Kokoris, M. S., Fitzgibbon, M., and Black, M. E. (1996) Cloning, characterization, and modeling of mouse and human guanylate kinase. *J. Biol. Chem.* **271,** 16,734–16,740.
35. Nitschmann, W. H. (1985) A firefly luciferase assay for determination of cytidine 5'-triphosphate in biological samples. *Anal. Biochem.* **147,** 186–193.

6

Bioluminescent Assay
of the Adenylate Energy Charge

Sharon R. Ford and Franklin R. Leach

1. Introduction

The "adenylate control hypothesis" proposed by Atkinson from UCLA *(1)* was further developed into the adenylate energy charge (AEC) concept in the 1960s *(2,3)* and thereafter codified *(4)*. This concept arose from making an analogy between the adenine nucleotide pool of ATP, ADP, and AMP, and a chemical storage battery or accumulator cell. The common elements include: (1) the total amount of material typically remains constant, and (2) chemical energy can be stored and recovered by alteration of the ratio of the components. The charge of the three-component adenylate system is given by the mole fraction of ATP plus $1/2$ of the mole fraction of ADP. It is shown in the following expression:

$$\text{Adenylate energy charge} = \frac{[ATP] + 0.5\,[ADP]}{[ATP] + [ADP] + [AMP]} \tag{1}$$

The adenylate energy charge is a linear measure of the metabolic energy stored in the adenine nucleotide system. When only AMP is present, the energy charge is zero, and when only ATP is present, the energy charge is one. This is still a fundamental concept taught in biochemistry courses *(5)*.

ATP has several metabolic roles: it is a phosphoryl and pyrophosphoryl carrier, it is incorporated into several cofactors, it serves as an allosteric effector, it is incorporated into RNA as a mononucleotide residue, it serves in the activation of methyl groups via the formation of S-adenosyl methionine, and it functions in control by covalent modifications (adenylations/deadenylations in addition to phosphorylations/dephosphorylations). ATP is the free energy currency of living systems, and its energy can be used to make feasible unfavorable reactions via coupling.

From: *Methods in Molecular Biology, Vol. 102: Bioluminescence Methods and Protocols*
Edited by: R. A. LaRossa © Humana Press Inc., Totowa, NJ

Energy in the form of ATP and reducing power in the form of NADPH are required to drive the synthetic reactions essential for life. A measure of the energy status of a cell or organism can provide insight into its health. Therefore, a method for determination of the individual adenylates would be useful. Wiebe and Bancroft *(6)* observed that actively growing and dividing microbial cells have energy charge ratios of 0.8–0.95; cells that are in the stationary phase of growth have a ratio of about 0.6; and resting or senescent cells have ratios below 0.5.

Figure 7-1 of Atkinson's monograph *(7)* plots the value of the energy charge observed vs the year of its measure. In 1948 the average energy charge had a value of 0.45, and by 1972 it had climbed to values of 0.85. The changes are owing to improvements in analytical techniques for measurement of the adenylates.

This chapter is concerned with techniques for the bioluminescent measurement of the adenylate energy charge. Determination of the adenylate energy charge reveals the metabolic status of the biological entity sampled. This can be individual cells, tissues, organisms, or even communities of organisms.

2. Materials

2.1. Water and Glassware

Water quality is of paramount importance. Because of the sensitivity of the technique, minute contamination of reagents (especially bacterial contamination) will cause high background luminescence. We routinely prepare the water used in all reagents as follows: The building's reverse osmosis- and UV-treated water is passed through two mixed-bed, ion-exchange resins (Barnstead/ Thermolyne D 8902 Ultrapure Cartridges, Dubuque, IA), glass-distilled, pressure-filtered through a sterile 0.45-μm Millipore® (Bedford, MA) filter into sterile bottles and then autoclaved. After opening, a bottle of water can be used for several days if handled using good sterile technique.

We recommend as a minimum standard that "Milli-Q-quality" water be additionally filtered through a sterile 0.45-μm filter and autoclaved before use. Backgrounds in the standard ATP assay containing 100 μL of Firelight® luciferase/ luciferin and no ATP in a 500 μL total volume should be <100 counts/10 s in a Lumac Model 2010A Biocounter. If backgrounds are high, the "Milli-Q" water should be distilled before filtering and autoclaving.

We recommend that all glassware used for reagents for these assays be washed in phosphate-free detergent, soaked in Pierce brand RBS-pf® (Rockford, IL), rinsed in reverse osmosis- (RO) treated or deionized water, and sterilized (*see* **Note 1**).

2.2. Enzymes

1. Firefly luciferase: Dissolve Firelight® (cat. no. # 2005; Analytical Luminescence Laboratory, Ann Arbor, MI) according to the manufacturer's directions in 50 m*M* Tricine, pH 7.8; containing 1 m*M* MgSO$_4$; 0.1 m*M* dithiothreitol; 0.1 m*M* EDTA;

and 0.1% bovine serum albumin. Age the enzyme solution on ice for at least 30 min before use (**Note 2**). If more than one vial of enzyme is needed, all vials should be pooled before beginning the assays. Unused enzyme can be stored overnight at 4°C and used again the next day, but the enzyme loses activity with storage.

2. Adenylate (myo) kinase: We use M 3003 from Sigma (St. Louis, MO). The enzyme preparation comes as an ammonium sulfate suspension. Pellet the enzyme by centrifuging for 5 min, at 10,000g in a microcentrifuge. Dissolve the pellet in 20 mM Tricine, pH 7.5; and 0.1% bovine serum albumin. Prepare only the amount of enzyme to be used in a single day.

3. Pyruvate kinase: We use P 1506 from Sigma. Prepare this enzyme in the same manner as the adenylate kinase.

2.3. Chemicals and Stock Solutions

We report the brand of chemicals that we used and found suitable. Many other manufacturers have equivalent preparations that could be used. Since the reagents from other manufacturers were not tested, we cannot report on their specific suitability. The chemicals used were Tris (T 3253, Sigma), Tricine (T 9784, Sigma), bovine serum albumin (A 2153, Sigma), MgSO$_4$ (ACS-reagent-grade), KCl (ACS-reagent-grade), dithiothreitol (D 5545, Sigma), EDTA (E 1644, Sigma), DMSO (EM Science, Cherry Hill, NJ), urea (Fisher, Pittsburgh, PA, or Sigma), phosphoric acid (J.T. Baker, Phillipsburg, NJ), ethanolamine (Eastman, Rochester, NY), phenol red (Fisher), antifoam A (Sigma), ATP (Sigma), and polyoxyethylene 10 lauryl ether (P9769, Sigma).

Stock solutions follow. Use a stirring hot plate to warm and stir the solutions during preparation. The procedure for preparation of the extractant is improved by preparing and storing two mixtures separately.

1. Ethanolamine stock: 5 and 10 N are prepared by diluting 1 mL of the liquid with 2.3 mL of water and 10 mL of the liquid with 6.67 mL of water, respectively.

2. Phenol red stock: Dissolve 0.1 g phenolsulfonphthalein in 20% ethanol (useful range pH 6.8 [yellow] to 8.2, [red]; desired pH 7.6 [orange]).

3. Mix A detergent: Dissolve polyoxyethylene 10 lauryl ether (0.2 g) in 34 mL of sterile "Milli-Q quality" water by warming to about 45°C on a hot plate; keep the solution at about 35°C. (We did this by keeping the solution at the edge of the hot plate.)

4. 10 M urea: Dissolve 600 g urea in 500 mL of water with heating and stirring. Then dilute the solution to 1000 mL and store at room temperature.

5. Dimethylsulfoxide: Use at room temperature.

6. Adenosine solution: Dissolve 2.5 g adenosine in 350 mL of water, with stirring and heating. Then add 150 mL of water. Dispense the solution into 100-mL prescription bottles and autoclave. Store at room temperature.

7. 1 M Ethylenediaminetetraacetic acid: Dissolve 37.22 g of Na$_2$EDTA·H$_2$O in 100 mL of water with heat and stirring. Adjust to pH 6.4, if necessary, to achieve solution. Store at room temperature. If using another salt or hydrate form of EDTA, be sure to correct mass used to prepare a 1 M solution.

8. Mix A: To prepare 80 mL of mix A, add in order: 34 mL of mix A detergent (polyoxyethylene 10 lauryl ether), 20 mL of 10 M urea, 20 mL of DMSO, 4 mL of adenosine solution, and 2 mL of 1 M EDTA.
9. Mix B: Mix phosphoric acid (228 mL of 85%, reagent-grade acid) with 772 mL of water; store at room temperature.
10. Extractant: To prepare the extractant, mix 80 mL of mix A with 20 mL of mix B reagent. A precipitate may occur as the extractant cools (*see* **Note 5**).
11. 10X Buffer mix contains 750 mM Tricine, pH 7.5; 50 mM MgCl$_2$; and 0.125 mM KCl.

2.4. Instrumentation

Any high-quality luminometer can be used for these assays. For best results, the reactions should be initiated within the counting chamber of the luminometer. We routinely use a Lumac Model 2010A Biocounter (Celsis, Cambridge, UK) with automatic injection of the Firelight® reagent to start the reaction (*see* **Note 3**).

3. Methods

The first problem the analyst faces is the rapid termination of all metabolism allowing extraction of the adenylates or other small molecules in an unadulterated, representative form and quantity.

3.1. Extraction of Adenylates (or Other Nucleotides)

Several procedures are available for extraction of nucleotides from living material; the choice should be based on properties of the sample. Stanley *(8)* has reviewed extraction methods.

3.1.1. Boiling Buffer Extraction

The simplest extraction procedure is treatment with a boiling buffer solution. This is especially useful for bacterial cultures. The procedure this laboratory uses follows:

1. A 0.5-mL aliquot of the cell suspension is taken with a 2-mL Cornwall-type automatic syringe (Becton Dickinson, Rutherford, NJ, no. 3052, Cornwall repeating dispenser used in a single dispense mode with a fixed volume—a Pipetman type automatic pipetter could be similarly used) and injected into 4.5 mL of buffer (50 mM Tricine, 10 mM MgSO$_4$, and 2 mM EDTA at pH 7.8) in a boiling water bath.
2. The tubes containing the buffer are heated for ≥5 min in a rapidly boiling and stirred water bath before the samples are added.
3. After addition of the samples, the mixture is heated for 3 min, during which the temperature inside the tube will reach 95–96°C.
4. The tubes are removed and chilled in ice for a minimum of 10 min.
5. The samples are warmed to room temperature before ATP determination.

3.1.2. Tricholoracetic Acid Extraction

Lundin and colleagues *(9)* found that 1.25–10% trichloroacetic acid (with the optimum concentration depending on the cell type and requiring experimental determination) gave the highest yield of ATP and the highest AEC among seven compared extractants. With the particular cell type they studied (LNCaP-r cells), 2.3% trichloroacebic acid was used.

1. 1 vol of 10% trichloroacetic acid is added to 2 vol of culture.
2. After 30 min, the extracts are transferred to an ice bath.

3.1.3. Extraction from Environmental Samples

For extraction of nucleotides from complex environmental samples, such as soil, a more rigorous treatment is required. This laboratory has developed *(10)* and improved *(11,12)* such an extraction procedure from soil, which is described below (taken from **ref.** *12* with permission).

1. Scrape the surface of the area to be sampled (about 2.5 cm deep) to remove vegetation and debris.
2. Dig the soil using a shovel, and remove visible roots, rocks, sticks, and other plant or animal material (worms).
3. Pulverize the soil (we use a four-pronged garden tool).
4. Sift through a flour sifter (we use a Bromwell three-cup two-wire agitator, 1.8 mm wire mesh, no. RCS-00039).
5. Weigh samples (100 g) of soil into a convenient vessel (we use paper cups), and immediately cover with aluminum foil.
6. Add 45 mL of the freshly prepared extractant mixture, 5 mL of M-9 medium *(13)*, 100 g soil sample, and 75 µL of Antifoam A to a stainless-steel semimicro jar of a Waring® blender.
7. Blend the mixture for 1 min in 20-s intervals. After each 20-s treatment, mix the contents of the jar using a sterile spatula.
8. Place the blended sample into a centrifuge tube, and centrifuge for 20 min at $30 \times 10^3 g$. (In the field we centrifuge at ambient temperature; in the laboratory at 4°C).
9. Remove the supernatant solution, and dilute it $^1/_{10}$ with 0.1 M Tricine buffer, pH 11.2.
10. Check the pH of a 200-µL aliquot of the supernatant solution using 50 µL of phenol red (orange color is pH 7.6).
11. Adjust the solution to between pH 7.0 and 8.0 with ethanolamine. Note the amount of ethanolamine used, and calculate the amount of ethanolamine needed to neutralize the sample from which the aliquot was taken.
12. Since phenol red inhibits firefly luciferase, neutralize the samples to be used for ATP determination by adding the experimentally determined amount of ethanolamine without the indicator.
13. Store the neutralized samples on ice until analyzed.
14. Warm samples to room temperature before assay.

3.2. Measurement of ATP

ATP is measured in a 500-μL mixture with 50 μL of sample; 100 μL of Analytical Luminescence Laboratory's firefly luciferase (Firelight®); 50 μL of Tricine buffer, pH 7.8; containing 250 mM Tricine, 50 mM MgSO$_4$, 5 mM EDTA, and 5 mM dithiothreitol; and 300 μL water. The bioluminescence is determined on a Lumac Model 2010A Biocounter for 10 s.

1. Prepare mixture containing all components except the firefly luciferase, vortex, and place in the luminometer.
2. Initiate the reaction by injecting the luciferase using the instrument's automatic injection system.
3. Assay a second aliquot of each sample containing an internal standard of 0.1 ng of ATP to determine the extent of inhibition, if any, of the assay itself (we have found about 40% inhibition).

See Webster and Leach *(14)* and Chapter 1 for optimization of the firefly luciferase ATP assay. Standards should be used to ascertain the efficiency of various procedures (*see* **Note 4**).

3.3. Determination of ATP, ADP, and AMP in Samples

AEC is determined by modifications of the procedures described by Ball and Atkinson *(15)* and Lundin and Thore *(16)*. Set up the reaction mixtures as described by Holm-Hansen and Karl *(17,18)* with the substitution of Tricine buffer for phosphate buffer and the inclusion of KCl (required for pyruvate kinase). There is less inhibition of light production from ATP with Tricine buffer than with phosphate buffer. Aliquots, 200 μL, of the samples to be analyzed are separately incubated in 13 × 100 mm test tubes with 50 μL of three different buffered enzyme and substrate solutions as described in **Subheadings 3.3.1.–3.3.3.** Experiments should be performed to optimize the assays under each laboratory's experimental conditions.

3.3.1. ATP Measurement (Reaction A)

1. Prepare reaction A buffer containing 75 mM Tricine, pH 7.5; 5 mM MgCl$_2$; and 0.0125 mM KCl by making a $^1/_{10}$ dilution of the 10X buffer mix.
2. Add 50 μL of reaction A buffer to each 200-μL sample to be assayed. (The final salt concentration is 15 mM Tricine, 1 mM MgC1$_2$, 2.5 μM KCl plus whatever salt is contributed by the sample.)
3. Incubate samples for 30 min at 30°C.
4. Place all tubes in a boiling water bath for 3 min to stop the reactions.
5. Chill the tubes on ice.
6. Bring to room temperature just before assaying for ATP as described in **Subheading 3.2.**

3.3.2. ADP + ATP Measurement (Reaction B)

1. Prepare reaction B buffer containing, in addition to the contents of reaction A buffer (*see* **Subheading 3.3.1., step 1**), 0.5 mM phosphoenolpyruvate (Sigma P 7002) and 0.40 µg/µL of pyruvate kinase. The phosphoenolpyruvate and pyruvate kinase convert ADP in the samples to ATP.
2. Add 50 µL of reaction B buffer to each 200-µL sample to be assayed.
3. Incubate this sample for 30 min at 30°C.
4. Place all tubes in a boiling water bath for 3 min to stop the reactions.
5. Chill the tubes on ice.
6. Bring to room temperature just before assaying for ATP as described **Subheading 3.2.**

3.3.3. AMP + ADP + ATP Measurement (Reaction C)

1. Prepare reaction C buffer containing, in addition to the contents of reaction B buffer (*see* **Subheading 3.3.2., step 1**), 0.5 µg/µL of adenylate (myo)kinase. The adenylate kinase converts AMP in the sample to ADP. The phosphoenolpyruvate and pyruvate kinase convert ADP in the sample to ATP.
2. Add 50 µL of reaction C buffer to each 200-µL sample to be assayed.
3. Incubate samples for 90 min at 30°C.
4. Place all tubes in a boiling water bath for 3 min to stop the reactions.
5. Chill the tubes on ice.
6. Bring to room temperature just before assaying for ATP as described **Subheading 3.2.**

3.4. Calculation of AEC

The AEC is calculated as described by Ball and Atkinson *(15)*. Multiply the sum of the counts measured in Reactions A and B by 0.5 and divide by the counts in Reaction C. We recommend the inclusion of mixtures of the adenylates yielding energy charges from 0.25–0.75 as controls. **Table 1** shows results for these control determinations, which establish the suitability of the modified procedure and document the standard deviations for AEC measurements.

3.5. Alternative One-Tube Automated Reaction

A protocol for automated determination of the adenylates has been developed and applied by Lundin and colleagues *(9)*. An appropriate system is now commercially available from BioTherma AB (Dalarö, Sweden). A BioOrbit 1251 Luminometer equipped with multiple injectors is the measuring instrument. The luminometer is controlled by software running on either an IBM-compatible PC or Macintosh computer via BioThema's communication kit. There is a support kit for determination of ATP/ADP/AMP that involves the use of a Microsoft Excel macro for the necessary calculations. Instructions are provided on how to prepare the reagents, how to operate the Model 1251

Table 1
AEC Determinations

Mixture of nucleotides	AEC	
ATP—ADP—AMP	Known	Found
1—1—0	0.75	0 75 ± 0.01
1—0—1	0.50	0 49 ± 0 00
0—1—1	0.25	0.26 ± 0.01
1—1—1	0.50	0.48 ± 0.02

instrument, and unlimited phone and fax support. A similar determination of ATP and phosphocreatine in a single human skeletal muscle fiber using sequential injections into a single cuvet illustrates the types of results that are obtained *(19)*.

3.6. Other Methods of Determination and Improvements

Protocols developed for the sensitive spectrophotometric or fluorometric determination of AMP, ADP, or ATP at several different levels have been described by Passonneau and Lowry *(20)*. Spielmarn et al. *(21)* enhanced the bioluminescent determination of the adenylates by excluding the contaminating enzymes of adenylate kinase and nucleoside diphosphate kinase by using purified commercial luciferase from LKB and removing ammonium sulfate from adenylate kinase by centrifugation. Contaminants in pyruvate kinase and adenylate kinase were the limiting factors in the sensitivity of the assay.

4. Notes

1. It has been our experience that whenever we have trouble with high backgrounds, the most likely problem is that the water is contaminated. Investigators who have had trouble repeating our results have usually not been as careful about the preparation and storage of water and other reagents as we think is necessary. We find that careful attention paid to preparation of reagents pays off in the quality and reproducibility of results.

2. Firefly luciferase: Three grades of firefly luciferase with different degrees of purity are commercially available. Crude lantern extracts contain sufficient pyrophosphatase so that PP_i does not accumulate *(22)*. These preparations also contain adenylate kinase and nucleoside diphosphate kinase, which enable nucleotides other than ATP to be enzymatically converted to ATP and thus produce light in the assay system. These preparations are not recommended for sensitive determination of ATP. Purification procedures have been developed that remove the adenylate kinase, pyrophosphatase, and nucleoside diphosphate kinase. These preparations can be used for the sensitive determination of ATP. Many are supplemented with sufficient luciferin so that no additional luciferin is required. Crys-

talline luciferase is purer, but is somewhat more difficult to handle. There is little difference between crystalline native and recombinant firefly luciferases. The slight differences in conformation and lability to proteolytic enzyme that exist for these two luciferases are not significant *(23)*.

Although firefly luciferase can be fairly stable when properly stored *(24)*, we recommend the use of a commercial preparation (such as Analytical Luminescence Laboratory's Firelight®) freshly made and pooled each day. The use of a commercial preparation with its stabilizers and quality control means that the individual laboratory does not have to have its own reagent quality control program. This laboratory has operated both systems and finds that the use of commercial kits is better for routine studies. The use of commercial kits is now much more accepted with the advent of molecular biology's cloning kits—it is more time-efficient to let the supplier provide the quality control. This means carefully selecting a supplier of reagents. This laboratory evaluated the commercially available reagents in 1986 *(25)*. Much progress has been made in commercial firefly luciferase reagent kits during the subsequent decade. Many of the suppliers listed in Table 1 of **ref.** *25* no longer supply the reagents, and there are also many new suppliers. The techniques and experiments used in the comparative evaluations are still appropriate to evaluate those products. The commercial firms whose products have survived probably have done so because of good quality. Beginning in 1993 Stanley has published lists of commercial firms providing luminescence kits based on information provided by the supplier *(26–30)*. There is no experimental comparison of the kits and reagents in Stanley's listing.

3. Instrumentation—Luminometer: Although relatively expensive and specialized, we recommend the use of an instrument designed for bioluminescent/chemiluminescent measurements. These instruments have a wide range of specific properties (such as geometry of the detector) and design criteria (temperature control and sample size). Some permit variation of the high voltage supplied to the photomultiplier, whereas others have fixed voltage; some allow temperature regulation, but others operate at room temperature. Ten commercially available instruments have been experimentally compared by Jago and associates *(31)*. The most sensitive instruments were the Lumac Model 2010A Biocounter and the Turner 20 TD photometers, which had actual limits of 0.09 and 0.12 pg ATP/ sample, respectively. Turner *(32)* presents a provocative assessment of instrument development from the viewpoint of a person trained in physics and electronics trying to obtain the most out of the instrument/reagent system. Van Dyke *(33)* reviewed the manufacturers' provided information for photometers that were available in 1985. Further review of the commercial instrumentation was made by Stanley in a continuing series of articles *(26–30,34–36)*.

If the investigator desires to construct a photometer, Anderson et al. *(37)* give complete instructions. These instructions were updated in 1985 *(38)* with "the strong recommendation that in most cases a researcher would be better served to purchase a commercial instrument."

For calibration of light production, please refer to the methods described by O'Kane and coworkers *(39)* and by Lee and Seliger *(40)*.

4. Standards: Spiking a sample with a known amount of ATP gives a rapid assessment of whether or not there is inhibition resulting from the components of the assay mixture. To determine the effectiveness of the overall procedure, an aliquot of a bacterial culture is added to the extractant solution and experimental sample. The result obtained is compared with that resulting from a direct extraction of the bacterial culture.

5. Extractant: The precipitate does not interfere with the extraction. The extractant is best if made fresh daily, but can be used for up to 1 wk.

6. Examples of the application of bioluminescent techniques for the measurement of nucleotides include:
   a. Determination of the adenylate and guanylate energy charges in a subsurface *Pseudomonas* sp. during growth in a rich medium *(41)*;
   b. Determination of metabolic potential in subsurface samples *(42)*;
   c. Determination of the AEC of *Bacillus stearothermophilus* during growth *(43)*; and
   d. Measurement of AEC in soil samples *(44)*.

   A typical application of bioluminescent determination of the adenylates is the measurement of the adenine nucleotides in bile by Cari et al. *(45)*. ATP acts as a messenger to activate the purinergic receptors of biliary epithelial cells to release Cl^-. They conclude that adenine nucleotides are present in sufficient concentrations in the bile for physiological effects and might function in the regulation of biliary secretions.

Acknowledgments

This research was supported in part by the Oklahoma Agricultural Experiment Station (Project 1806) and is published with the approval of the director. E. C. Nelson and Vanitha Thulasiraman read the manuscript and suggested improvements.

References

1. Atkinson, D. E. (1966) Regulation of enzyme activity. *Ann. Rev. Biochem.* **35,** 85–124.
2. Atkinson, D. E. and Walton, G. M. (1967) Adenosine triphosphate conservation in metabolic regulation. Rat liver citrate cleavage enzyme. *J. Biol. Chem.* **242,** 3239–3241.
3. Atkinson, D. E. and Fall, L. (1967) Adenosine triphosphate conservation in biosynthetic regulation. *Escherichia coli* phosphoribosylpyrophosphate synthase. *J. Biol. Chem.* **242,** 3241,3242.
4. Atkinson, D. E. (1968) The energy charge of the adenylate pool as a regulatory parameter. Interaction with feedback modifiers. *Biochemistry* **7,** 4030–4034.
5. Garrett, R. H. and Grisham, C. M. (1995) Metabolic integration and the unidirectionality of pathways, in *Biochemistry,* Saunders College Publishers, Fort Worth, TX, pp. 816–819.
6. Wiebe, W. J. and Bancroft, K. (1975) Use of the adenylate energy charge ratio to measure growth state of natural microbial communities. *Proc. Natl. Acad. Sci. USA* **72,** 2112–2115.

7. Atkinson, D. E. (1977) *Cellular Energy Metabolism and Its Regulation.* Academic, New York.
8. Stanley, P. E. (1986) Extraction of adenosine triphosphate from microbial and somatic cells. *Methods Enzymol.* **133,** 14–22.
9. Lundin, A., Hasenson, M., Persson, J., and Pousette, A. (1986) Estimation of biomass in growing cell lines by adenosine triphosphate assay. *Methods Enzymol.* **133,** 26–42.
10. Webster, J. J., Hampton, G. J., and Leach, F. R. (1984). ATP in soil: a new extractant and extraction procedure. *Soil Biol. Biochem.* **16,** 335–342.
11. Vaden, V. R., Webster, J. J., Hampton, G. J., Hall, M. S., and Leach, F. R. (1987) Comparison of methods for extraction of ATP from soil. *J. Microbiol. Methods* **7,** 211–217.
12. Pangburn, S. J., Hall, M. S., and Leach, F. R. (1994) Improvements in the extraction of bacterial ATP from soil with field application. *J. Microbiol. Methods* **20,** 197–209.
13. Anderson, E. H. (1940) Growth requirements of virus-sensitive mutants of *Escherichia coli* and strain "B." *Proc. Natl. Acad. Sci. USA* **32,** 120–128.
14. Webster, J. J. and Leach, F. R. (1980) Optimization of the firefly luciferase assay for ATP. *J. Appl. Biochem.* **2,** 469–479
15. Ball, W. J. and Atkinson, D. E. (1975) Adenylate energy charge in *Saccharomyces cerevisiae* during starvation. *J. Bacteriol.* **121,** 975–982.
16. Lundin, A. and Thore, A. (1975) Comparison of methods for extraction of bacterial adenine nucleotides determined by firefly assay. *Appl. Microbiol.* **30,** 713–721.
17. Karl, D. M. and Holm-Hansen, O. (1978) Methodology and measurement of adenylate energy charge ratios in environmental samples. *Marine Biol.* **48,** 185–197.
18. Holm-Hansen, O. and Karl, D. M. (1978) Biomass and adenylate energy charge determination in microbial cell extracts and environmental samples. *Methods Enzymol.* **57,** 73–85.
19. Wibom, R., Soderlund, K., Lundin, A., and Hultman, E. (1991) A luminometric method for the determination of ATP and phosphocreatine in single human skeletal muscle fibres. *J. Biolumin. Chemilumin.* **6,** 123–129.
20. Passonneau, J. V. and Lowry, O. H. (1993) *Enzymatic Analysis: A Practical Guide.* Humana, Totowa, NJ.
21. Spielman, H., Jacob-Muller, U., and Schulz, P. (1981) Simple assay of 0.1–1.0 pmol of ATP, ADP, and AMP in single somatic cells using purified luciferin luciferase. *Anal. Biochem.* **113,** 172–178.
22. DeLuca, M. and McElroy, W. D. (1978) Purification and properties of firefly luciferase. *Methods Enzymol.* **57,** 3–15.
23. Ford, S. R., Hall, M. S., and Leach, F. R. (1992) Comparison of properties of commercially available crystalline native and recombinant firefly luciferases. *J. Biolumin. Chemilumin.* **7,** 185–193.
24. Hall, M. S. and Leach, F. R. (1988) Stability of firefly luciferase in tricine buffer and in a commercial enzyme stabilizer. *J. Biolumin. Chemilumin.* **2,** 41–44.

25. Leach, F. R. and Webster, J. J. (1986) Commercially available firefly luciferase reagents. *Methods Enzymol.* **133,** 51–70.
26. Stanley, P. E. (1993) A survey of some commercially available kits and reagents which include bioluminescence or chemiluminescence for their operation. *J. Biolumin. Chemilumin.* **8,** 51–63.
27. Stanley, P. E. (1993) Commercially available luminometers and imaging devices for low-light measurements and kits and reagents utilizing chemiluminescence or bioluminescence: Survey update 1. *J. Biolumin. Chemilumin.* **8,** 234–240.
28. Stanley, P. E. (1993) Commercially available luminometers and imaging devices for low-light measurements and kits and reagents utilizing chemiluminescence or bioluminescence: Survey update 2. *J. Biolumin. Chemilumin.* **9,** 51–53.
29. Stanley, P. E. (1993) Commercially available luminometers and imaging devices for low-light measurements and kits and reagents utilizing chemiluminescence or bioluminescence: Survey update 3. *J. Biolumin. Chemilumin.* **9,** 123–125.
30. Stanley, P. E. (1993) Commercially available luminometers and imaging devices for low-light measurements and kits and reagents utilizing chemiluminescence or bioluminescence: Survey update 4. *J. Biolumin. Chemilumin.* **11,** 175–191.
31. Jago, P. H., Simpson, W. J., Denyer, S. P., Evans, A. W., Griffiths, M. W., Hammond, J. R. M., Ingram, T. P., Lacey, R. F., Macey, N. W., McCarthy, B. J., Salusbury, T. T., Senior, P. S., Sidorowicz, S., Smithers, R., Stanfield, G., and Stanley, P. E. (1989) An evaluation of the performance of ten commercial luminometers. *J. Biolumin. Chemilumin.* **3,** 131–145.
32. Turner, G. K. (1985) Measurement of light from chemical or biochemical reactions, in *Bioluminescence and Chemiluminescence: Instruments and Application,* vol. I (Van Dyke, K., ed.), CRC, Boca Raton, FL, pp. 43–78.
33. Van Dyke, K. (1985) Commercial instruments, in *Bioluminescence and Chemiluminescence: Instruments and Application,* vol I (Van Dyke, K., ed.), CRC, Boca Raton, FL, pp. 83–128.
34. Stanley, P. E. (1985) Characteristics of commercial radiometers. *Methods Enzymol.* **133,** 587–603.
35. Stanley, P. E. (1992) A survey of more than 90 commercially available luminometers and imaging devices for low light: measurement of chemiluminescence and bioluminescence, including instruments for manual, automatic and specialized operation for HPLC, LC, GLC and microplates. Part 1 descriptions, *J. Biolumin. Chemilumin.* **7,** 77–108.
36. Stanley, P. E. (1992) A survey of more than 90 commercially available luminometers and imaging devices for low light measurement of chemiluminescence and bioluminescence, including instruments for manual, automatic and specialized operation for HPLC, LC, GLC and microplates. Part 1 photographs. *J. Biolumin. Chemilumin.* **7,** 157–169.
37. Anderson, J. M., Faini, G. J., and Wampler, J. E. (1978) Construction of instrumentation for bioluminescence and chemiluminescence assays. *Methods Enzymol.* **57,** 529–540.

38. Wampler, J. E. and Gilbert, J. C. (1985) The design of custom radiometers, in *Bioluminescence and Chemiluminescence: Instruments and Application,* vol I (Van Dyke, K., ed.), CRC, Boca Raton, FL, pp. 129–150.
39. O'Kane, D. J., Ahmad, M., Matheson, I. B. C., and Lee, J. (1986) Purification of bacterial luciferase by high-performance liquid chromatography. *Methods Enzymol.* **133,** 109–127.
40. Lee, J. and Seliger, H. H. (1972) Quantum yields of the luminol chemiluminescence reaction in aqueous and aprotic solvents. *Photochem. Photobiol.* **15,** 109–127.
41. Ford, S. R., Hall, M. S., Vaden, V. R., Webster, J. J., and Leach, F. R. (1994) Adenylate and guanylate energy charges in a subsurface *Pseudomonas* sp. *Proc. Okla. Acad. Sci.* **74,** 31–36.
42. Webster, J. J., Hall, M. S., and Leach, F. R. (1992) ATP and adenylate energy charge determinations on core samples from an av-fuel spill site at the Traverse City, Michigan airport. *Bull. Environ. Contam. Toxicol.* **49,** 232–237.
43. Webster, J. J., Walker, B. G., and Leach, F. R. (1988) ATP content and adenylate energy charge of *Bacillus stearothermophilus* during growth. *Curr. Microbiol.* **16,** 271–275.
44. Vaden, V. R., Webster, J. J., Hampton, G. J., Hall, M. S., and Leach, F. R. (1987) Comparison of methods for extraction of ATP from soil. *J. Microbiol. Methods* **7,** 211–217.
45. Chari, R. S., Schultz, S. M., Haebig, J. E., Shimokura, G. H., Cotton, P. B., Fitz, J. G., and Meyers, W. C. (1996) Adenine nucleotides in bile. *Am. J. Physiol.* **270,** G246–G252.

III

MOLECULAR BIOLOGY TOOLS

7

Photorhabdus luminescens luxCDABE Promoter Probe Vectors

Tina K. Van Dyk and Reinhardt A. Rosson

1. Introduction

The discovery and characterization of *cis*-acting promoter regions and *trans*-acting regulatory proteins is often aided by use of reporter gene fusions. In bacterial systems, frequently used reporters include *lacZ* (β-galactosidase), *phoA* (alkaline phosphatase), *cam* (chloramphenicol transacetylase), and *lux* (bacterial luciferase). Among these, the *lux* reporter is distinct, because the reporter genes' ultimate product, light, can be measured without disrupting the cell. Furthermore, utilization of the five-gene *luxCDABE* reporter system allows continuous monitoring of light production. An additional advantage of the *lux* reporter is the sensitivity and large dynamic range of light detection equipment, which enables very weak to very strong transcriptional activity to be measured.

Bioluminescent bacteria are found in marine and terrestrial environments. The *lux* gene products from marine organisms often exhibit thermolability, such that they do not efficiently function in standard media at growth temperatures typical for many other bacteria *(1–3)*. In contrast, the *lux* gene products from a terrestrial microorganism, *Photorhabdus luminescens* (formerly called *Xenorhabdus luminescens*), produce light at temperatures up to 45°C *(1)*.

Two promoter probe vectors that use the *P. luminescens luxCDABE* reporter are described here. Plasmid pJT205 is of high copy number, whereas plasmid pDEW201 has a moderate copy number. These vectors can be used to discover and characterize promoter sequences. Briefly, isolated chromosomal DNA is partially digested with a restriction enzyme to give overlapping fragments, which are subsequently ligated within a multiple cloning site upstream of the *lux* genes. Those DNA sequences containing promoters will result in tran-

From: *Methods in Molecular Biology, Vol. 102: Bioluminescence Methods and Protocols*
Edited by: R. A. LaRossa © Humana Press Inc., Totowa, NJ

scription and subsequent translation of the *luxCDABE* gene products and, hence, light production from cells containing the fusion plasmid. Quantitation of this bioluminescence results in identification of sequences with promoter activity. The details of methodology to discover promoters from *Eschorichia coli* DNA are given. These methods could also be adapted to find promoters in DNA isolated from other microorganisms.

2. Materials

2.1. Chromosomal DNA Isolation

1. *E. coli* strain W3110 [F⁻ prototroph] *(4)*.
2. LB medium: per liter, 10 g Bacto-tryptone, 5 g Bacto-yeast extract, 10 g NaCl. Sterilize by autoclaving *(5)*.
3. Proteinase K solution: 5 mg/mL (Boehringer Mannheim, Indianapolis, IN) in autoclaved water. Store at –20°C.
4. RNase A solution: 10 mg/mL (Boehringer Mannheim) in water. Boil for 10 min after preparation. Cool to room temperature. Store at –20°C.
5. Phenol/chloroform/isoamyl alcohol: Prepare at 25:24:1 and store at 4°C.
6. Chloroform/isoamyl alcohol: Prepare at 24:1 and store at room temperature.
7. TE: 10 m*M* Tris, 1 m*M* EDTA, pH 8.0. Store at –20° or 4°C.
8. Strataclean™ resin (Stratagene, La Jolla, CA).

2.2. Partial Sau3A I Digestion

1. 10X buffer A: 330 m*M* Tris acetate, 100 m*M* magnesium acetate, 660 m*M* potassium acetate, 5 m*M* dithiothreitol, pH 7.9. Store at –20°C.
2. Restriction enzyme *Sau*3A I (Promega, Madison, WI).
3. 6X gel-loading solution: 0.25% bromophenol blue, 30% glycerol in water. Store at room temperature.
4. SeaKem GTG agarose (FMC, Rockland, ME).
5. Spin-X 0.45 µ*M* cellulose acetate filter units (Costar, Cambridge, MA).
6. Spin-X UF concentrators, 30,000 mol-wt cutoff (Costar).

2.3. Promoter Probe Vectors and Their Preparation

1. pJT205: Plasmid pJT205 was made by cloning the *P. luminescens luxCDABE* genes from pCGLS1 *(6)* into pUC18 *(7)* followed by nuclease *Bal*31-generated deletion of DNA 5' to the *lux* operon. Plasmid pCGLS1 was digested with *Eco*RI, and the 6.9-kbp fragment containing the *lux* genes was isolated after agarose gel electrophoresis. This fragment was ligated to pUC18 DNA that had been linearized by digestion with *Eco*RI. The resultant recombinant plasmid, pCGLS200, contained the *lux* genes oriented opposite to the direction of transcription of P*lac*. Plasmid pCGLS200 was linearized by digestion with *Sst* I, treated with nuclease *Bal*31 for various amounts of time, and then sized on agarose gels. Digestions that resulted in loss of about 500 bp were chosen and circularized by ligation. After transformation into *E. coli*, dimly luminescent colonies were

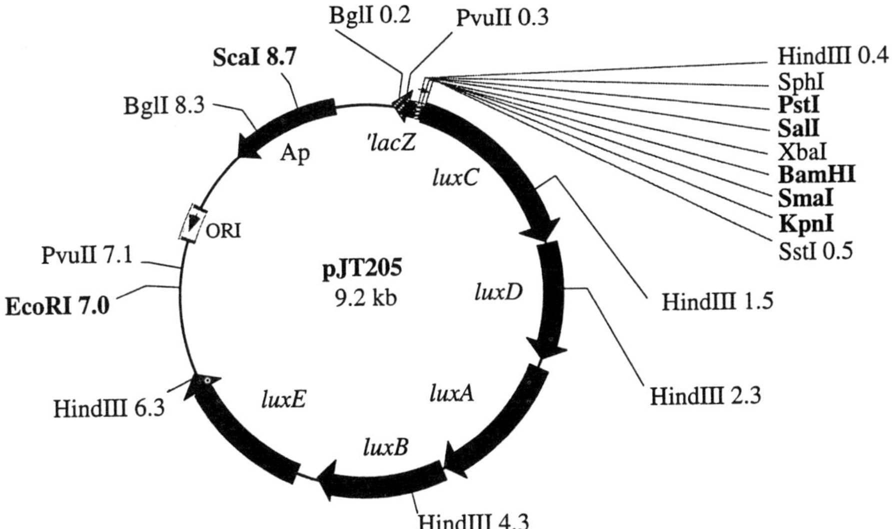

Fig. 1. Map of high copy number plasmid pJT205. The map is drawn approximately to scale. Unique restriction sites are shown in bold.

selected, and plasmid DNA was isolated from these transformants. Such a plasmid was linearized with *Kas*I, which cut approx 165 bp upstream of the 5'-end of the *lux* operon, the ends were made blunt by digestion with Mung Bean nuclease, *Sst*I linkers were attached, and the fragment was religated. The resultant plasmid, pJT202, was digested with *Eco*RI and *Sst*I, and the 6.4-kbp fragment containing the *lux* genes was isolated after agarose gel electrophoresis. This fragment was directionally ligated into pUC18 DNA that had been digested with *Eco*RI and *Sst*I to yield pJT205. DNA sequence analysis showed that all but 8 bp of the *P. luminescence* upstream sequence proximal to *luxC* were deleted.

A map of plasmid pJT205 is shown in **Fig. 1.** Features of this plasmid are:
a. Unique *Pst*I, *Sal*I, *Bam*H I, *Xma*I, and *Kpn*I sites in the multiple cloning site;
b. Promoterless *P. luminescence luxCDABE* genes downstream of the multiple cloning site;
c. Low light production from *E. coli* cells containing this vector without promoter DNA in the multiple cloning site;
d. An ampicillin resistance gene allowing selection for maintenance of the plasmid; and
e. High copy number owing to the deletion of *rop* present in pUC18.
2. pDEW201: Plasmid pDEW201 was made by replacing the promoterless *lacZ* reporter of pRS415 *(8)* with the promoterless *luxCDABE* cassette from pJT205. Plasmid pRS415 contains an origin of replication from pBR322, a *bla* gene, and four tandem transcription terminators from phage T1 upstream of a multiple cloning site. This plasmid DNA was digested with *Bam*HI, *Nru*I and *Eco*RV, and ligated with pJT205 DNA that had been digested with *Bam*HI, *Pvu*II, and *Pst*I.

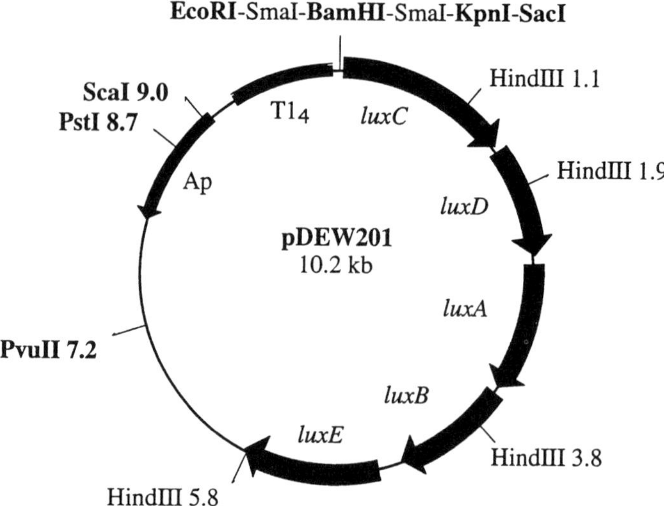

Fig. 2. Map of moderate copy number plasmid pDEW201. The map is drawn approximately to scale. Unique restriction sites are shown in bold.

The ligation mixture was used to transform *E. coli* strain DH5, selecting for ampicillin resistance. Light production from overnight cultures of transformant colonies that had been grown in LB medium containing 150 µg/mL of ampicillin was quantitated in an ML3000 luminometer in the presence and absence of 0.0033% nonanal, a substrate for bacterial luciferase. Light production in the presence of nonanal was, thus, indicative of *luxAB* expression independent of the other *lux* genes. Plasmid DNA was isolated from three transformants that had very low levels of light production (0.0022–0.0029 RLU) in the absence of nonanal and moderate light production (0.91–0.97 RLU) in the presence of nonanal. The background of an empty well was 0.0000–0.0004 RLU. Restriction digestion analysis with various enzymes and combinations of enzymes showed that each of these three plasmids had the expected structure, except that a portion (~900 bp) near the 3'-terminus of the *luxCDABE* operon containing an *Eco*RI site was deleted. One of these plasmids was saved and named pDEW201. The entire *luxCDABE* operon was shown to be intact in pDEW201, because placement of DNA with promoter activity in the multiple cloning site resulted in light production in the absence of exogenously added aldehyde. Furthermore, DNA sequence analysis of the 3'-region of the *lux* operon in pDEW201 showed that the joint of the sequences derived from pJT205 to sequences derived from pRS415 was one nucleotide beyond the termination codon of *luxE*.

A map of plasmid pDEW201 is shown in **Fig. 2.** Features of this plasmid include:
a. Unique *Eco*RI, *Bam*HI, *Kpn*I, and *Sac*I sites in the multiple cloning site;
b. Promoterless *P. luminescens luxCDABE* genes downstream of the multiple cloning site;

   c. Transcription terminators upstream of the multiple cloning site resulting in a very low level of read-through transcription and, hence, very low light production from cells containing this plasmid without promoter DNA cloned into the multiple cloning site;

   d. Ampicillin resistance selection for maintenance of the plasmid; and

   e. Moderate copy number owing to the replication origin and *rop* gene of pBR322.

3. 10X buffer B: 100 m*M* Tris-HCl, 50 m*M* MgCl$_2$, 1 *M* NaCl, 10 m*M* 2-mercaptoethanol, pH 8.0. Store at –20°C.
4. Restriction enzyme *Bam*HI (Promega).
5. 10X CIP buffer: 0.5 *M* Tris-HCl, 10 m*M* MgCl$_2$, 1 m*M* ZnCl$_2$, 10 m*M* spermidine, pH 9.3. Store at –20°C.
6. Calf intestinal alkaline phosphatase (Promega, Madison, WI).

2.4. Library Generation

1. 10X ligase buffer: 300 m*M* Tris-HCl, 100 m*M* MgCl$_2$, 100 m*M* dithiothreitol, 5 m*M* ATP, pH 7.8. Store at –20°C in small aliquots.
2. T4 DNA ligase (Promega).
2. XL2Blue ultracompetent cells (Statagene): Store at –80°C.
3. LB and LBAmp plates: per liter, 10 g Bacto-tryptone, 5 g Bacto-yeast extract, 10 g NaCl, 15 g agar/L. Sterilize by autoclaving. Add ampicillin to a final concentration of 150 µg/mL if required, after autoclaving and cooling the medium to 50–60°C.

2.5. Plasmid Library Isolation

1. Disposable plastic spreaders.
2. Qiagen tip-20 and solutions (Qiagen, Valencia, CA).

2.6. Transformation of the Testing Strain

1. *E. coli* strain DPD1675 *(ilvB2101 ara thi Δ[pro-lac] tolC*::miniTn*10)*: This strain was constructed by P1*vir (5)* mediated transduction selecting for tetracycline resistance using phage grown on strain DE112 *(rpsL200, galK2, lac74 tolC*::miniTn*10)* as the donor and *E. coli* strain CU847 *(ilvB2101 ara thi Δ[pro-lac])* as the recipient. The presence of the *tolC* mutation was confirmed by the transductant's bile salt sensitivity that was scored regarding the inability to grow on MacConkey agar *(9)*.
2. Medium A: Sterile LB broth with 10 m*M* MgSO$_4$·7H$_2$O and 0.2% glucose.
3. Storage solution B: 36% glycerol, 12% PEG8000, 12 m*M* MgSO$_4$·7H$_2$O in LB broth at pH 7.0. Stir over low heat and filter-sterilize.
4. Falcon 2059 tubes, 14 mL vol.

2.7. Toothpicking

1. Microtest III ™ tissue-culture plates, 96-well, flat-bottom with low evaporation lid, sterile (Falcon®, Lincoln Park, NJ).
2. 50X E (Vogel-Bonner minimal medium): Add in order to water at 45°C: 670 mL distilled water, 10 g MgSO$_4$·7H$_2$O, 100 g citric acid·1H$_2$O, 500 g K$_2$HPO$_4$ anhydrous, 175 g NaHNH$_4$PO$_4$·4H$_2$O. Store over chloroform at room temperature *(9)*.

3. E₁GluUraPro medium: Per liter: 20 mL of 50 × E, 10 mL of 40% glucose, 10 mL of 0.25% uracil, 8.3 mL of 2% proline, 0.2 mL of 0.1% thiamine. Filter-sterilize, and store at 4°C. On the day used, add ampicillin to the desired concentration by diluting from a sterile 10 mg/mL solution stored at –20°C.

2.8. Luminometry

1. Sterile, white, flat-bottom microplates: Microlite™ (Dynatech Laboratories, Chantilly, VA).
2. Sterile covers for microplates (Dynatech Laboratories).
3. ML3000 microplate luminometer (Dynatech Laboratories).

3. Methods

3.1. Chromosomal DNA Isolation

1. Grow *E. coli* strain W3110 overnight in 10.0 mL LB medium at 37°C.
2. Pellet the cells and resuspend in 2.0 mL of 50 m*M* Tris-HCl, pH 8.0, 20 m*M* EDTA.
3. Transfer aliquots of 0.5 mL to four sterile 13 × 100 glass tubes, and add 0.1 mL of proteinase K solution + 10 μL of 10% SDS + 6 μL of RNase solution to each tube.
4. Mix well, and then incubate at 37°C for 1–5 h, until lysed.
5. Transfer aliquots to four microfuge tubes, and extract three times with phenol/chloroform/isoamyl alcohol by adding 600 μL, vortexing, spinning 1 min in a microfuge, and removing the top, aqueous layer. Then extract once with 600 μL chloroform/isoamyl alcohol.
6. Precipitate nucleic acids by adding ¹/₁₀ vol 3 *M* sodium acetate and 2 vol of cold ethanol. Leave at –20°C for 1 h. Spin in microfuge for 30 min. Wash with cold 70% ethanol, and dry briefly at room temperature.
7. Resuspend with total of 200 μL TE, pH 8.0. Check the quality of the preparation by agarose gel electrophoresis of an aliquot. If a low-mol-wt smear of RNA is present, treat with RNase again, followed by Strataclean™ treatment.

3.2. Partial Sau3A I Digestion of Chromosomal DNA

1. Digest 10 μg of *E. coli* W3110 chromosomal DNA with 0.67 U of *Sau*3A I in 1X buffer A in a total volume of 100 μL at 37°C for 15 min. Stop the reaction by adding 2.5 μl of 500 m*M* EDTA and heating at 70°C for 10 min (*see* **Note 2**).
2. Add 20 μL of 6X gel load. Load the entire sample onto 0.6% SeaKem GTG agarose in 1X TBE gel. Also run mol-wt markers in another lane.
3. Following electrophoresis, cut off the marker lane and stain it with ethidium bromide. Using the stained markers as a guide, cut out the 1000–3500 bp region and mince the gel slice into fine pieces with a razor blade.
4. Place the gel pieces into a Spin-X filter unit, add 100–150 μL of TE, pH 8.0, and mix with a micropipet tip. Allow the DNA to diffuse out of the gel for 1–2 h at 4°C.
5. Spin the filter units for 5 min in a microfuge at 12,000*g*.

6. Add an additional 100 µL of TE, pH 8.0, to the agarose pieces in the top of the
   filter unit, and spin as in **step 5**. Remove the DNA sample from the bottom of the
   filter unit.
7. Concentrate and desalt the DNA using a Spin-X UF filter unit. Spin at 12,000*g* in
   a microfuge for 10 min. Twice, add 400 µL autoclaved water to top of the filter
   unit and spin. Then add 200 µL of autoclaved water and spin for the fourth time.
   Remove the DNA sample from the top of the filter unit.

3.3. lux *Vector Preparation*

1. Digest 20 µg of pDEW201 DNA with 30 U of *Bam*HI in 1X buffer B in a total
   volume of 100 µL at 37°C for 60 min. Heat-inactivate the restriction enzyme at
   75°C for 10 min. Take a 10-µL sample to analyze completeness of the digestion
   by agarose gel electrophoresis.
2. Add 10 µL of 10X CIP buffer and 0.5 U of calf intestinal alkaline phosphatase to
   the digested pDEW201. Incubate at 37°C for 30 min. Inactivate the alkaline
   phosphatase by incubating at 75°C for 15 min. Use Strataclean™ treatment to
   remove the enzymes.
3. Desalt and concentrate with a Spin-X UF unit (*see* **Subheading 3.2.7., step 7**).

3.4. *Plasmid Library Generation by Ligation and Transformation*

1. In a total volume of 10 µL, 1X ligase buffer, ligate for 5–7 h at 15°C the *Bam*HI-
   digested, alkaline phosphatase-treated pDEW201 vector DNA (400 ng) with the
   partially *Sau*3AI-digested chromosomal DNA (64 ng) using 3 U of T4 ligase.
   These quantities yield a molar ratio of approx 10:7 vector: insert.
2. Use 8 µL of the ligation mixture to transform ultracompetent *E. coli* XL2Blue by
   the manufacturer's protocol.
3. Plate the transformation mix on LB ampicillin (150 µg/mL) plates as follows:
   2 µL in 200 µL LB broth on one plate, 20 µL in 200 µL LB broth on a second
   plate, and then four 200-µL samples of the transformation mixture individually
   onto four additional plates.
4. Incubate the plates at 37°C overnight. Use the 2- and 20-µL plates to estimate the
   total number of transformant colonies.

3.5. *Isolation of Plasmid Library DNA*

1. Add 2.5 mL LB medium to each of the four plates that had been spread with
   200 µL of the transformation mix. Using a sterile spreader, resuspend the
   colonies. Combine the resuspended colonies from the four plates and mix
   well.
2. Store the cells containing the plasmid library at –80°C by taking four 0.5-mL
   aliquots of the resuspended cells and mixing with 0.5 mL 24% glycerol.
3. Recover the mixed population plasmid library by taking four 1.0-mL aliquots
   of the resuspended cells and isolating plasmid DNA using four Qiagen tip20
   columns and the manufacturer's miniprep protocol.
4. Resuspend each miniprep in 25 µL TE, and store at 4° or –20°C.

3.6. Transformation of the Testing Strain for Bioluminescence

1. Prepare competent *E. coli* strain DPD1675 by growing a culture to midlog phase at 37°C in 50-mL medium A, chilling on ice for 10 min, pelleting cells at 4500*g* for 10 min at 4°C, and resuspending in 0.5 mL of medium A, which had been precooled on ice. Add 2.5 mL of precooled storage solution B, and mix well without vortexing. Divide into 100-µL aliquots, and store at –80°C. A quick freeze is not necessary *(10)*.
2. Thaw cells on ice.
3. Transfer three aliquots of 100 µL to a precooled Falcon 2059 tubes. One will be used for the plasmid library transformation, another for the parental vector, pDEW201, and the third for the cells alone control.
4. Add 1.0 µL of a 1:50 dilution of plasmid library DNA isolated from the mixed population (*see* **Note 5**) to a 100-µL cell suspension. To the second tube, add 1 µL of pDEW201. Add nothing to the third tube. Mix by gentle tapping. Incubate on ice for 30 min (30 min are the minimum, but longer is fine).
5. Heat-pulse at 42°C for 1 min.
6. Add 1 mL of LB medium, and shake at 37°C for 30 min for phenotypic expression of ampicillin resistance (timing is important; *see* **Note 6**).
7. To ≈20 plates (LB Amp150 µL/mL; no more than 3 wk old; stored at 4°C) spread 50 µL each of the mixed plasmid population transformation mixture. Spread 1 LB Amp plate with 50 µL of the pDEW201 transformation mixture. Spread 1 LB Amp and 1 LB plate with 50 µL each of the cells alone control.
8. Incubate overnight at 37°C.

3.7. Toothpicking Transformants to Microplates

1. Label plates appropriately and fill all wells of a clear 96-well microplate with 190 µL of E₁GluUraProAmp (25 µg/mL) medium (*see* **Note 7**).
2. Using sterile toothpicks, inoculate the control wells first: pDEW201 transformants of DPD1675 to wells H9 and H10; sterile (toothpick only) in wells H11 and H12. Leave these toothpicks in the wells until the whole plate is inoculated.
3. Select well-isolated, but otherwise random, transformant colonies, and using sterile toothpicks, inoculate each of the remaining 92 wells of the microplate. Leave the toothpicks in the wells to mark those that have been inoculated.
4. Carefully remove all toothpicks, so that well-to-well crosscontamination is avoided.
5. Incubate covered plates overnight at 37°C, stationary.
6. Visually check for growth in all wells except the sterile controls.

3.8. Screening Bioluminescence as a Measure of Promoter Activity

1. Dilute the fresh overnight cultures in microplates by transferring 15 µL from each well to 150 µL E₁GluUraProAmp (10 µg/mL) in the wells of a white luminometer plate, prewarmed to 37°C.
2. Incubate plate, covered at 37°C, for 3 h to allow cells to become actively growing (*see* **Note 8**).

3. Read the uncovered plate in an ML3000 luminometer, using the cycle mode with the following settings: MEDIUM gain; data ALL; 1 cycle; 2-s pause; 20 A/D reads/well; auto gain ON; mix ON; temperature 37°C (*see* **Note 9**).

3.9. Sample Data

Table 1 shows sample bioluminescence data from transformants of *E. coli* DPD1675 prepared and tested as described. Under these conditions, the bioluminescence from cells containing the promoterless vector, pDEW201, (wells H9 and H10) are indistinguishable from sterile controls (wells H11 and H12). Using 10-fold over the highest RLU reading of the sterile and promoterless vector controls as a minimal arbitrary cutoff, we assigned a measurement of 0.05 RLU for putative promoter activity. The instrumentation gives a linear response to light production between 10^{-3} and 10^3 RLU. Thus, these data demonstrate that >1000-fold differences in transcriptional activity were observed: from the low of 0.050 RLU in well C9 to the high of 85.338 RLU in well A1. The *P. luminescens luxCDABE* reporter provides a quick, sensitive assay for promoters that is useful over a wide range of transcriptional activity.

4. Notes

1. Either described promoter probe vector pDEW201 or pJT205 may be used for discovering promoters. The higher copy number of pJT205 may be useful to amplify the signal from weakly expressed promoters. In general, pDEW201 may be superior, because the transcription terminators upstream of the *lux* operon reduce background transcription of the reporter genes.
2. Trial partial *Sau* 3A1 digestions will likely be necessary to determine the precise conditions required to generate incomplete digestions with fragments in the desired size range.
3. The use of ultracompetent cells with very high transformation frequency allows generation of a large number of fusion-containing strains at one time. Other types of competent cells may be used, but more transformations may have to be combined to yield a library representative of the chromosome.
4. The host *E. coli* strain for the bioluminescence testing can vary with the application. We are interested in further screening the plasmid library transformants for alterations of gene expression owing to the presence of xenobiotic compounds. For this purpose, the *tolC* mutation is useful in that it renders *E. coli* hypersensitive to hydrophobic compounds *(11,12)*.
5. After transformation of the strain for testing bioluminescence, it is desirable to obtain about 100 colonies on each plate. This facilitates subsequent toothpicking of isolated single colonies. The amount of library DNA to add to the competent cells to yield this number needs to be empirically determined for each batch of frozen competent cells.
6. To minimize generation of siblings, which will minimize the duplicates in a transformation, it is critical to limit the phenotypic expression time to 30 min prior to plating on selective medium.

Table 1
**Sample Bioluminescence Data from Cells
with Random *E. coli* DNA Fused to *P. luminescens luxCDABE*[a]**

	1	2	3	4	5	6	7	8	9	10	11	12
A	85.338	0.198	22.596	0.003	0.004	2.901	0.007	0	0.002	0.013	0.001	
B	0.029	0.005	0.016	0.013	2.155	43.779	0.005	0.008	0	0.017	41.335	0.095
C	0.133	0.012	33.103	0.013	0.017	1.491	0.008	3.211	0.05	0	0.006	7.892
D	0.584	0.036	0.007	0.006	0.027	0.003	0	3.525	0.011	1.114	12.829	0.005
E	0.006	0.005	0.537	0.072	79.463	0.003	0.039	0.032	0.029	0.011	69.621	0.204
F	0.188	6.898	4.674	0.003	0.236	9.869	0	0.167	0	0	0	1.546
G	14.784	18.439	0.028	0.015	0.254	0.013	0	0.008	1.806	4.773	0.029	0.007
H	6.125	0.001	13.994	0.021	0	0.042	6.532	0.006	0	0.002	0	0.005

[a]*E. coli* strain DPD1675 was transformed, and individual transformants were picked, grown, and bioluminescence measured 3 h after inoculation, as described in **Subheading 3**. Control transformants with pDEW201, the promoterless vector, are in wells H9 and H10. Sterile controls are in wells H11 and H12.

7. Either a rich medium, such as LB, or a defined medium, as described here, can be used to screen for promoter activity. The relative activity of many promoters is expected to be dependent on the growth medium.

8. We have found that, in general, bioluminescence decreases as cultures go into stationary phase. Thus, it is best to screen actively growing cells for promoter activity.

9. Very bright cultures may result in spurious light reading in neighboring wells. We find such crosstalk to be <0.01% by measurement of a bright culture and a neighboring uninoculated well. Nevertheless, because light production is quantifiable over a very large range, one should consider potential crosstalk before low bioluminescence values are assumed to be significant, if they came from wells neighboring strongly bioluminescent cultures.

References

1. Szittner, R. and Meighen, E. (1990) Nucleotide sequence, expression, and properties of luciferase coded by *lux* genes from a terrestrial bacterium. *J. Biol. Chem.* **265**, 16,581–16,587.

2. Rupani, S. P., Gu, M. B., Konstantinov, K. B., Dhurjati, P. S., Van Dyk, T. K., and LaRossa, R. A. (1996) Characterization of the stress response of a bioluminescent biological sensor in batch and continuous cultures. *Biotechnol. Prog.* **12**, 387–392.

3. Hill, P. J., Rees, C. E. D., Winson, M. K., and Stewart, G. S. A. B. (1993) The application of *lux* genes. *Biotechnol. Appl. Biochem.* **17**, 3–14.

4. Ernsting, B. R., Atkinson, M. R., Ninfa, A. J., and Matthews, R. G. (1992) Characterization of the regulon controlled by the leucine-responsive regulatory protein in *Escherichia coli. J. Bacteriol.* **174**, 1109–1118.

5. Miller, J. H. (1972) *Experiments in Molecular Genetics.* Cold Spring Harbor Laboratory, Cold Spring Harbor, NY.

6. Frackman, S., Michael, A., and Nealson, K. H. (1990) Cloning, organization, and expression of the bioluminescence genes of *Xenorhabdus luminescens. J. Bacteriol.* **172**, 5767–5773.

7. Yanisch-Perron, C., Vieira, J., and Messing, J. (1985) Improved M13 phage cloning vectors and host strains: nucleotide sequences of the M13mp18 and pUC19 vectors. *Gene* **33**, 103–119.

8. Simons, R. W., Houman, F., and Kleckner, N. (1987) Improved single and multicopy *lac*-based cloning vectors for protein and operon fusions. *Gene* **53**, 85–96.

9. Davis, R. W., Botstein, D., and Roth, J. R. (1980) *Advanced Bacterial Genetics.* Cold Spring Harbor Laboratory, Cold Spring Harbor, NY.

10. Nishimura, A., Morita, M., Nishimura, Y., and Sugino, Y. (1990) A rapid and highly efficient method for preparation of competent *Escherichia coli* cells. *Nucleic Acids Res.* **18**, 61–69.

11. Schnaitman, C. (1991) Improved strains for target-based chemical screening. *ASM News* **57**, 612.

12. Van Dyk, T. K., Majarian, W. R., Konstantinov, K. B., Young, R. M., Dhurjati, P. S., and LaRossa, R. A. (1994) Rapid and sensitive pollutant detection by induction of heat shock gene-bioluminescence gene fusions. *Appl. Environ. Microbiol.* **60**, 1414–1420.

8

Insertion of Promoter Region::*luxCDABE* Fusions into the *Escherichia coli* Chromosome

David A. Elsemore

1. Introduction

Transcriptional reporter systems have been extensively used to measure gene expression. Facile measurement of a reporter signal enables the study of gene expression in cases where protein product quantification is problematic. Reporters, such as ß-galactosidase, provide convenient chromogenic assays *(1)*. More recently, bioluminescence has been developed as a monitoring system *(2)*. Reporter gene systems may be located extrachromosomally or integrated into the chromosome.

Integration of transcriptional fusions offers several advantages over plasmid-borne systems. These advantages include avoidance of repressor molecule titration, read-through from plasmid-borne promoters, construct stability, and ability to grow the reporter strain in the absence of antibiotic. These may be important considerations depending on the application of the transcriptional fusion.

Experiments designed to identify promoter regions traditionally use mobile genetic elements to randomly deliver a cassette containing a particular reporter gene into the chromosome. However, introduction of previously constructed, plasmid-based promoter::reporter fusions into the *Escherichia coli* chromosome is problematic. Transposon systems usually depend on a transposase supplied in trans and therefore require introduction of two replicons where one component carries the engineered transposon while a second component carries the transposase gene. These systems offer a few convenient cloning sites and may be suitable for some applications.

Recombination with a chromosomal locus is another means of integrating an engineered fragment of DNA into a particular locus of the *E. coli* chromosome.

From: *Methods in Molecular Biology, Vol. 102: Bioluminescence Methods and Protocols*
Edited by: R. A. LaRossa © Humana Press Inc., Totowa, NJ

The ability to render *E. coli* cells competent for DNA transformation through chemical treatment and the mutational elimination of exonuclease and endonuclease systems allow chromosomal integration of engineered DNA to proceed efficiently.

In this chapter, the integration of the *E. coli recA* promoter region:: *luxCDABE*_{Photobacterium luminescens} fusion into the chromosome of *E. coli* strain W3110 is described. The source of the *recA::lux* operon fusion is a pJT205-based plasmid (*see* Chapter 7) into which a 300-bp fragment of DNA containing the *recA* promoter region has been inserted upstream of *luxCDABE*. This specific example is presented for ease of description, but the protocol may be adapted to successfully integrate any promoter::*luxCDABE* fusion into the *E. coli* chromosome at the *lacZ* locus.

In general, the process involves several intermediate constructions (*see* **Fig. 1**). First the promoter::*luxCDABE* fusion is inserted into the multiple cloning site of a pBRINT plasmid *(3)*. *E. coli* strain JC7623 is transformed with the pBRINT-based plasmid. The inability of pBR322-based plasmids to replicate in a *recB21*, *recC22*, *sbc15* background of strain JC7623 allows the selection of chromosomal integrants produced by recombination between the chromosomal *lacZ* locus and the plasmid's *lacZ* DNA that flanks the multiple cloning site. Finally, P1_{vir}-mediated generalized transduction is used to move the *luxCDABE* integration into a phage-sensitive *E. coli* working strain.

2. Materials
2.1. Strains and Plasmids

1. pBRINT plasmid (*see* **Note 1**).
2. Promoter region::*luxCDABE* fusion construct.
3. *E. coli* JC7623 (ATCC47002) F⁻, *recB21*, *recC22*, *sbcC201*, *sbcB15*, *thr-1*, *leu-6*, *ara-14*, *his-4*, λ⁻, Δ(*gpt-proA*)62, *lacY1*, *tsx-33*, *supE44*, *galK2*, *rac*, *rfbD1*, *mgl-51*, *rpsL31*(Str^R), *kdgK51*, *xyl-5*, *mtl-1*, *argE3*, *thi* (*see* **Note 2**).
4. P1_{vir}-sensitive recipient strain.

2.2. Preparation of Competent E. coli Cells

1. Solution A: LB broth supplemented with 10 m*M* MgSO₄·7H₂O and 0.2% (w/v) glucose. Filter-sterilize using 0.22-µm filter. Store at room temperature.
2. Solution B: LB broth supplemented with 36% glycerol (v/v), 12% PEG 8000 (w/v), and 12 m*M* MgSO₄·7H₂O. Stir over low heat to dissolve. Filter-sterilize as above. Store at room temperature.

2.3. Selection of Integrants

1. LB agar plates supplemented with 50 or 25 mg/L of kanamycin.
2. LB agar plates supplemented with 100 mg/L of ampicillin.
3. LB agar plates supplemented with Bromo-4-chloro-3-indolyl-β-D-galacto-pyranoside (X-Gal) and IPTG (*see* **Note 3**).

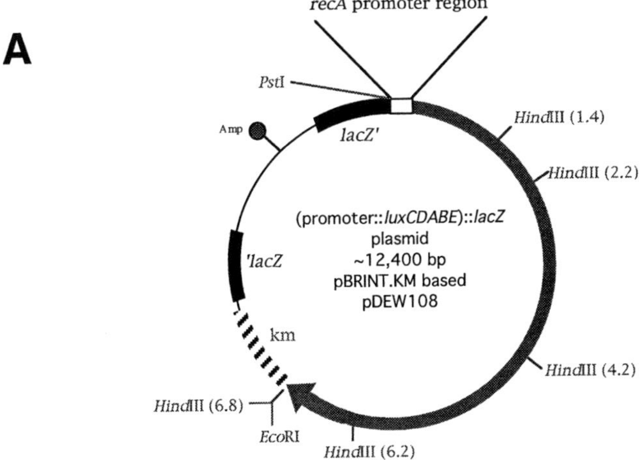

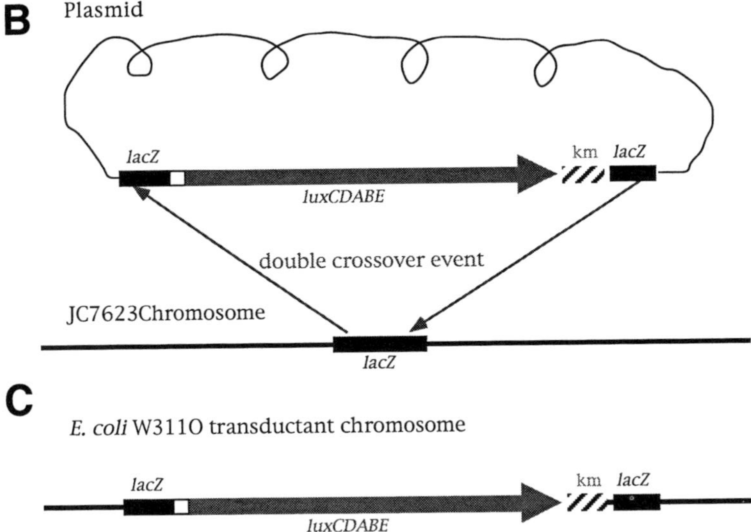

Fig. 1. Intermediate constructions. **(A)** A previously constructed *recA* promoter region::*lux* operon fusion is cloned into the multiple cloning site of pBRINT.Km plasmid. **(B)** Transformation of *E. coli* strain JC7623 allows integration of the fusion at the chromosomal *lacZ* locus via homologous recombination. **(C)** P1$_{vir}$ generalized transduction allows movement of the construct to wild-type *E. coli* strain W3110.

2.4. P1$_{vir}$ Transduction

1. P1$_{vir}$ stock.
2. Appropriate phage-sensitive recipient *E. coli* strain (*see* **Note 4**).

3. $CaCl_2$: 1 M in distilled water. Autoclave or filter-sterilize. Store at room temperature.
4. Glucose: 20% (w/v) in water. Filter-sterilize. Store at room temperature.
5. Top R-agar: 0.5% agar in water supplemented with 2.0 mM $CaCl_2$, and 0.4% glucose (v/v). Store at room temperature (*see* **Note 5**).
6. Bottom R-agar: 1.0% agar in water supplemented with 2.0 mM $CaCl_2$ and 0.4% glucose (v/v). Store at room temperature (*see* **Note 6**).
7. MC buffer: 100 mM $MgSO_4 \cdot 7H_2O$ and 5.0 mM $CaCl_2$ in water. Filter-sterilize. Store at room temperature.

3. Methods

3.1. Construction of recA *promoter::*luxCDABE *pBRINT Plasmids*

1. The previously constructed, plasmid-borne *recA* promoter::*luxCDABE*$_{Photobacterium\ luminescens}$ fusion was used as the source of *recA* promoter::*lux* operon DNA *(4)* (*see* Chapter 12, and **Note 7**).
2. Digest the *recA* promoter::*luxCDABE* fusion plasmid with appropriate restriction enzymes, such that one enzyme cuts upstream of the promoter region (*Pst*I) and the second enzyme cuts downstream of the *lux* operon (*Eco*RI).
3. Digest pBRINT.Km plasmid with compatible restriction enzymes.
4. Ligate fragments (*see* **Note 8**).
5. Transform competent subcloning efficiency *E. coli* cells (*see* **Note 9**).
6. Select transformants on LB agar with kanamycin (50 mg/L) overnight at 37°C.
7. Screen transformant colonies for light production by exposing colony-bearing plates to X-ray film for several hours (*see* **Note 10**).
8. Isolate plasmids from kanamycin resistant, light producing transformants.
9. Digest plasmid DNA with *Hind*III (*see* **Fig. 1A**). The internal *Hind*III sites within the *lux* operon result in unique restriction digest patterns (pDEW108 yeilds fragments with the following kbp sizes: 6.9, 2.0 [doublet], 0.8, and 0.9) that allow verification of the construction.

3.2. Making E. coli *Strain JC7623 Competent for DNA Transformation*

This method is from the published procedure of Nishimura et al. *(5)*.

1. Inoculate 0.5 mL of an overnight culture of JC7623 grown in LB into 50 mL of prewarmed (37°C) medium A in a 150-mL flask.
2. Grow the cells at 37°C with shaking until the cells have reached midlog (OD_{600} = 0.3–0.5) (*see* **Note 11**).
3. Chill cells on ice for 10 min.
4. Pellet cells at 4500*g* for 10 min at 4°C.
5. Remove supernatant from cells, and gently resuspend the cells in 0.5 mL of pre-cooled medium A (*see* **Note 12**).
6. Add 2.5 mL of precooled storage solution B, and mix well. Do not vortex.
7. Aliquot 0.1 mL of cells/chilled 1.5-mL microcentrifuge tube. Cells are now competent for DNA transformation, and may be used immediately or frozen at –80°C (*see* **Note 13**). Frozen cells remain competent for months.

3.3. Transformation of JC7623 and Identification of Integrants

1. Thaw competent cells on ice.
2. Add promoter::*lux* operon fusion containing pBRINT plasmid DNA (total amount = 1–2 µg/µL DNA) to 0.1 mL of cells. Mix gently, and incubate on ice for 30 min.
3. Heat-shock the cells at 42°C for 1 min.
4. Return the cells to ice for an additional 2 min.
5. Add 0.9 mL of LB media, and incubate at 37°C with shaking (*see* **Note 14**) for 1 h.
6. Plate on LB with kanamycin (25 mg/L). Incubate overnight at 37°C (*see* **Note 15**).
7. Screen kanamycin-resistant colonies for light production (*see* **Note 10**).
8. Pick kanamycin-resistant colonies onto LB plates containing ampicillin (100 mg/L) and LB plates containing X-Gal and IPTG (*see* **Note 16**).

3.4. Preparation of P1$_{vir}$ Stock

1. Grow an overnight culture of the donor integrant strain prepared in **Subheading 3.3.** in LB supplemented with kanamycin (25 mg/L) at 37°C.
2. Reinoculate 50 µL of the overnight culture into 5 mL of fresh LB media supplemented with 5 mM CaCl$_2$.
3. Grow the cells for 2 h at 37°C.
4. Preabsorb P1$_{vir}$ at various concentrations to 0.1 mL of cells (*see* **Note 17**) in 13 × 100 mm glass culture tubes.
5. Incubate the cells and phage at 37°C for 20 min without shaking.
6. Add 2.5 mL of melted R-top agar, cooled to 46°C, to cells and phage mix. Vortex quickly, and plate entire contents onto an R-bottom agar plate (*see* **Note 18**).
7. Incubate plates, agar side up, at 37°C overnight.

3.5. Harvesting Phage

Select the phage dilution that results in confluent lysis, which is recognized as a lacy-looking plate. Too high a concentration of phage results in mucoid, phage-resistant colonies, which are difficult to separate from the phage. Too low a concentration of phage will result in a low titer.

1. Scrape off R-top agar into a chloroform-resistant centrifuge tube containing five drops of chloroform (*see* **Note 19**).
2. Vortex tube for 30 s. Let the tube sit for 10 min.
3. Add 1 mL of LB to plate to wash off remaining phage. Add this volume to the centrifuge tube.
4. Vortex tube for 30 s. Let the tube sit for 10 min.
5. Centrifuge at 3000g for 10 min.
6. Pour off supernatant into an microcentrifuge tube containing five drops of chloroform.
7. Vortex for 1 min and allow to settle for 10 min. This phage stock is best stored at 4°C. The titer will diminish over time.

3.6. P1$_{vir}$ Transduction

1. Grow an overnight culture of appropriate phage-sensitive host strain. Strain W3110 (F$^-$, prototroph) was used in this example.
2. Centrifuge 1 mL of grown culture in a 1.5-mL microcentrifuge tube. Decant and discard supernatant.
3. Resuspend cells in 1 mL of MC buffer.
4. Shake at 37°C for 15 min.
5. Add 0.1 mL of the MC buffer-resuspended cells to 10 µL of various dilutions of the phage preparation (10^0, 10^{-1}, 10^{-2}, and 10^{-3}) in a microcentrifuge tube (*see* **Note 20**). Include a cells-alone and a phage-alone control.
6. Preabsorb the P1$_{vir}$ phage and cells at 37°C for 15 min without shaking.
7. Add 1.0 mL of LB media supplemented with 10 m*M* sodium citrate (chelates excess calcium). Incubate the cells at 37°C for 1 h.
8. Centrifuge cells in a microfuge for 4 min at maximum speed. Remove 800 µL of supernatant. Resuspend the cell pellet in the remaining 200 µL of media.
9. Plate the entire volume on an LB plate supplemented with 25 mg/L kanamycin.
10. Incubate plates at 36°C for *Photorhabdus luminescens lux* integrants.
11. Purify kanamycin-resistant colonies from the lowest phage dilution that produces transductants.
12. Screen for light production (*see* **Note 21**).

4. Notes

1. The pBRINT plasmid series is available from the laboratory of F. Valle, Instituto de Biotecnología, UNAM, Apdo. Postal 510-3, Cuernavaca, Morelos, México. Three individual pBRINT plasmids containing different selectable markers (kanamycin, chloramphenicol, and gentamycin) are available. The pBRINT.Km plasmid was used in this example.
2. *E. coli* strain JC7623 is streptomycin-resistant. This marker may be used to verify the final strain construction.
3. Stock solutions of X-Gal and IPTG are spread on individual LB plates. Spread 100 µL of 10% (w/v) X-Gal solution dissolved in DMSO (wear gloves) and a 100 µL of a 100-m*M* solution of IPTG dissolved in water/plate. Allow the plate to dry before using. Both solutions are stable at –20°C. X-Gal is light-sensitive.
4. Strains containing mutations in *galE, galU, lpcA,* or *rfaD* locus **(6)** are not P1$_{vir}$-phage-sensitive.
5. Aliquots of 0.5% agar are prepared and sterilized in glass bottles. When the molten agar has cooled to 50°C, the CaCl$_2$ and glucose can be added. When needed, carefully melt solidified top R-agar in a microwave oven. Cool to 50°C before using.
6. Autoclave 1.0% agar and allow to cool to 50°C. Supplement with CaCl$_2$ and glucose. Pour the plates, and allow to solidify. Plaque formation is best when the plates are fresh. Spreading CaCl$_2$ and glucose on an LB plate can substitute for R-bottom agar.
7. Stability of the protein products responsible for bioluminescence depends on the source of the *luxCDABE* operon. *P. luminescens lux* operon products are

thermotolerant and produce light at 37°C, whereas *Vibrio fisheri lux* operon proteins are more sensitive to heat and produce bioluminescence at temperatures <30°C under normal conditions.

8. Typical ligation mixes included 1–2 µg of total DNA and 1.0 U of T4 ligase in a 20–30 µL volume.

9. Subcloning efficiency *E. coli* cells is commercially available or alternatively, can be prepared by the procedure of Nishimura et al. as described in **Subheading 3.2.**

10. Empty X-ray film boxes (20.3 x 25.4 cm) make convenient containers to expose Petri plates to film. Four to five Petri plates, sealed shut with parafilm, can be attached to a thin piece of cardboard. Attach the bottoms of the plates to the cardboard with doubled over labeling tape. Put a piece of X-ray film on top of the plates. A second piece of cardboard is added to the top to provide stability. The entire stack can then be placed in a light-tight bag and inserted into the film box. All manipulations of the film should be done in a darkroom. Place the box in an incubator at the appropriate temperature.

11. At this point, prechill medium A, storage buffer B, and microcentrifuge tubes on ice.

12. Decant supernatant. Resuspend by gently pipeting cells with medium A.

13. Quick-freezing the cells in a dry-ice/ethanol bath is not necessary.

14. The microcentrifuge tubes may be taped to the platform of a shaking incubator.

15. Typically, 200 µL of transformed cells are plated/Petri plate. If the *luxCDABE* operon is from *V. fisheri*, incubate the plates at 26°C.

16. The desired *recA::luxCDABE* integrant of strain JC7623 is kanamycin^R, streptomycin^R, ampicillin^S, LacZ$^-$, and bioluminescent. Ampicillin^R in the genetic cross is indicative of insertion of the entire plasmid into the chromosome. Such recombinants should be avoided because they may be unstable.

17. A good P1$_{vir}$ phage stock will have a titer of 10^{9-10} PFU/mL. Set up separate mixtures of cells and phage that contain 100 µL of cells and 1 µL of P1$_{vir}$ phage at several dilutions (10^0, 10^{-1}, and 10^{-2}). Typically, the 10^{-2} dilution of a fresh phage stock produces a good phage preparation.

18. Pour entire mixture of top agar, cells, and phage into a puddle at the edge of the R-bottom agar plate. Spread the mixture across the entire R-bottom agar plate by gently tipping the plate back and forth. Allow the overlay to solidify.

19. The R-top agar overlay can be scraped off using the tip of a disposable 1-mL micropipeter. Make sure the plastic centrifuge tube is not sensitive to chloroform.

20. The multiplicity of infection (MOI) can approach a ratio of 1:1 of cells to phage. Infections with serial dilutions of phage eliminate the need to experimentally determine the titer of the freshly prepared P1$_{vir}$ phage stock.

21. The W3110 transductant is kanamycin^R, streptomycin^S, ampicillin^S, and bioluminescent.

References

1. Silhavy, T. J., Berman, M., and Enquist, L. (1984) *Experiments with Gene Fusions.* Cold Spring Harbor Laboratory, Cold Spring Harbor, NY.
2. Stewart, G. S. A. B. and Williams, P. (1992) *lux* genes and the applications of bacterial bioluminescence. *J. Gen. Microbiol.* **138,** 1289–1300.

3. Balbas, P., Alexeyev, M., Shokolenko, I., Bolivar, F., and Valle, F. (1996) A pBRINT family of plasmids for integration of cloned DNA into the *Escherichia coli* chromosome. *Gene* **172,** 65–69.

4. Vollmer, A. C., Belkin, S., Smulski, D. R., Dyk, T. V., and LaRossa, R. A. (1996) Detection of DNA damage using *Escherichia coli* carrying *recA'::lux, uvrA'::lux* or *alkA'::lux* reporter plasmids. *Appl. Environ. Microbiol.* **63,** 2566–2571.

5. Nishimura, A., Morita, M., Nishimura, Y., and Sugino, Y. (1990) A rapid and highly efficient method for preparation of competent *Escherichia coli* cells. *Nucleic Acids Res.* **18,** 6169.

6. LaRossa, R. A. (1996) Mutant selections linking physiology, inhibitors, and genotype, in *Escherichia coli and Salmonella* (Neidhardt, F. C., Curtiss, R., III, Ingraham, J. L., Lin, E. C. C., Low, K. B., Magasanik, B., Reznikoff, W. S., Riley, M., Schaechter, M., and Umbarger, H. E., eds.), ASM, Washington, DC, pp. 2527–2587.

9

Probing for Promoters
with Luciferase-Transposons

David C. Alexander and Michael S. DuBow

1. Introduction

The union of transposon mutagenesis and reporter gene fusion technologies has created a very powerful tool for analyzing gene expression. Transposable elements, such as phage Mu and Tn*5*, are able to move within a DNA molecule or from one DNA molecule to another, and can randomly integrate into the bacterial chromosome or plasmids. Transposition into a coding sequence can disrupt gene function and produce a null mutant phenotype. If the null mutant exhibits an obvious phenotype, then it is possible to screen for specific mutations on selective media. Once a mutant is isolated, the responsible gene can be identified by mapping the transposon insertion site. However, not all genes encode an obvious phenotype or readily assayable product. By coupling transcription or translation to an easily assayed reporter gene, it is possible to monitor expression of any gene, even if its phenotype is not known *a priori.*

The ideal reporter gene should provide a sensitive, real-time measurement of gene expression and encode a product that is not native to the organism of interest. The assay for the ideal reporter gene product should be simple and inexpensive to perform, and should not require cell disruption. Bacterial luciferases, such as those encoded by the *luxAB* genes of *Vibrio fischeri* and *Vibrio harveyi,* satisfy these criteria. The luciferase reaction requires oxygen and a reduced flavin mononucleotide, which are readily available in viable cells, and a linear aldehyde, which can be produced in vivo by the *luxCDE* genes or exogenously supplied *(1,2).* The product, photons of light at 490 nm, can be easily, inexpensively, and noninvasively measured in realtime with photographic film, a scintillation counter (with coincidence turned off), a luminometer, or the naked eye. In addition, light emission can be localized spatially

From: *Methods in Molecular Biology, Vol. 102: Bioluminescence Methods and Protocols*
Edited by: R. A. LaRossa © Humana Press Inc., Totowa, NJ

to individual cells and is not blocked by cell pigments that can mask colorimetric reporter gene products.

By incorporating a reporter gene into a transposon (*see* **Table 1**), it is possible to create reporter gene fusions within any gene and mutagenize all nonessential regions of a chromosome or plasmid. An effective transposon-mediated reporter gene fusion strategy involves:

1. A method to deliver the transposon to the target organism;
2. A procedure for selection of successful transposon insertion events;
3. An assay to identify reporter fusions to the genes of interest; and
4. A method to identify the site of transposon insertion.

Engebrecht et al. *(3)* were the first to combine transposon and *lux* gene fusion technologies. They replaced the *lac'ZYA* genes of a mini-Mu transposon, Mud II 1681 *(4,5)* with the *luxCDABE* genes from *V. fischeri,* oriented such that *luxC* was closest to the Mu right end. The resultant mini-Mu*lux* transposon contained the Mu *A* and *B* transposase genes, a thermosensitive *c* repressor, an antibiotic resistance marker (Tc^R or Km^R), as well as the genes for luciferase (*luxAB*) and aldehyde synthesis (*luxCDE*) that are necessary for light production. Mini-Mu*lux* lacked the *lux* operon promoter and the *luxR* and *luxI* control elements. As such, expression of the *luxCDABE* reporter genes was regulated by promoters located outside of the Mu right end. The mini-Mu*lux*(Tc^R) construct could be packaged by a helper Mu phage for transduction into chromosomal or plasmid DNA in susceptible strains of *Escherichia coli.* Selecting for Tc^R led to the identification of clones with insertions in a plasmid-borne *galK* gene, as well as insertions in the chromosomal *lac* and *ara* loci. Expression of these genes resulted in increased light emission in approx 50% of the clones. This demonstrated that the mini-Mu*lux* construct was capable of inserting at random, and in either orientation, into chromosomal or plasmid DNA. Subsequently, phage P1 *clr-100*CM was used to package and transduce mini-Mu*lux* (Km^R) into *Vibrio parahaemolyticus (3,6).* Of the 9800 Km^R transductants, 106 mutants exhibited a nonswarming, Laf-negative phenotype when grown on solid media. Approximately 40% of these clones contained insertions in the proper orientation to produce light. Light expression increased when cells were grown in viscous medium, which was suitable for induction of swarming. When cells were returned to dilute medium, the rate of light expression slowed and eventually decreased. These experiments demonstrated that transposon mutagenesis with mini-Mu*lux* could be effectively used to identify genes involved in cell differentiation and monitor expression nondestructively in realtime.

DeLorenzo et al. *(7)* described a series of plasmid-borne mini-Tn*5*-based promoter-probe transposons that could be mobilized by conjugation. Mini-Tn*5luxAB* contains the promoterless *V. harveyi luxAB* genes and a Tc^R marker,

Table 1
Features of the lux-based Promotor Probe Transposons[a]

Construct	*lux* Genes	Marker	Features	Refs.
Mini-Mu*lux*	*V. f luxCDABE*	Tc^R or Km^R	Temperature sensitive *c* repressor, inducible at 40°C; delivery via transduction with P1 or Mu helper phage; endogenous aldehyde production	3, 6
Mini-Tn*5 luxAB*	*V. h luxAB*	Tc^R	Donor plasmid, pUT (Km^R), contains RP4 *oriT* for efficient conjugal transfer; P protein-dependent R6K origin of DNA replication for increased rate of transposition in *pir-* host strains; transposase supplied, in cis, by donor plasmid; transposon does not produce transposase, and secondary transposition events cannot occur; unique site for restriction enzyme *NotI* facilitates isolation of transposon-flanking DNA	7
Tn*5*-1063a	*V. f luxAB*	Nm^R/Km^R Sm^R, Bm^R	Donor plasmid contains RK2 *oriT* for conjugal tranfer;transposon with flankingDNA can be excised by restriction enzymes *Eco*RI, *Eco*RV, *Dra*I, *Cla*I, *Sca*I or *Spe*I; owing to p15 origin of DNA replication within transposon, recircularized fragment is self-replicating	8
Tn*5*-1113a	same as Tn*5*-11063a except *V. fischeri luxCDABE* genes			10
Tn*5*-1406	same as Tn*5*-1063a except *P. leiognathi luxCDABEG rib*BAH genes			10
Tn*5*-1407	same as Tn*5*-1063a except *V. harveyi luxCDABE* genes			10
Tn*5*-1419	same as Tn*5*-1063a except *X. luminescens luxCDABE* genes			10

(continued)

Table 1
(continued)

Construct	*lux* Genes	Marker	Features	Refs.
Tn*4431*	*V. f luxCDABE*	Tc^R	Delivery by donor plasmids pDS1 or pUCD622; conjugal plasmid pDS1 (Gm^R) has thermosensitive origin of DNA replication, inactive at 40°C, for increased ease of selection in enteric hosts; conjugal plasmid pUCD622 does not replicate in nonenteric host strains	*14* *13* *14*
Tn*5351*	*V. h luxAB*	Viomycin^R	Delivery to *Streptomyces* spp. by transformation or conjugal transfer of donor plasmid (*tsr*+); unique sites for restriction enzymes *Ase*I and *Ssp*I facilitate mapping of insertion site by pulse-field gel electrophoresis.	*16,17*
Tn*5353*	*V. h luxAB*	Nm^R/Km^R	Delivery to *Streptomyces* spp. by transformation or conjugal transfer of donor plasmid (*tsr*+) or by transduction with fC31; unique sites for restriction enzymes *Ase*I and *Ssp*I	*16,17*
Tn*5::lux*	*V. h luxAB*	Tc^R	Delivery by transformtaion with donor pFUSLUX; to prevent pFUSLUX replication in enteric hosts, cotransform with RNA1 overproducing plasmid pTF421 (Ap^R); sites for restriction enzymes *Hind*III and *Bam*HI facilitate isolation of DNA flanking insertion site	*19,23–26*

[a]Abbreviations used: *V. f = V. fischeri, V. h = Vibrio harveyi,* Tc = tetracycline, Nm = neomycin, Km= kanamycin, Sm = streptomycin, Bm = bleomycin, Gm = gentamicin, Ap = ampicillin, *tsr* = thiostrepton.

flanked by 19-bp Tn*5* O and I ends. Transcription from exogenous promoters can continue through the O element and into *luxAB*. Alternatively, DNA fragments can be inserted into a polylinker between the O end and *luxAB*. The transposon has a modular construction, such that the Tc^R marker can be replaced with other antibiotic resistance genes. The donor plasmid supplies transposase in *cis* and contains the RP4 *ori*T for efficient conjugal transfer. The plasmid also has a π protein-dependent R6K origin of replication. When transferred to a π protein-deficient recipient cell, the transposon-containing donor plasmid is unable to replicate, and antibiotic-resistant colonies cannot form until mini-Tn*5luxAB* has inserted within the chromosome. After the initial transposition event, the transposase-supplying donor plasmid is lost. Without transposase, the chromosomally inserted mini-Tn*5luxAB* is unable to transpose again. When used to mutagenize *Pseudomonas putida,* 90% of Tc^R recipients arose from authentic transposition events, whereas 10% contained the complete delivery plasmid integrated into the chromosome. The transposition efficiency was less for *E. coli* K-12, because the donor strain, *E. coli* SM10(λ *pir*), harbored a π protein-supplying prophage, which was induced during conjugation. The phage could lysogenize the recipient cells and supply them with π protein, which enabled the mini-Tn*5luxAB* vector to replicate. In these recipient *E. coli* strains, up to 90% of the exconjugants arose from chromosomal integration of the delivery plasmid, and not authentic transposition.

Wolk et al. *(8)* used a different Tn*5*-based promoter probe to study environmentally responsive genes in the filamentous cyanobacterium, *Anabaena.* This probe employs a modified Tn*5* transposon, in which the IS*50*R sequence and transposase activity are intact, but the IS*50*L sequence is truncated to 53 bp. Between IS*50*L and IS*50*R are the promoterless *luxAB* genes from *V. fischeri,* the origin of replication (*ori*V) from plasmid p15A, and genes for neomycin, bleomycin, and streptomycin resistance. This construct, Tn*5*-1063a, was contained within the conjugal plasmid, pRL1063a. Tn*5*-1063a has several intriguing features. The truncated IS*50*L supports transposition, and its diminutive size ensures that transcription from exogenous promoters can continue, through the left end, into *luxAB*. The *luxCDE* genes were omitted because it had been reported that it was difficult to transfer the entire *luxCDABE* unit *(9)*. Plasmid pRL1063a was unable to replicate in *Anabaena,* and antibiotic-resistant colonies developed only after integration of the transposon into the genome. Neomycin, bleomycin, and streptomycin were more appropriate markers for the photosynthetic cyanobacteria, because, unlike tetracycline, they are not photoinactivated. To identify the site of transposon integration, total genomic DNA was digested with a restriction enzyme (e.g., *Cla*I or *Eco*RI) that cleaved the chromosome, but not the transposon, and the restriction fragments were recircularized. Because the transposon included *ori*V, the recircularized fragment containing the transposon and flanking DNA was self-replicating and could be transformed, selected, and manipulated in *E. coli.* Wolk et al. *(8)* used this strategy to generate and isolate *Anabaena* mutants that demonstrated increased light emission after nitrogen or phosphate deprivation. Southern blot analysis indicated that each mutant contained a single copy of the transposon at a random position in the chromosome.

Because Tn5-1063a did not contain *luxCDE,* it was necessary to add aldehyde exogenously when assaying for luminescence. Such treatment tended to lyse the *Anabaena* cells. To solve this problem, Fernandez-Pinas and Wolk *(10)* developed a set of promoter-probe transposons that used luciferase as a reporter enzyme and produced aldehyde endogenously. These constructs were identical to Tn5-1063a, except that they contained *V. fischeri luxCDABE* (Tn5-1113a), *V. harveyi luxCDABE* (Tn5-1407), *Xenorhabdus luminescens luxCDABE* (Tn5-1419), or *Photobacterium leiognathi luxCDABEG ribBAH* (Tn5-1406) in place of *luxAB*. These constructs could be transferred to *Anabaena* and randomly transpose into the genome. Although a few clones were self-luminescent, most of the transposon-derived colonies still did not emit light unless aldehyde was added exogenously. In short, this strategy did not permit protracted, nontoxic transcriptional reporting, but aldehyde-limited luminescence. A system to deliver aldehyde endogenously was achieved with pRL1472, a plasmid containing the *X. luminescens luxCDE* genes under control of their native promoter. This plasmid could be stably maintained in *Anabaena* and produced enough endogenous aldehyde to support self-luminescence of clones containing expressed chromosomal *luxAB* fusions.

Shaw et al. *(11)* constructed Tn*4431,* a promoter probe that uses the promoterless *luxCDABE* genes from *V. fischeri* and a TcR marker. The Tn*4431* construct is contained on a conjugal plasmid, pUCD623. This donor-plasmid cannot replicate in nonenteric hosts and Tc^R exconjugate colonies form only after integration of Tn*4431* into the host chromosome. On transposon mutagenesis of the nonenteric plant pathogen *Xanthomonas campestris* pv. campestris 2D520 (*X.c.c.*) with Tn*4431,* a nonpathogenic mutant, JS111, was identified *(12)*. Like the pathogenic parental strain, the *X.c.c.* JS111 mutant produced exopolysaccharide and a variety of degradative enzymes. Even so, this *hrp* (hypersensitivity response and pathogenicity) mutant grew poorly on radish, cauliflower, and other host plants, did not induce black rot, and was unable to induce a hypersensitive response on nonhost plants. However, JS111 exhibited wild-type behavior when coinoculated with a wild-type strain. By monitoring luminescence, it was determined that the *lux*-interrupted locus, *hrpXc,* was strongly induced when cells were grown on radish leaves.

Steinmann et al. *(13)* used Tn*4431* to mutagenize plasmids. Plasmid pDS1 has a thermosensitive origin of replication, which is inactive at 40°C, and contains Tn*4431* and a gentamycin resistance (Gm^R) marker. By propagating *E. coli* at 40°C and screening for a Gm^S, Tc^R phenotype, it was possible to identify clones that maintained Tn*4431,* but lost pDS1. For mutagenesis of plasmid DNA, pDS1 was introduced into an *E. coli* strain transformed previously with pBK2, a plasmid containing a 10-kbp fragment of *X.c.c.* DNA and a Km^R marker. These cells were grown at 40°C, and Gm^S, Tc^R and Km^R colonies, which maintained pBK2 and Tn*4431,* but not pDS1, were isolated. Tn*4431* transposes at random into either genomic or plasmid DNA. To identify pBK2 derivatives containing Tn*4431,* plasmid DNA from the Tc^R Km^R clones was collected *en masse* and transformed into antibiotic-sensitive *E. coli.* Seventy transformants containing pBK2::Tn*4431* were identified by selection for Km^R and Tc^R. Characterisation of 24 plasmids confirmed that the position and orientation of the transposon insertion was random. In 21 plasmids, the site of insertion mapped to the 10-kbp X.c.c. fragment. Light emission varied with the location and orientation of the transposon. Steinmann et al. *(13)* also demonstrated that pBK2::Tn*4431* plasmids could be mobilized, via conjugation, into *X.c.c.* Plasmid pBK2 is unable to replicate in *X.c.c.* However, by screening for Km^R and Tc^R, it was possible to select for recombination of pBK2::Tn*4431* into the genome. One such recombinant was used to infect cauliflower. Infection by the self-luminescent pathogen could be followed with autophotography.

King et al. *(14)* used Tn*4431* to monitor expression of naphthalene degradation genes in *Pseudomonas fluorescens.* A strain of *P. fluorescens,* carrying a naphthalene catabolic plasmid, was mutagenized with Tn*4431* and screened for increased bioluminescence in response to naphthalene exposure. One luminescent clone contained a Tn*4431* insertion at the plasmid *nah*G locus *(15).* A culture containing this plasmid was grown in a chemostat and exposed to naphthalene at various intervals. Luminescence increased 15 min after addition of napthalene and decreased to background levels 15 min after removal of naphthalene. In addition, light emission by this culture could be used to differentiate between naphthalene contaminated and uncontaminated soil slurries.

Sohaskey et al. *(16,17)* described several plasmid- and transposon-based promoter-probe vectors. The plasmid-based promoter probes were derivatives of pRS1105, a *Streptomyces* replicon that contains a thiostrepton resistance marker and the promoterless *luxAB* genes from *V. harveyi.* The pRS1105 construct was originally used to identify novel *Streptomyces coelicolor* genes associated with aerial mycellium development *(18).* The transposon-based probes were derivatives of Tn*4556,* a Tn*3*-type transposon. Both Tn*5351* and Tn*5353* contain promoterless *V. harveyi luxAB* genes near the right inverted repeat, transposase and resolvase functions near the left inverted repeat, and sites for the restriction enzymes *Ase*I and *Ssp*I. The plasmid-borne transposons can be mobilised via conjugation. Tn*5351* is larger (11.1 Kb,112 bp) and contains a viomycin resistance marker. Tn*5353* contains a *neo* gene for kanamycin resistance and is small enough (8062 bp) to be packaged and delivered by bacteriophage φC31. These constructs were capable of transposing from a multicopy donor plasmid to the *S. coelicolor* genome. Southern blot analysis indicated that the transposons inserted randomly and only once per chromosome. Tn*5353*-derived mutants of *S. coelicolor* displayed wide variations in the intensity and temporal pattern of light emission. On screening of these clones, a variety of *bld* (no aerial hyphae) and *whi* (no gray-pigmented spores) mutants were collected. The site of each transposon insertion was mapped by isolating total genomic DNA and comparing the mutant *Ase*I and *Ssp*I restriction pattern to established *S. coelicolor* restriction maps.

Guzzo and DuBow *(19)* also developed a Tn*5*-based promoter probe. The left end of this modified Tn*5* element is truncated to 23 bp, but the right end is intact and encodes transposase. The promoterless *V. harveyi luxAB* genes and a Tc^R marker are contained between the left and right ends. The Tn*5::lux* transposon is contained within the ColE1-based multicopy plasmid, pFUSLUX. This plasmid can replicate in enteric hosts. To prevent replication of pFUSLUX and thus select for isolation of transposition events, the plasmid was transformed into *E. coli* containing pTF421, a p15A-based plasmid that expresses ColE1 RNA 1 from a *trp* promoter *(20)*. RNA 1 inhibits correct RNA primer formation and negatively regulates DNA replication from the ColE1 origin *(21)*. The majority of Tc^R colonies did not form until the Tn*5::lux* transposon integrated within the host genome (or, rarely, into pTF421). Over 90% of the Tc^R colonies were found to contain single inserts of the Tn*5-luxAB* element in random chromosomal locations *(19)*. Total genomic DNA was isolated and digested with either *Hind*III or *Bam*HI. Each of these enzymes cleave the transposon once and result in DNA fragments containing the Tc^R marker and flanking upstream (*Hind*III) or downstream (*Bam*HI) DNA. By shotgun cloning the digested DNA into a similarly cleaved pUC vector *(22)* and selecting for tetracycline resistance, DNA at the site of transposon insertion was isolated. Single-stranded DNA was produced from the pUC vector M13 origin, and the flanking DNA was sequenced. In this manner, environmentally regulated genes that show increased expression when exposed to increasing concentrations of heavy metal ions *(23,24)*, arsenic *(25)*, organotin *(26)*, and organic compounds, such as xylose *(19)*, were identified.

Transposons containing promoterless *lux* reporter genes constitute an exquisite promoter-probe system. Because of their ability to move between and within DNA molecules, transposable elements are able to deliver the *lux* reporter genes, at random, to nonessential sites in bacterial chromosomes and plasmids. When integrated within an operon, expression of the promoterless *lux* genes is effectively coupled to that of an upstream promoter. This coupling allows sensitive, inexpensive, real-time monitoring of promoter activity. As described above, numerous *lux*-transposon promoter-probe strategies have been developed. Although they employ a variety of transposable elements, different sets of *lux* genes, and are targeted to a diverse array of host organisms, each strategy has been successfully used to identify often novel genes whose expression is mediated by environmental stressors. The use of these promoter probes, and subsequent characterisation of the genes they identify, will greatly enhance our understanding of genetically programmed physiological changes and responses to environmental agents.

The following sections explain in detail how to generate a library of Tn*5luxAB* fusion mutants by transposon mutagenesis with Tn*5::lux,* how to screen this library for environmentally regulated genes, and how to isolate these gene::*luxAB* fusions.

2. Materials

1. Bacterial strains: *E. coli* strain NM522 (*supE, thi,* Δ[*lac-proAB*], Δ*hsd5* (*rk⁻, mk⁻*), [F', *proAB, lac*I^{q}ZΔM15]) was used for the construction of plasmid pFUSLUX *(27)*. *E. coli* strain DH1 (F⁻, *rec*A1, *end*A1, *gyr*A96, *thi, hsd*R17 *rk⁻, mk⁺*] *sup*44, *rel*A1) was used for construction of the luciferase-fusion library *(28)*.
2. Plasmids: Plasmid pTF421 was constructed by Fitzwater et al. *(20)*. This 6263-bp plasmid confers ampicillin resistance and has a p15A origin of replication such that it can coexist with ColE1-based vectors. RNA1, which inhibits ColE1 plasmid replication, is expressed from a *Serratia marscesens trp* promoter.
   Plasmid pFUSLUX was constructed by Guzzo and DuBow *(19)*. This 18.8-kbp vector is derived from pRZ341-21::Tn*5lac*, a ColE1-based plasmid that contains a Tn*5lacZYA* promoter probe *(29)*. The intact Tn*5* right end (IS*50*R) encodes the transposase. The left end (IS*50*L) is truncated to 23 bp, supports transposition, and allows transcriptional read-through from external promoters. The Tn*5* element also contains the Tn*10* tetracycline resistance marker. To create a *luxAB* promoter probe, pRZ341-21::Tn*5lac* was digested with *Bam*HI. This removed the *lacZYA* genes and allowed the insertion of a 3.25-kbp *Bam*HI fragment containing the promoterless *luxAB* genes from *V. harveyi. (30)* A modified pFUSLUX vector also exists. Plasmid pAG3 is identical to pFUSLUX, except that the *Bam*HI site between *luxAB* and the Tc^R marker was eliminated, allowing the cloning of *Bam*HI-compatible fragments (e.g., *Bgl*, II, *Sav*3Δ) upstream of the *lux*ΔB genes in pFUSLUX.
3. Media: Bacterial strains were routinely propagated at 32 or 37°C in LB broth (1% NaCl, 1% tryptone, 0.5% yeast extract, adjust to pH 7.0 by adding 1.5 mL 2 *N* NaOH/L) or on LB plates containing 1.5% agar. Ampicillin was used at a final concentration of 40 µg/mL. Tetracycline was used at a final concentration of 10 µg/mL in broth and 20 µg/mL in plates.
4. Cracking Buffer: This consists of 100 m*M* Tris-HCl, pH 7.5, 4 m*M* EDTA, 0.2% (w/v) sodium dodecyl sulfate, 0.8 *M* sucrose, and 0.2 mg/mL bromophenol blue.

5. Decyl Aldehyde: This can be obtained from Aldrich Chemical Company (Milwaukee, WI).
6. Film: Film should be Kodak XAR-5 or Agfa Curix RPI, 35 cm × 43 cm sheets.

3. Methods
3.1. Preparing the Library

1. Transform *E. coli* strain DH1 with pTF421, and select for transformants on LB agar containing ampicillin.
2. Transform DH1:pTF421 with pFUSLUX.
3. Select for Tc^R Ap^R clones on LB agar containing tetracycline and ampicillin. Grow at 37°C for 24 h and then incubate the plates, at room temperature, for a further 7 d (*see* **Note 1**).
4. Pick colonies, and master onto LB agar containing tetracycline and ampicillin. We master 48 colonies/plate, a spacing that allows optimal differentiation between luminescent clones.
5. Identify the plasmid content of the colonies to confirm the loss of pFUSLUX, and the maintenance of pTF421. We use the following "cracking" procedure.
6. Using a sterile toothpick, pick part of a colony and disperse it in 25 µL cracking buffer.
7. Incubate at room temperature for 10 min.
8. Centrifuge at 15,000*g* for 15 min at 4°C.
9. Remove the supernatant fluid, and subject it to agarose gel electrophoresis. Colonies with a chromosomally integrated Tn*5*::*lux* transposon will contain plasmid pTF421 (6203 bp), but not plasmid pFUSLUX (18.8 kbp).

3.2. Measuring Light Emission with Film

1. Prepare LB agar containing tetracycline and ampicillin.
2. Pour thin plates (i.e., 12.5 mL/10 cm Petri plate).
3. Replica plate (e.g., with velvets) the collection (i.e., library) of *lux* gene fusion clones on to these plates, allow them to grow for at least 6 h.
4. In a fume hood, add 50 µL of aldehyde (we use decyl aldehyde [Aldrich]) to the plastic lid, and invert the Petri dish. To reduce the pungent, unpleasent odor of the aldehyde, seal the plates with Parafilm (American National Can, Neenah, WI).
5. In a darkroom, cover the inverted plates with film (we use 35 cm x 43 cm Kodak XAR-5 or Agfa Curix RPI film). It is important that the film is flat on top of the plates, such that a uniform colony-to-film distance is maintained (*see* **Note 2**).
6. Expose for up to 4 h.
7. Develop the X-ray film. Transposon-derived clones will exhibit a variety of light-emitting phenotypes. After a 4-h exposure, any clones containing free plasmid pFUSLUX exhibit a medium to high light-emitting phenotype. For our studies of genes that are induced by environmental agents, we use cells that contain chromosomally integrated Tn*5*::*lux* transposons, and have a non- or low light-emitting phenotype.

3.3. Screening the Library

1. Prepare three sets of thin LB agar plates containing tetracyline, ampicillin, and the chemical of interest. Add the chemical agent to an appropriate final concentrations (*see* **Note 3**). For metals, such as iron and nickel, we had success with final concentrations of 0, 1, and 10 µg/mL *(23,24)*.
2. Replica plate the library onto these plates, and grow for at least 6 h.
3. Invert the plates (bottoms up), add 50 µL of aldehyde to the lid of each plate, and seal with Parafilm.
4. In a darkroom, cover the plates with film, and expose for various times (up to 4 h).
5. Develop film and look for "spots" of increasing size or intensity (i.e., "responsive" clones that exhibit increasing light emission with increasing concentration of compound).
6. Pick the "responsive" clones and streak for single colonies.
7. Retest the single colonies, and isolate clones that exhibit a consistent pattern of light emission. That is, on exposure to a particular concentration of compound, the X-ray spots are always a particular size, shape, and intensity (*see* **Note 4**).

3.4. Measuring Light Emission with a Luminometer

Although film is an excellent method for monitoring light emission, a more sophisticated technique employs a luminometer. A luminometer uses a photomultiplier tube, which converts photons into electrical pulses, to measure light emission. Many luminometers include injection systems that allow the rapid and precise addition of aldehyde or other reagents. In addition, by connecting the luminometer to a computer, data can be rapidly documented and analyzed. We use a Tropix Optocomp I Luminometer (MGM Instruments, Hamden, CT), which holds single samples in disposable cuvets.

1. Grow an overnight culture of the "responsive" clone to be tested in LB broth containing ampicillin and tetracycline.
2. Dilute cells 1:20 in fresh LB plus antibiotics and grow to midlog phase (i.e., A_{600} = 0.4–0.5).
3. Dilute cells 1:100 in fresh LB broth (final volume at least 150 mL) and grow until A_{600} = 0.05 (*see* **Note 5**).
4. Split cells into three 50-mL aliquots.
5. To measure light emission, remove three 1-mL samples from each 50-mL aliquot of cells (*see* **Note 6**).
6. Set the luminometer to inject 10 µL of aldehyde (diluted 1:100 in LB broth or sterile distilled water). We use decyl or dodecyl aldehyde (Aldrich). Record the photon emissions for an appropriate time (e.g., 10–60 s).
7. Add the chemical agent to an appropriate series of final concentrations (e.g., 0, 1, or 10 µg/mL). and measure light emission after a variety of time-points (e.g., 5, 10, 15, 30, 45, and 60 min after addtion of the chemical agent). Keep the cells at a constant temperature. Light emission varies with cell density, so before counting a sample, measure the optical density (*see* **Note 5**). If necessary, dilute cells

to A_{600} = 0.05 with LB broth (plus chemical agent and antibiotic) kept at the same temperature as the cells.

3.5. Verifying the Copy Number of the Integrated Transposon

1. To obtain a *luxAB* probe, digest pFUSLUX with *Bam*HI and isolate the 3.25-kbp fragment.
2. Label the probe (e.g., by nick-translation *[31]* or random priming *[32]*).
3. Isolate total chromosomal DNA from the responsive clone *(19)*.
4. Digest 10 µg chromosomal DNA with *Sal*I and 10 µg with *Eco*RI.
5. Separate chromosomal digests by agarose gel electrophoresis.
6. Transfer chromosomal digests to a Hybond-N membrane (Amersham, Oakville, Canada), and hybridize (e.g., by the bidirectional method of Smith and Summers *[33]*). If the Tn*5-lux* transposon is present in single copy, the probe will hybridize to a single *Sal*I band or two *Eco*RI bands. One of the two *Eco*RI bands should be an internal 5.1-kb *Eco*RI fragment from the Tn*5::lux*-transposon (*see* **Fig. 1**).

3.6. Isolating the Insert/Sequencing the Fusion Junction

1. Isolate total chromosomal DNA from the "responsive" clone *(19)*.
2. Digest 3–5 µg with *Bam*HI and an equal amount with *Hind*III. *Bam*HI allows cloning of the Tc marker, IS*50*R, and the adjacent, downstream chromosomal DNA. *Hind*III allows cloning of the Tc^R marker, *luxAB,* IS*50*L, and the adjacent, upstream chromosomal DNA (*see* **Fig. 2**).
3. Digest 1.0 µg pUC119 (or pUC120) with *Bam*HI and an equal amount with *Hind*III.
4. Ligate (i.e., shotgun clone) the pUC and chromosomal DNA digests *(32)* (*see* **Note 7**).
5. Transform *E. coli* DH1 with the ligation mixture. Plate onto LB agar supplemented with tetracycline and ampicillin, and grow overnight.
6. Restreak and isolate single colonies.
7. Check the plasmid content of the isolated colonies. Recircularized pUC vector is 3.2 kb in size. Plasmids containing the *Hind*III insert and upstream genomic DNA are larger than 11.9 kb (3.2 kb for pUC plus 8.7 kb for the Tc^R marker and the *luxAB* genes), whereas plasmids containing the *Bam*HI insert and downstream genomic DNA are larger than 11.5 kb (3.2 kb for pUC plus 8.3 kb for the Tc^R marker and IS*50*R) (*see* **Fig. 2**).
8. Prepare single-stranded DNA for dideoxy DNA sequencing *(22)*.
9. We use a Sequenase Kit (US Biochemical Corp., Cleveland, OH) for dideoxy sequencing. To determine the sequence at the junction between IS*50*R and the chromosome, we use an oligonucleotide (5' AAGGTTCCGTTCAGGAC 3') that corresponds to bp 1497–1513 of IS*50*R *(34)*. To determine the sequence at the *Bam*HI site in the chromosome, we use the "–40" primer from within pUC119/120.
10. Compare sequence to GeneBank, EMBL, or other nucleotide sequence data base (*see* **Note 8**).

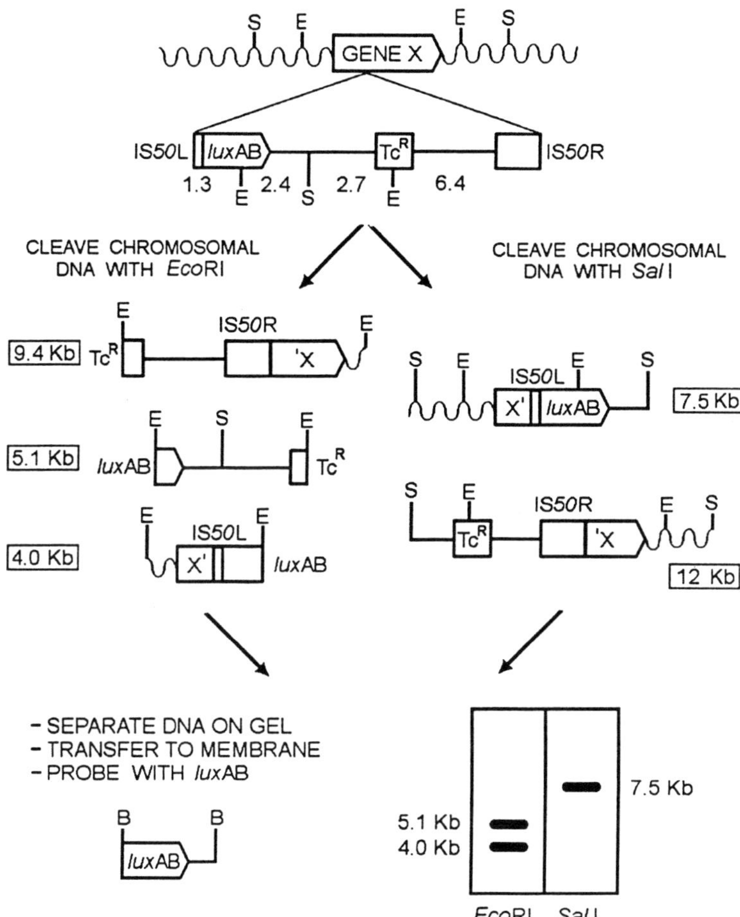

Fig. 1. Scheme to verify the copy number of the integrated Tn*5::lux* transposon. Integration of the Tn*5::lux* transposon into an unknown gene (Gene X) and cleavage sites for restriction enzymes *Eco*RI (E) and *Sal*I (S) are shown. Cleavage of total bacterial DNA with *Eco*RI generates a variety of restriction fragments, two of which contain portions of *luxAB*. One of these is an internal 5.1-kb fragment from the Tn*5::lux* transposon. Cleavage of total bacterial DNA with *Sal*I also generates a variety of restriction fragments, one of which contains the *luxAB* genes. The chromosomal digests are separated by gel electrophoresis. Then, a Southern blot is performed using the *luxAB* genes (i.e. a 3.25-kb *Bam*HI [B] fragment from plasmid pFUSLUX) as a probe. If the Tn*5::lux* transposon is present in single copy, the probe will hybridize to a single *Sal*I fragment and two *Eco*RI fragments. One of the *Eco*RI fragments should be the internal 5.1-kb fragment.

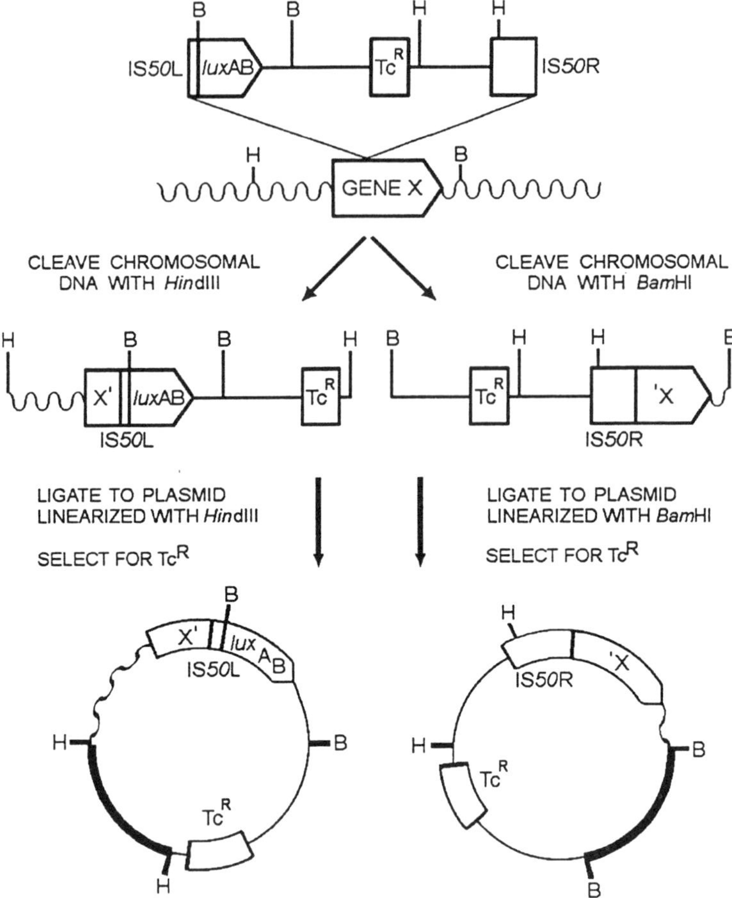

Fig. 2. Scheme for the isolation of DNA flanking the Tn*5::lux* transposon integration site. Integration of the Tn*5::lux* transposon into an unknown gene (Gene X) and cleavage sites for restriction enzymes *Hin*dIII (H) and *Bam*HI (B) are shown. Cleavage of chromosomal DNA with *Hin*dIII generates a fragment containing the tetracycline resistance marker (Tc^R), *lux*AB, and IS*50*L from the transposon, the 5'-end of Gene X (X'), and upstream chromosomal DNA (squiggly line). Cleavage of chromosomal DNA with *Bam*HI generates a fragment containing Tc^R and IS*50*R from the transposon, the 3'-end of Gene X ('X), and downstream chromosomal DNA (squiggly line). Total *Hin*dIII (or *Bam*HI) cleaved chromosomal DNA is then ligated to plasmid DNA (thick line), which has been linearized with *Hin*dIII (or *Bam*HI). By selecting for Tc^R transformants, plasmid vectors containing a portion of the transposon and upstream (or downstream) flanking DNA is isolated.

3.7. Other Applications of the pFUSLUX System

Although we have only used pFUSLUX in *E. coli,* our system should work in other species of bacteria. In enteric organisms, the above strategy should be satisfactory. In organisms that do not permit the replication of ColE1 plasmids, pFUSLUX would become a "suicide plasmid" and the requirement for pTF421 would be eliminated. In addition, pFUSLUX could be used for saturation mutagenesis of plasmids in a manner similar to that of Steinmann et al. *(13).* Indeed, we have observed integration of the transposon onto pTF421. A suitable target plasmid should not have a p15A or ColE1 origin of replication, or encode ampicillin or tetracycline resistance. Alternatively, target DNA can be cloned into pTF421. To isolate mutagenized plasmids, one should transform cells containing the target plasmid with pFUSLUX and then select for Ap^R, Tc^R, and the target plasmid marker. Plasmid DNA from all transformants should then be isolated and used to transform an antibiotic sensitive strain. Plasmids containing *lux* transposon insertions can be isolated by selection for Tc^R and target plasmid marker.

4. Notes

1. RNA 1 production from pTF421 should prevent pFUSLUX replication. In theory, this should prevent colony formation on media supplemented with tetracycline until after transposition of the Tc^R transposon has occurred. However, pFUSLUX-containing colonies can be found within 24–48 h, whereas new colonies continue to appear for more than 7 d. Nonetheless, more than 90% of the clones that appear on or after 4 d were found to be the product of authentic transposition events *(19).* These are picked and used for further study.

2. For best results, we have created a simple device known affectionately as the "Flat-o-Matic." Briefly, the Petri plates are taped, bottoms up, to a sheet of stiff cardboard or scrap X-ray film. In a darkroom, unexposed X-ray film is placed on top of the inverted Petri plates. Then, the plates and film are sandwiched between stiff cardboard such that this sandwich slides snuggly into a box (we use empty X-ray film boxes). The "Flat-o-Matic" prevents the plates and film from moving around, and keeps the X-ray film flat on top of the Petri plates such that the colony-to-film distance is uniform.

3. The appropriate concentrations vary with each test compound. However, if the results are to be relevant, the concentrations used should reflect what is available in the environment. Avoid cytotoxic concentrations. Dead cells do not form colonies or emit light.

4. At this stage, it is especially important to inoculate the plates with approximately the same number of cells. Different numbers of cells emit different amounts of light. Instead of replica plating, grow cells for at least 6 h in 10 mL LB broth (with antibiotics). Then, dilute cells 1:20 in fresh LB broth and grow to midlog phase (i.e., A_{600} = 0.4–0.5). Dilute this culture 1:100 in fresh LB broth (with

antibiotics) and grow until A_{600} = 0.05. Finally, deposit 5 µL samples on to labeled Petri plates containing the appropriate media. After drying of the aliquots and incubation, the plates are suitable for light testing.

5. Light emission measurements are affected by cell density. Results are most reliable when the absorbance at 600 nm (i.e., A_{600}) is between 0.001 and 0.05. At higher cell densities (i.e., $A_{600} > 0.1$), results are inconsistent, perhaps because cells block the light emitted by their neighbors from reaching the photomultiplier tube of the luminometer. When counted, all samples should be at the same optical density. During a time-course experiment, cells will grow and, if samples are not diluted before counting, there will be an apparent increase in light emission.

6. Multiple (at least three) samples are required to allow calculation of mean values and standard deviations from the mean values. These values are important for determining if differences in light emission (e.g., increasing light emission on exposure to increasing concentration of some compound) are statistically significant. Within any set of three samples, readings should not be significantly different.

7. We use the phagemid vectors pUC119/120 because they allow rapid generation of single-stranded DNA, which can be used for dideoxy DNA sequencing *(22)*.

8. Initially, the Kohara et al. *(35)* phage library of the *E. coli* chromosome was used to map the site of transposon insertion *(19)*. Briefly, the Kohara et al. phage library was propagated on *E. coli* strain NM621 *(36)* and transferred to nitrocellulose filters via a plaque lift method *(37)*. A probe was prepared by digesting plasmid DNA from the Tc^R colony with *Hind*III and isolating 1 µg of the fragment that contains 1.1 kb of IS*50*R and downstream chromosomal DNA. This probe was labeled and hybridized with a Digoxygenin DNA Labeling and Detection Kit (Boehringer-Mannheim, Laval, Canada) according to the manufacturer's directions.

Acknowledgments

The authors wish to thank C. Diorio for her indispensable help. This work was supported by a grant (97043) from the Center for the Alternatives to Animal Testing (USA). DCA is supported by a Medical Research Council of Canada Studentship (#ST 46660 AP007480).

References

1. Hastings, J. W. (1968) Bioluminescence. *Ann. Rev. Biochem.* **37,** 597–630.
2. Engebrecht, J., Nealson, K., and Silverman, M. (1983) Bacterial bioluminescence: isolation and genetic analysis of functions from *Vibrio fischeri. Cell* **32,** 773–781.
3. Engebrecht, J., Simon, M., and Silverman, M. (1985) Measuring gene expression with light. *Science* **227,** 1345–1347.
4. Casadaban, M. J. and Cohen, S. N. (1979) Lactose genes fused to exogenous promoters in one step using a Mu-*lac* bacteriophage: *in vivo* probe for transcriptional control sequences. *Proc. Natl. Acad. Sci. USA.* **76,** 3530–3533.

5. Castilho, B. A., Olfson, P., and Casadaban, M. J. (1984) Plasmid insertion mutagenesis and *lac* gene fusion with mini-Mu bacteriophage transposons. *J. Bacteriol.* **158,** 488–495.

6. Belas, R., Simon, M., and Silverman, M. (1986) Regulation of lateral flagella gene transcription in *Vibrio parahaemolyticus. J. Bacteriol.* **167,** 210–218.

7. DeLorenzo, V., Herrero, M., Jakubzik, U., and Timmis, K. N. (1990) Mini-Tn*5* transposon derivatives for insertion mutagenesis promoter probing, and chromosomal insertion of cloned DNA in Gram-negative eubacteria. *J. Bacteriol.* **172,** 6568–6572.

8. Wolk, C. P., Cai, Y., and Panoff, J. M. (1991) Use of a transposon with luciferase as a reporter to identify environmentally responsive genes in a cyanobacterium. *Proc. Natl. Acad. Sci. USA* **88,** 5355–5359.

9. Meighen, E. A. (1991) Molecular biology of bacterial luminescence. *Microbiol. Rev.* **55,** 123–142.

10. Fernandez-Pinas, F. and Wolk, P. C. (1994) Expression of *luxCD-E* in *Anabaena* sp. can replace the use of exogenous aldehyde for *in vivo* localisation of transcription by *luxAB. Gene* **150,** 169–174.

11. Shaw, J. J., Settles, L. G., and Kado, C. I. (1988) Transposon Tn4431 mutagenesis of *Xanthomonas campestris* pv. campestris: characterization of a nonpathogenic mutant and cloning of a locus for pathogenicity. *Mol. Plant. Microbe. Interact.* **1,** 39–45.

12. Kamoun, S. and Kado, C. I. (1990) A plant-inducible gene of *Xanthomonas campestris* pv. campestris encodes an exocellular component required for growth in the host and hypersensitivity on nonhosts. *J. Bacteriol.* **172,** 5165–5172.

13. Steinmann, D., Wiggerich, H.-G., Klauke, B., Schramm, U., Puhler, A., and Priefer, U. B. (1993) Saturation mutagenesis in *Escherichia coli* of a cloned *Xanthomonas campestris* DNA fragment with the *lux* transposon Tn*4431* using the delivery plasmid pDS1, thermosensitive in replication. *Appl. Microbiol. Biotechnol.* **40,** 356–360.

14. King, J. M. H., DiGrazia, P. M., Applegate, B., Burlage, R., Sanseverino, J., Dunbar, P., Larimer, F., and Sayler, G. S. (1990) Rapid, sensitive, bioluminescent reporter technology for naphthalene exposure and biodegradation. *Science* **249,** 778–781.

15. Menn, F.-M., Applegate, B. M., and Sayler, G. S. (1993) NAH plasmid-mediated catabolism of anthracene and phenanthrene to naphthoic acids. *Appl. Environ. Microbiol.* **59,** 1938–1942.

16. Sohaskey, C. D., Im, H., and Schauer, A. (1992) Construction and application of plasmid- and transposon-based promoter-probe vectors for *Streptomyces* spp. that employ a *Vibrio harveyi* luciferase reporter cassette. *J. Bacteriol.* **174,** 367–376.

17. Sohaskey, C. D., Im, H., Nelson, A. D., and Schauer A. (1992) Tn*4556* and luciferase: synergistic tools for visualising transcription in *Streptomyces. Gene* **115,** 67–71.

18. Schauer, A., Ranes, M., Santamaria, R., Guijarro, J., Lawlor, E., Mendez, C., Chater, K., and Losick, R. (1988) Visualising gene expression in time and space in the filamentous bacterium *Streptomyces coelicolor. Science* **240,** 768–772.

19. Guzzo, A. and DuBow, M. S. (1991) Construction of stable, single-copy luciferase gene fusions in *Escherichia coli. Arch. Microbiol.* **156,** 444–448.

20. Fitzwater, T., Tamm, J., and Polisky, B. (1984) RNA1 is sufficient to mediate plasmid ColE1 incompatibility *in vivo. J. Mol. Biol.* **175,** 409–417.

21. Polisky, B. (1988) ColE1 replication control circuitry: sense from antisense. *Cell* **55,** 929–932.

22. Vieira, J. and Messing, J. (1987) Production of single-stranded plasmid DNA. *Methods Enzymol.* **153,** 3–11.

23. Guzzo, A., Diorio, C., and DuBow, M. S. (1991) Transcription of the *Escherichia coli fli*C gene is regulated by metal ions. *Appl. Environ. Microbiol.* **57,** 2255–2259.

24. Guzzo, A. and DuBow, M. S. (1994) A *luxAB* transcriptional fusion to the cryptic *cel*F gene of *Escherichia coli* displays increased luminescence in the presence of nickel. *M. G.G.* **242,** 455–460.

25. Cai, J. and DuBow, M. S. (1996) Expression of the *Escherichia coli* chromosomal *ars* operon. *Can. J. Microbiol.* **42,** 662–671.

26. Briscoe, S. F., Diorio, C., and DuBow, M. S. (1996) Luminescent biosensors for the detection of tributyltin and dimethyl sulfoxide and the elucidation of their mechanisms of toxicity, in *Environmental Biotechnology: Principles and Applications* (Moo-Young, M., Anderson, W. A., and Chakrabarty, A. M., eds.), Kluwer Academic Publishers, The Netherlands, pp. 645–655.

27. Gough, J. and Murray, N. (1983). Sequence diversity among related genes for recognition of specific targets in DNA molecules. *J. Mol. Biol.* **166,** 1–19.

28. Hanahan, D. (1983) Studies on transformation of *Escherichia coli* with plasmids. *J. Mol. Biol.* **166,** 557–580.

29. Johnson, R. C. and Reznikoff, W. S. (1983). Sequences at the ends of transposon Tn5 required for transposition. *Nature* **304,** 280–282.

30. Miyamoto, C. M., Graham, A. D., Boylan, M., Evans, J. F., Hasel, K. W., Meighen, E. A., and Graham, A. F. (1985) Polycistronic mRNAs code for polypeptides of the *Vibrio harveyi* luminescence system. *J. Bacteriol.* **161,** 995–1001.

31. Rigby, P. W. J., Dieckmann, M., Rhodes, C., and Berg, P. (1977) Labelling deoxyribonucleic acid to high specific activity *in vitro* by nick translation with DNA polymerase I. *J. Mol. Biol.* **113,** 237–251.

32. Sambrook, J., Fritsch, E., and Maniatis, T. (1989) *Molecular Cloning. A Laboratory Manual.* 2nd ed. Cold Spring Harbor Laboratory, Cold Spring Harbor, NY.

33. Smith, E. G. and Summers, M. D. (1980) The bidirectional transfer of DNA and RNA to nitrocellulose or diazobenzyloxymethyl-paper. *Anal. Biochem.* **109,** 123–129.

34. Auerswald, E. A., Ludwig, G., and Schaller, H. (1980) Structural analysis of Tn5. *Cold Spring Harbor Symp. Quant. Biol.* **45,** 107–113.

35. Kohara, Y., Akiyama, K., and Isono, K. (1987) The physical map of the whole *E. coli* chromosome: application of a new strategy for rapid analysis and sorting of a large genomic library. *Cell* **50,** 495–508.

36. Whittaker ,P. A., Campbell, A. J. B., Southern, W. M., and Murray, N. E. (1988) Enhanced recovery and restriction mapping of DNA fragments cloned in a new λ vector. *Nucleic Acids Res.* **16,** 6725–6736.

37. Maniatis, T., Fritsch, E., and Sambrook, J. (1982) *Molecular Cloning. A Laboratory Manual.* Cold Spring Harbor Laboratory, Cold Spring Harbor, NY.

10

Cryopreservation and Reawakening

L. Winona Wagner and Tina K. Van Dyk

1. Introduction

Laboratories like the American Type Culture Collection (Rockville, MD) have been using freeze-drying or lyophilization techniques as a means of preserving microorganisms for more than 50 years. There are numerous variations of the method, and elaborate equipment is available to improve environmental control and to allow large numbers of samples to be processed.

Lyophilization of cultures is a process that removes water from frozen cultures by sublimation under reduced pressure *(1)*. The method chosen will depend on such factors as cell viability, genetic mutations, frequency of culture use, maximum viability of the cells, maximum storage time, age of culture, and the selection of a suitable cryoprotectant, such as horse serum, skim milk, sucrose, dextran, inositol, or others *(2)*.

Bioluminescent microorganisms have been lyophilized, and when reconstituted retain bioluminescence *(3)*. Such lyophilized and reconstituted cultures are useful for toxicity assays that quantitate decreases in bioluminescence *(4)*. It was not certain whether lyophilized and reconstituted bacteria would be useful without an outgrowth period for assays that look for an increased bioluminescence response. In such tests, increased bioluminescence results from increased transcription initiated at the promoter controlling *lux* gene expression. Therefore, regulated transcriptional initiation, transcription, translation, protein folding and assembly, as well as biochemical functioning of the *lux* gene products are all required for an increase in bioluminescence. Here, we describe a method for lyophilization of a recombinant *Escherichia coli* strain containing a fusion of an *E. coli* heat-shock promoter to the *Vibrio fischeri luxCDABE* genes *(5)*. Following lyophilization and reconstitution, these cells were found to be useful without requiring outgrowth for assays

From: *Methods in Molecular Biology, Vol. 102: Bioluminescence Methods and Protocols*
Edited by: R. A. LaRossa © Humana Press Inc., Totowa, NJ

that monitor increased bioluminescence induced by sublethal concentrations of toxicants.

2. Materials

2.1. E. coli *Strain and Growth Media*

1. *E. coli* strain TV1061, a transformant of plasmid pGrpELux5 into host strain RFM443 *(5)*.
2. LBG with kanamycin *(6)*: 10 g/L tryptone, 5 g/L yeast extract, 10 g/L sodium chloride. Autoclave for 15 min at 121°C. After cooling the medium to approx 60°C, add sterile glucose to 10 g/L and kanamycin to 2.5 g/L.
3. 15-mL sterile conical centrifuge tubes, such as Corning (VWR Scientific Products, West Chester, PA).
4. Incubator set at 26°C containing a shaker platform, such as New Brunswick Scientific Model G24 (Edison, NJ).
5. Lyophilization medium: 0.3 g/L NH_3SO_4, 0.45 g/L $MgSO_4$, 0.047 g/L $NaCitrate·2 H_2O$, 0.025 g/L $FeSO_4·7H_2O$, 0.06 g/L thiamine-HCl, 1.95 g/L $K_2HPO_4·2H_2O$, 0.9 g/L NaH_2PO_4, 0.005 g/L biotin, 20 g/L casamino acids, 1 mL trace element solution, 0.1 g/L uracil, 20 g/L glucose, 0.026 g/L $CaCl_2·2H_2O$. Autoclave for 15 min at 121°C.
6. Trace element solution: 8 g/L $ZnSO_4·7H_2O$, 3 g/L $CuSO_4·5H_2O$, 2.5 g/L $MnSO_4·H_2O$, 0.15 g/L boric acid, 0.1 g/L $NH_4MoO_4·4H_2O$, 0.06 g/L $CoCl_2·6H_2O$. Autoclave for 15 min at 121°C.

2.2. Lyophilization

1. Cryoprotectant solution: 24 g/100 mL sucrose. Autoclave for 15 min at 121°C.
2. Lyophilization vials sterilized by autoclaving for 15 min at 121°C (*see* **Note 1**).
3. Filters for lyophilizer sterilized by autoclaving for 15 min at 121°C. (Pall Emflon II 0.2 µ absolute, Cortland, NY).
4. Lyophilizer used suitable for vials selected (*see* **Note 1**), for example, FTS System, Model FD-14–84, Stone Ridge, NY.

2.3. Reawakening

1. LB medium *(6)*: per liter, 10 g Bacto-tryptone, 5 g Bacto-yeast extract, 10 g salt. Adjust to pH 7.0 with 5 *N* NaOH. For agar plates, add 15 g/L of Bacto-agar. Sterilize by autoclaving.
2. Sterile, white, flat-bottom microplates: Microlite™ (Dynatech Laboratories, Waltham, MA).
3. ML3000 microplate luminometer (Dynatech Laboratories).

3. Methods

3.1. Cell Growth

1. Grow *E. coli* strain TV1061 in LBG plus kanamycin to an OD_{600} of 2 to ensure viability of culture.

2. Cells grown in **step 1**, are used to inoculate lyophilization medium (100 mL for 2-L fermenter) at 5% v/v.
3. Cells in lyophilization medium are grown at 26°C, pH 7.0, DO_2 50%. Dissolved oxygen is controlled by automated feedback regulation of agitation and aeration. Initial parameters of 0.6 vvm for aeration and 300 rpm for agitation are set (*see* **Note 4**).
4. Grow cells to OD_{600} of 1.8. (*see* **Notes 3** and **4**).
5. Pellet cells by centrifugation (Sorvall Superspeed RC5B [Newtown, CT], SS34 rotor, 10,352g, 4°C for 20 min). Decant medium and store cells on wet ice.

3.2. Freezing and Lyophilization

1. Resuspend pellet in one-half volume of starting culture with fresh lyophilization medium and an equal half volume of 24% sucrose solution (sterile).
2. Dispense into sterile lyophilization vials and freeze at –70°C. Typically, vials are filled to 20% of their total volume (*see* **Note 6**).
3. Place frozen vials on lyophilizer. Process for at least 3 h at ≤20 mtorr and –100°C.
4. Seal vials and store at either refrigerated (4°C) or freezer temperatures (–20°C) until rehydrated.

3.3. Reconstitution and Stress Induction of Lyophilized Cells

1. Resuspend lyophilized *E. coli* TV1061 cells in a volume of sterile water equal to the volume of the samples prior to lyophilization. These reconstituted cells can be used immediately for a stress induction test (*see* Chapter 13) or may be incubated at room temperature or 26°C for up to 60 min prior to use.
2. Determine viable cell counts by plating serially diluted reconstituted cells on LB plates and incubating overnight at 37°C.
3. Add 20 µL reconstituted cells to 80 µl LB medium with or without a chemical or sample at 1.25X the desired final concentration in a white 96-well microplate. For example, 2.5% (v/v) ethanol in LB was used to test the stress induced by a final concentration of 2% ethanol.
4. Quantitate bioluminescence in a Dynatech ML3000 microtiter plate luminometer with temperature controlled at 26°C using the cycle mode and the following settings: Gain—High; Data—All; Cycles—20; Pause—300 s; Auto gain—On; Mixing—On.
5. Plot data as RLU of treated and untreated cells as a function of time after chemical addition, and examine for increased bioluminescence caused by stress.

3.4. Sample Data

E. coli strain TV1061 was lyophilized and reconstituted as described above. Aliquots of this culture, which had 1.0×10^9 viable cells/mL, were immediately challenged with a final concentration of 2% ethanol. **Figure 1** shows the results of this test. The kinetics of induction, consisting of a lag time of 15–20 min followed by rapid increase in bioluminescence from the stressed cells, was similar to previous results with actively growing cells in LB medium (*5*).

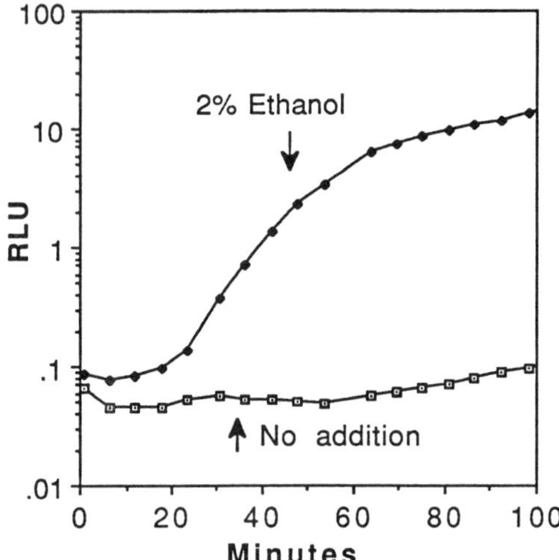

Fig. 1. Induction of increased bioluminescence by 2% ethanol from *E. coli* strain TV1061 that had been lyophilized and reconstituted.

4. Notes

1. We expect that this protocol would be useful for other recombinant *E. coli* strains containing stress-inducible *lux* genes.
2. Types of vials and lyophilizer equipment selected are not expected to alter the responses obtained.
3. Cells must be in early logarithmic phase rather than late logarithmic or early stationary phase for best results following lyophilization. This parallels the results using actively grown cells *(7)*.
4. Cells grown to late logarithmic phase were difficult to work with and showed greater variation in viable counts after rehydration.
5. A KLF 2000 fermenter (Bioengineering AG, Wald, Switzerland) was used to grow large quantities of cells for lyophilization under controlled dissolved oxygen and pH. Other types of fermenters can also be used. Good control over pH and dissolved oxygen is the determining factor for choice of fermenter.
6. Once cells have been dispensed and frozen, do not allow to thaw.
7. The lyophilization medium gave better results than LBG.
8. Sucrose was the best cryoprotectant tested. Glucose at 1% concentration in rich medium (LB) was also an acceptable cyroprotectant. Skim milk was unsatisfactory.

References

1. Kirsop, B. E. and Snell, J. J. S. (1984) *Maintenance of Microorganisms.* Academic, Orlando, FL.

2. Gherna, R. L. (1994) Culture Preservation, in *Methods for General and Molecular Bacteriology* (Gerhardt, P., Murray, R. G. E., Wood, W. A., and Krieg, N. R., eds.), American Society for Microbiology, Washington, DC, pp. 278–292.
3. Janda, I. and Opekarová, M. (1989) Long-term preservation of active luminous bacteria by lyophilization. *J. Biolum. Chemilum.* **3,** 27–29.
4. Bulich, A. A. (1982) A practical and reliable method for monitoring the toxicity of aquatic samples. *Process Biochem.* **17,** 45–47
5. Van Dyk, T. K., Majarian, W. R., Konstantinov, K. B., Young, R. M., Dhurjati, P. S., and LaRossa, R. (1994) Rapid and sensitive pollutant detection by induction of heat shock gene–bioluminescence gene fusions. *Appl. Environ. Microbiol.* **60,** 1414–1420.
6. Sambrook, J., Fritsch, E. F., and Maniatis, T. (eds.) (1989) *Molecular Cloning A Laboratory Manual,* 2nd ed., vol. 3, appendix A.1. Cold Spring Harbor Laboratory, Cold Spring Harbor, NY.
7. Rupani, S. P., Gu, M. B., Konstantinov, K. B., Dhurjati, P. S., Van Dyk, T. K., and LaRossa, R. A. (1996) Characterization of the stress response of a bioluminescent biological sensor in batch and continuous cultures. *Biotechnol. Prog.* **12,** 387–392.

11

Luciferase Renaturation Assays of Chaperones and Chaperone Antagonists

Vanitha Thulasiraman and Robert L. Matts

1. Introduction

Firefly luciferase has been widely used as a model substrate to study folding *(1–3)* and renaturation *(4–6)* of protein because of its rapid and sensitive bioluminescenct activity. Although normally localized in peroxisomes, luciferase folds to the native state on expression in bacteria, in mammalian cells in culture *(7)*, and in rabbit reticulocyte lysate (RRL) *(1–3)*. RRL also efficiently facilitates the renaturation of thermally *(5,6)* or chemically *(1,4,8–10)* denatured luciferase. Efficient luciferase renaturation requires optimal ATP, Mg^{2+}, and K$^+$ concentrations *(5)*. Thermal denaturation of luciferase produces unfolded intermediates that mimic denatured protein produced in a cell subjected to heat stress, whereas chemical denaturation of luciferase with guanidinium·HCl was thought to generate unfolded luciferase, which closely mimics nascent unfolded luciferase. However, recent results indicate that the folding pathway of chemically denatured luciferase is not identical to that followed by newly synthesized luciferase *(1)*.

To define the components involved in facilitating luciferase folding and renaturation several laboratories are using purified heat-shock proteins (hsps) and hsp cohorts to reconstitute the ability to renature luciferase *in vitro*. Components that have been identified to be required for or to stimulate luciferase renaturation in vitro include: hsp90 *(3,5,6)*; hsc70 *(4–6)*; DnaJ *(11)*; Hip *(12)*; and p60 *(13)*. However, these reconstituted systems are very inefficient and require hsps to be present in large excesses compared to the amount of denatured luciferase that is renatured.

Recently, we have demonstrated that the rate at which the renaturation of thermally denatured luciferase is catalyzed in RRL obeys Michelis-Menten kinetics *(6)*. In addition, we have used the kinetics of luciferase renaturation in

From: *Methods in Molecular Biology, Vol. 102: Bioluminescence Methods and Protocols*
Edited by: R. A. LaRossa © Humana Press Inc., Totowa, NJ

RRL to analyze the mechanism by which the hsp90 binding drug, geldanamycin *(14)*, inhibits hsp90 function. This chapter describes the use of the luciferase renaturation assay for analysis of the effects of pharmacological agents on hsp (chaperone) function in RRL.

2. Materials
2.1. Reagents

1. Rapamycin (Calbiochem, La Jolla, CA).
2. Cyclosporin (Calbiochem).
3. L-683-590, an FK506 derivative (Merck, Sharp and Dohme, Rahway, NJ).
4. Rabbit reticulocyte lysate is available from Green Hectares (Oregon, WI), Ambion (Austin, TX), and Promega (Madison, WI), or it can be prepared as previously described *(5,15–18)*.

2.2. Equipment

1. Water bath.
2. Lumac (3 *M*) bioluminometer.
3. Lumacuvet from Celsis (Monmouth Junction, NJ).
4. Vortex.
5. Micropipeters (2-μL, 20-μL, 200-μL and 1-mL capacity).
6. Eppendorf tubes (500 μL to 1.5 mL), since all the assays are done in small volumes.

2.3. Buffers

1. Assay buffer (AB): 25 mM Tricine-HCl, pH 7.8, 8 mM MgSO$_4$, 0.1 mM EDTA, 33 μM dithiothreitol (DTT), 470 μM D-luciferin, 240 μM coenzyme A, and 0.5 mM ATP. Assay buffers are quick-frozen in liquid nitrogen and stored at −70°C as 1.3-mL aliquots.
2. Stability buffer (SB): 25 mM Tricine-HCl, pH 7.8, 8 mM MgSO$_4$, 0.1 mM EDTA, 10 mg/mL bovine serum albumin (BSA), 10% glycerol, and 1% Triton X-100. Stability buffer is stored at 4°C and can be used over a month.
3. Dialysis buffer (DB): 10 mM Tris-HCl, pH 7.4, 100 mM KCl, 3 mM Mg(OAc)$_2$, and 2 mM DDT.
4. Cold mix (CM): 200 μM of the 20 amino acids commonly found in proteins, 100 mM Tris-HCl, pH 7.7, 2 mM GTP, 10 mM Mg(OAc)$_2$, and 750 mM KCl. Cold mix is stored as 250-μL aliquots at −20°C.
5. Deletion mix (DM): CM without amino acids and GTP.

2.4. Stock Solutions

2.4.1. Stocks for Preparing Assay Buffer AB and Stability Buffer SB

1. 1 M Tricine-HCl, pH 7.8, stored at −20°C as 1-mL aliquots.
2. 1 M MgSO$_4$ stored at 4°C.
3. 0.5 M EDTA stored at room temperature.

4. 250 mM DTT dissolved in water is made fresh on the day of use.
5. 0.1 M ATP in water stored at –20°C as 100-μL aliquots.
6. 40 mg/mL of coenzyme A in water stored as 500-μL aliquots at –20°C.
7. 6.5 mg/mL of luciferin (Sigma, St. Louis, MO) in water stored as 1-mL aliquots at –70°C. Luciferin is quick-frozen in liquid nitrogen.
8. 10% Triton X-100 (Sigma) in water stored at 4°C.
9. 50 mg/mL acetylated BSA (Sigma) in water stored at 4°C (*see* **Note 7**).

2.4.2. Stocks for Preparing CM and Refolding Mix

1. 1 M Tris-HCl, pH 7.7, is stored at –20°C as 5-mL aliquots.
2. 20 mM amino acid dissolved in water is stored as 15-mL aliquots at –20°C.
3. 0.1 M GTP dissolved in water is stored as 250-μL aliquots at –70°C.
4. 1 M Mg(OAc)$_2$ stored at 4°C.
5. 2 M KCl stored at 4°C.
6. 0.1 M creatine phosphate (Sigma) in water, stored as 1-mL aliquots at –20°C.
7. 1600 U/mL of creatine phosphokinase (Sigma) in 50% glycerol, stored as 50-μL aliquots at –20°C.

2.4.3. Stock Solutions of the Different Drugs

1. 50-mM stock of the peptide pepN (NIVRKKK, Sarkeys Biotechnology Research Laboratory, Oklahoma State University, OK) in water, can be stored at 4°C for a week.
2. 50-mM stock of the peptide pepF (FYQLALT, Sarkeys Biotechnology Research Laboratory) in DMSO, stored at –70°C.
3. 1 mg/mL geldanamycin (GA, provided by the Drug Synthesis and Chemistry Branch, Developmental Therapeutics Program, Division of Cancer Treatment, National Cancer Institute) in DMSO stored as 10-μL aliquots at –20°C. Geldanamycin is light-sensitive and is stored in an appropriate container to protect it from exposure to light.
4. 5 mg/mL geldampicin (Kenneth L. Rinehart, University of Illinois, Urbana) in DMSO stored as 10-μL aliquots at –20°C.
5. 300 mM clofibric acid (Sigma) dissolved in water is stored at 4°C and can be used up to a week (*see* **Note 9**).
6. 200 mM ibuprofen (Sigma) dissolved in water is preferably prepared fresh on the day of use (*see* **Note 9**).
7. 200 mM indomethacin (Sigma) dissolved in water is prepared fresh on the day of use (*see* **Note 9**).
8. 400 mM salicylic acid (Sigma) dissolved in water is stored at 4°C and can be used up to a week (*see* **Note 9**).

3. Methods

3.1. Thermal Denaturation of Luciferase

1. Firefly luciferase (Sigma) is dissolved in SB lacking Triton X-100 and glycerol. When the luciferase is completely dissolved, Triton X-100 and glycerol are added (*see* **Note 1**).

2. Luciferase stock (0.5 mg/mL) is centrifuged to remove any undissolved luciferase and stored at 4°C (*see* **Note 2**).
3. Desired aliquot of the luciferase stock is incubated in 41°C water bath for 10 min. The thermally denatured luciferase is transferred to 4°C (*see* **Notes 3** and **4**).

3.2. Preparation of Refolding Mix

1. Rabbit reticulocyte lysate from Green Hectares can be purchased in a variety of volumes. We aliquot the lysate when it is first unthawed and store as 150-µL aliquots in liquid nitrogen (*see* **Note 14**).
2. Just before the experiment, the lysate is thawed by incubation in a 30°C water bath for 1 min or until it is fully thawed. The thawed lysate is placed on ice.
3. The lysate is diluted 1:1 by the addition of 30 µL of CM, 15 µL of creatine phosphate (10 mM), 3.0 µL of creatine phosphokinase (16 U/mL), and the appropriate amount of water or other additions (*see* **Note 5**).

3.3. Renaturation of Luciferase

1. The denatured luciferase (*see* **Notes 5**, **11**, and **12**) is diluted 20-fold into RRL refolding mix supplemented with the different drugs or appropriate drug vehicle as a control (*see* **Note 8**).
2. The assay mix is then transferred to a 30°C water bath.
3. At desired times, an 1.5-µL aliquot is mixed with 50 µL of AB, vortexed for 2 s, and assayed in the luminometer for 10 s (*see* **Note 10**).

3.4. Results and Discussion

RRL is an ideal system to screen for pharmacological agents that inhibit chaperone function. Recently, we have demonstrated that the hsp90 binding drugs, geldanamycin and herbimycin A, inhibit the rate of luciferase renaturation in RRL by 50% at concentrations of 0.2 and 1.3 µM, respectively *(6)*. A kinetic analysis indicated that geldanamycin inhibited luciferase renaturation noncompetitively with respect to luciferase concentration and uncompetitively with respect to ATP concentration (*see* **Note 13**). This analysis indicated that geldanamycin bound to the hsp90 chaperone machinery after the binding of ATP, but that it can bind either before or after the binding of the luciferase substrate to the hsp90 chaperone machine/ATP complex. The binding of geldanamycin to hsp90 markedly decreased the K_{app} of the hsp90 chaperone machinery for ATP, indicating that its binding either increased the binding affinity of hsp90 chaperone machinery for ATP or slowed the rate of ATP hydrolysis.

3.4.1. Effect of Chaperone Antagonists on Luciferase Renaturation

Clofibric acid is a pharmacological agent that causes peroxisomal proliferation and has been demonstrated to specifically bind hsc70 *(19)*. To determine whether the binding of clofibric acid to hsc70 inhibits hsc70 function, we examined the effect of clofibric acid on luciferase renaturation in RRL. The rate of

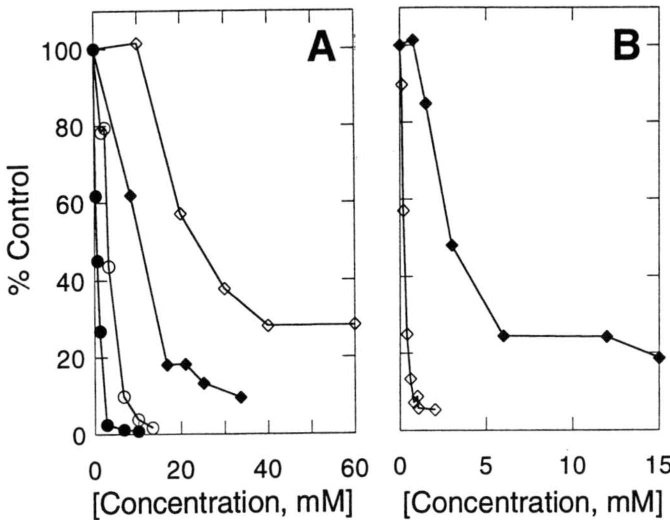

Fig. 1. Effect of drugs on luciferase renaturation in RRL. Refolding mixes were incubated at 28°C and **(A)** with no additions (controls) or with varying concentrations of clofibric acid (◆), salicylic acid (◇), ibuprofen (○), and indomethacin (●); or **(B)** with DMSO or water (controls) or with varying concentration of pepN (◆) and pepF (◇). One microliter of denatured luciferase (1 mg/mL) was diluted 20-fold into the mix. After 10 min, the amount of luciferase activity present in a 1.5 μL of aliquot of each reaction was measured as described in **Subheadings 2.** and **3.**

luciferase renaturation was inhibited by clofibric acid in a concentration-dependent manner **(Fig. 1A)**. Luciferase renaturation was analyzed in the presence of saturating ATP and varying concentrations of luciferase and clofibric acid **(Fig. 2A)**. The Eadie-Hofstee plot of the data gave parallel lines, indicating that clofibric acid inhibited luciferase renaturation noncompetively. This result indicates that clofibric acid binds to both the hsp70 chaperone machine complex and the hsp70 chaperone machine/luciferase complex.

Screening of phage display libraries has indicated that hsc70 can bind with high affinity to short polypeptides that are primarily composed of either hydrophobic (pepN; FYQLALT) or hydrophilic (pepN; NIVRKKK) amino acids *(20)*. Therefore, we examined whether the addition of either of these polypeptides to RRL would inhibit luciferase renaturation. Both pepF and pepN inhibited luciferase renaturation in a concentration-dependent manner **(Fig. 1B)**, with their relative inhibitory potencies reflecting the binding affinity of hsc70 for the polypeptides *(20)*. Kinetic analysis of the inhibition of luciferase renaturation at saturating ATP and varying luciferase concentrations indicated that pepF also inhibited luciferase renaturation by a noncompetitive mechanism **(Fig. 2B)**.

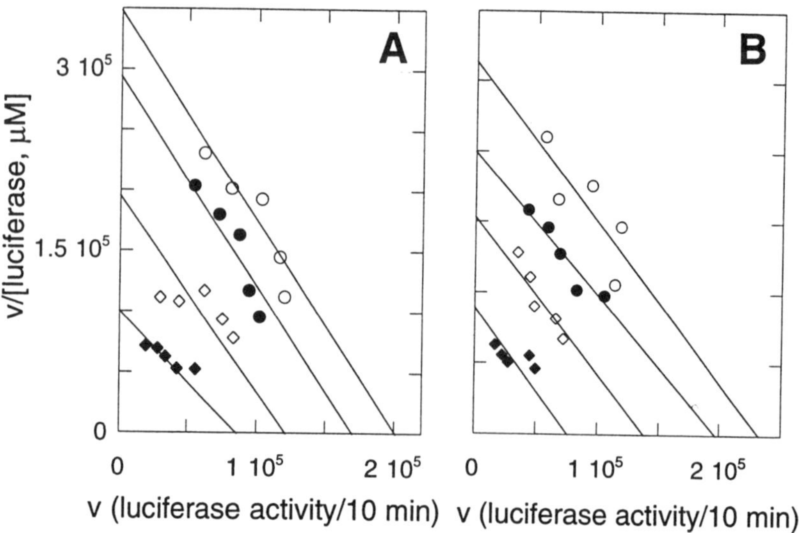

Fig. 2. Effect of clofibric acid and pepF on the kinetics of luciferase renaturation in the presence of saturating ATP and varying luciferase concentrations. **(A)** Eadie-Hofstee plot of the kinetics of luciferase renaturation measured at saturating ATP (1.7 mM endogenous ATP) and varying concentrations of luciferase and clofibric acid. The rate of luciferase renaturation was measured in 1.5 µL of refolding mix with no addition (○), or 5 mM (●), 10 mM (◇), and 15 mM (◆) clofibric acid. **(B)** Eadie-Hofstee plot of the kinetics of luciferase renaturation measured at saturating ATP and varying concentrations of luciferase and pepF. The rate of luciferase renaturation was measured in 1.5 µL of refolding mix with DMSO (0.1 µL/100 µL, ○), or 0.19 mM (●), 0.28 mM (◇), and 0.37 mM (◆) pepF.

hsp70 has been suggested to negatively regulate the DNA binding affinity of heat-shock transcription factor-1 (HSF1) *(21,22)*. It has been proposed that stress agents cause the accumulation of denatured protein in cells, which sequesters hsc70 and activates HSF1. The nonsteroidal anti-inflammatory drugs (NSAD) salicylate and indomethacin induce the DNA binding activity of HSF1, and potentiate the effect of heat shock by maintaining HSF1 in the activated DNA binding state for a prolonged period of time *(23)*. Since salicylate, ibuprofen, and indomethacin have structural similarity to clofibric acid, we examined their effect on luciferase renaturation in RRL. The NSAD strongly inhibited luciferase renaturation in RRL (**Fig. 1A**). Similar to the order of potency for NSAD-induced inhibition of cyclooxygenase and activation of HSF1 DNA binding activity, the order of potency of NSAD-induced inhibition of luciferase renaturation was: indomethacin > iboprofen > salicylate.

The hsp90 chaperone machine is associated with a number of individual immunophilins *(24,25)*, proteins that specifically bind to either the immunosuppressant cyclosporin A, such as CBP, or the immunosuppressants FK506 and/or rapamycin, such as FKBP. Immunophilins have been demonstrated to have peptidyl-prolyl *cis-trans* isomerase activity. Therefore, we examined the effect of these immunosuppresant on luciferase renaturation in RRL. Rapamycin (10 µg/mL) and cyclosporin A (1 mg/mL) were found to have little reproducible inhibitory effect on the rate of luciferase renaturation in RRL, whereas 125 µg/mL of FK506 derivative L-683-590 reproducibly inhibited luciferase renaturation by 20%. This result suggests that an FKBP containing chaperone complex may be involved in facilitating the renaturation of a subpopulation of denatured luciferase.

3.4.2. Effect of Chaperone Antagonists on Luciferase Stability

Luciferase is thermally labile at temperatures above 28°C in vitro. However, native luciferase displays no apparent instability when incubated in RRL at 37°C (*3*; *see* **Note 15**). An alternative method by which to screen for the antichaperone activity of pharmacological drugs is to study the effect of the drugs on the rate of loss of the activity of luciferase incubated at elevated temperature in RRL. At 42°C, luciferase is unstable in normal RRL, but the addition of either of the chaperone inhibitors, geldanamycin or clofibric acid, markedly accelerated the rate at which luciferase denatured (**Fig. 3A**). Similarly, the presence of salicylate (**Fig. 3A**), pepF, or pepN (**Fig. 3B**) accelerated the rate at which luciferase activity was lost when incubated in RRL at 42°C. These results indicate that the RRL luciferase assay can also be useful in identifying agents that interfere with the ability of chaperone to interact reiteratively with proteins and maintain their structure.

The importance of chaperones in facilitating protein folding is underscored by the fact that the hsp90 inhibitor, geldanamycin, is currently in phase one clinical trials because of its demonstrated tumoricidal activity *(26)*. The RRL luciferase renaturation assay represents a simple, fast, and reproducible method to assay for chaperone function. The findings presented above demonstrate the utility of the RRL luciferase renaturation assay in identifying pharmacological agents with antichaperone activity. This assay represents a means by which a large number of agents could be rapidly screened to identify new potentially important drugs that act by inhibiting chaperone function. Moreover, using kinetic analyses one can study the mechanism of action of these drugs on chaperone function.

4. Notes

1. For complete solubilization of luciferase, it is important to add luciferase to the stability buffer before the addition of Triton X-100 and glycerol.
2. Because of the expense of the luciferase, we store the luciferase stock (nondenatured) at 4°C for use over a period of a month. If the luciferase stock is to be

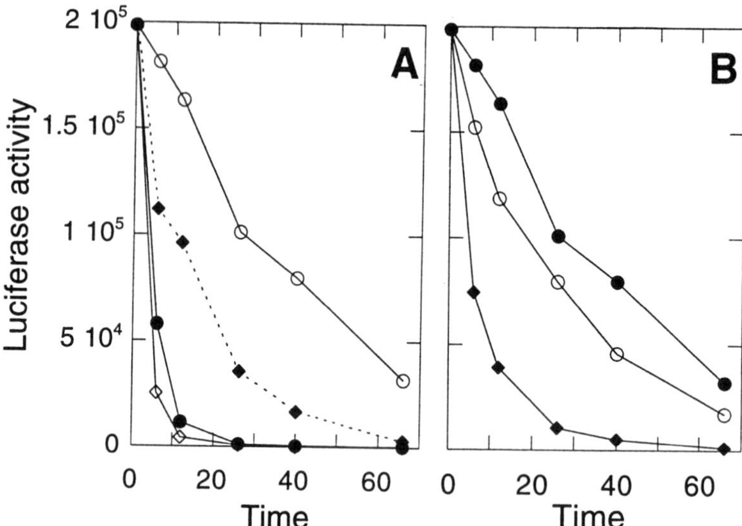

Fig. 3. Effect of pharmacological agents on the rate of luciferase denaturation at 42°C. (**A**) Purified luciferase was added to rabbit reticulocyte lysate containing an ATP-regenerating system and DMSO or water (○), or clofibric acid (15 m*M*, ◇), salicylic acid (15 m*M*, ●), and geldanamycin (8 μ*M*, ◆) as indicated. Reactions were incubated at 42°C, and luciferase activity in 1.5-μL aliquots was assayed at times indicated in the figure. (**B**) Purified luciferase was added to rabbit reticulocyte lysate containing an ATP-regenerating system and DMSO or water (●), or pepN (2 m*M*, ◆) and pepF (0.1 m*M*, ○) as indicated. Reactions were incubated at 42°C and luciferase activity in 1.5-μL aliquotes was assayed at times indicated in the figure.

stored for a long period of time, it is important to add Triton X-100 and glycerol to storage solution. However, the luciferase slowly denatures and aggregates on prolonged storage, such that the concentration of soluble luciferase in the stock progressively decreases. Severely aggregated luciferase does not renature.
a. Prior to each experiment, an aliquot of the stock is centrifuged to remove any aggregated luciferase.
b. The best alternative is to snap-freeze aliquots of the luciferase stock in a dry ice /ethanol bath and stored at –70°C. Thawed aliquots give reproducible results over a period of 6 mon.
3. Luciferase denatures and aggregates at a much faster rate, when there is no Triton X-100 and glycerol in the stability buffer. Glycerol gradient analysis indicates that this luciferase is aggregated to a greater degree than luciferase that is denatured in the presence of Triton X-100 and glycerol. The kinetics of renaturation of partially aggregated luciferase are slower and are inhibited to a greater degree by such agents as RCM (reduced, carboxymethylated)-BSA *(27)*:
a. RCM-BSA inhibits renaturation of luciferase that has been denatured in the presence of Triton X-100 and glycerol in the stability buffer only after the refold-

ing mix is preincubated with RCM-BSA for 20 min at 30°C, before the addition of denatured luciferase;

b. RCM-BSA inhibits renaturation of luciferase that has been denatured in the absence of Triton X-100 and glycerol even without prior incubation with RCM-BSA; and

c. Similarly, the effect of preincubation on the ability of as yet uncharacterized agents to inhibit luciferase renaturation in RRL should be examined.

4. Temperature of denaturation of luciferase is very important. We have observed that the accuracy of different thermometers significantly varies. Variations in denaturation temperature of as little as 1°C leads to different populations of unfolded intermediates whose renaturation can:

a. Occur at different rates;

b. Give different yields in the amount of overall activity recovered; and

c. Be more or less inhibited by pharmacological agents.

5. If one is doing renaturation assays under protein-synthesizing conditions, one should be aware that addition of hemin reduces the renaturation capacity of reticulocyte lysate. This effect of hemin is specific, since other protein synthesis initiation inhibitors do not show the same stimulatory effect as the lack of hemin. To study the renaturation of luciferase without any interference of chaperones involved in protein synthesis, one can replace CM with DM and not add hemin.

6. Since preparations of RRL markedly vary in the level of different heat-shock proteins that are present *(15)*, different preparations of RRL will renature luciferase at different rates *(5)*.

7. The rate of luciferase renaturation varies depending on the type of BSA that is used to make the SB. Luciferase spontaneously aggregates extensively in the presence of some lots of BSA. For best results, use molecular-biology-grade acetylated BSA.

8. Proper vehicle and/or buffer controls should be done while studying the effect of different drugs on luciferase renaturation.

a. When possible, use an inactive analog of the drug as the control (e.g., geldampicin as a control for geldanamycin *[6]*).

b. If an inactive analog is not available, then the vehicle or buffer in which the drug is dissolved should be used as the control: e.g., DMSO was the vehicle control for geldanamycin; refolding assay mixes can tolerate up to 1% DMSO without affecting rates of protein synthesis or renaturation of luciferase.

c. The effect of an agent on luciferase activity should be assayed in vitro to verify that the agent is not simply acting by inhibiting the activity of luciferase directly.

9. To dissolve clofibric acid and nonsteroidal anti-inflammatory drugs in water, one has to slowly add 0.1 *M* KOH.

10. The rate of luciferase renaturation in our hands is linear with time up to around 25 min. We have routinely measured the rate of luciferase renaturation by determining the amount of luciferase renatured in an assay mix in 10 min. However, if one is planning on carrying out a kinetic analysis using a fixed time-point

to measure the rate of luciferase renaturation, preliminary experiments should be carried out to verify that the rate of luciferase renaturation is linear over the time frame of the assay. This is necessary because of the variations in renaturation rates that occur between lots of lysate and preparations of denatured luciferase.

11. For studies in which the overall extent to which luciferase can be renatured is to be measured, one should use a luciferase stock with a concentration of $3.2 \times 10^{-11} - 3.2 \times 10^{-10}$ *M*. Under these conditions, the activity of 60 to nearly 100% of the denatured luciferase can be renatured *(5)*.

12. For kinetic analyses, we use a stock solution containing 30 µ*M* luciferase, and luciferase at a final concentrations from 0.25–1.0 µ*M* in RRL assay mix, so that <1% of the denatured luciferase is renatured over the time-course of the assay.
    a. We find that a 1.0-µ*M* final concentration of luciferase is a near-saturating concentration for the chaperone machinery in most RRLs *(6)*.
    b. When renaturing luciferase at concentrations that nearly saturate the RRL chaperone machine, one needs to verify that the amount of luciferase activity, which is renatured in the aliquot of the RRL mix that is assayed, does not exceed the linear response range of the instrument being used to measure light production.

13. RRL can be dialyzed against DB to determine the optimum concentration of ATP and to carry out kinetic analyses of the effect of agents on the rate of luciferase renaturation at near-saturating luciferase concentrations and varying concentrations of ATP as described *(6)*.
    a. We use Slide-A-Lyzer Cassettes™ (Pierce: 10,000 MWCO; 0.5–3.0 mL sample volume) to dialyze RRL against 250 vol of DB at 4°C. For efficient removal of bound nucleotide, DB is changed every 1.5 h. At the end of 6 h, the RRL is aliquoted into microfuge tubes, and then frozen and stored in liquid nitrogen. ATP remaining in the RRL can be quantitated using the luciferase luminescence assay as described *(6)*.
    b. A creatine kinase/creatine phosphate ATP-regenerating system is used with varying concentrations of ATP to generate RRL that maintains constant ATP levels over the time-course of the experiment *(6)*.
    c. Ions (i.e., Mg^{2+}, or K^+) or other small molecules can be added or omitted from the DB to examine their effects on luciferase renaturation. Optimization of the requirements for Mg^{2+}, and K^+ for luciferase renaturation was done in this manner *(5)*.

14. For reproducible results, if a given lot of RRL will be used over an extended period of time, it should be stored in liquid nitrogen. Although manufacturers claim that RRL remains active for translation when stored for up to 6 mon at −80°C, it is our experience that RRL begins to loose activity in as little as 3 wk under these conditions. Loss of vigor in translation of proteins occurs concomitantly with loss in vigor of the chaperone machinery, and is probably due to the accumulation of denatured protein.

15. A portion of the firefly luciferase population in the luciferase stock solution is inactive when dissolved in the stability buffer. This inactive luciferase regains its

activity when incubated in RRL at 37°C, if the concentration of luciferase is kept below some critical concentration. The renaturation of this inactive population of luciferase is inhibited in the presence of geldanamycin or RCM-BSA. However, the enzymatically active luciferase population appears to be stable under these conditions, and no loss of activity is observed on the addition of geldanamycin or RCM-BSA to RRL at 37°C.

The ability of RRL to renature the inactive population of luciferase is progressively lost as the concentration of luciferase added to the RRL is increased. At even higher concentrations, RRL can no longer maintain the population of enzymatically active luciferase at 37°C, and luciferase activity decreases with time of incubation. Under these conditions, geldanamycin and RCM-BSA accelerate the rate at which luciferase activity is lost. These observations imply that RRL can maintain and renature luciferase at 37°C when the concentration of denatured luciferase is low and does not saturate the chaperone machinery of the RRL. However, when the concentration of denatured luciferase exceeds the capacity of the RRL to renature the luciferase, luciferase activity is lost as it spontaneously denatures at 37°C , and the presence of chaperone antagonists accelerates the rate of loss of luciferase activity.

Acknowledgments

This work was supported by grant number ES-04299 from the National Institute of Environmental Health Sciences, NIH, and by the Oklahoma Agricultural Experiment Station (Project No. 1975).

References

1. Frydman, J. and Hartl, F. U. (1996) Principles of chaperone-assisted protein folding: differences between in vitro and in vivo mechanisms. *Science,* **272,** 1497–1502.
2. Frydman, J., Nimmesgern, E., Ohtsuka, K., and Hartl F. U. (1994) Folding of nascent polypeptide chains in a high molecular mass assembly with molecular chaperones. *Nature* **370,** 111–117.
3. Thulasiraman, V., Hartson, S. D., and Matts, R. L. (1997) Conditional involvement of molecular chaperones in the folding of luciferase. Manuscript submitted.
4. Freeman, B. C., Myers, M. P., Schumacher, R., and Morimoto, R. I. (1995) Identification of a regulatory motiff in hsp70 that affects ATPase activity, substrate binding and interaction with HDJ-1. *EMBO J.* **14,** 2281–2292.
5. Schumacher, R. J., Hurst, R., Sullivan, W. P., McMahon, N. J., Toft, D. O., and Matts, R. L. (1994) ATP-dependent chaperoning activity of reticulocyte lysate. *J. Biol. Chem.* **269,** 9493–9499.
6. Thulasiraman, V. and Matts, R. L. (1996) Effect of geldanamycin on the kinetics of chaperone-mediated renaturation of firefly luciferase in rabbit reticulocyte lysate. *Biochemistry* **35,** 13,443–13,450.
7. Pinto, M., Morange, M., and Bensaude, O. (1991) Denaturation of proteins during heat shock: in vivo recovery of solubility and activity of reporter enzymes. *J. Biol. Chem.* **266,** 13,941–13,946.

8. Frydman, J., Nimmersgern, E., Erdujment-Bromage, H., Wall, J. S., Tempst, P., and Hartl, F. U. (1992) Function in protein folding of TRiC, a cytosolic ring complex containing TCP-1 and structurally related subunits. *EMBO. J.* **11,** 4767–4778.

9. Buchberger, A., Schroder, H., Buttner, M., Valencia, A., and Bukau, B. (1994) A conserved loop in the ATPase domain of the DnaK chaperone is essential for stable binding of GrpE. *Struct. Biol.* **1,** 95–101.

10. Nimmesgern, E. and Hartl, F. U. (1993) ATP-dependent protein refolding activity in reticulocyte lysate: Evidence for the participation of different chaperone components. *FEBS Lett.* **331,** 25–30.

11. Schumacher, R. J., Hansen, W. J., Freeman, B. C., Alnmri, E., Litwack, G., and Toft, D. O. (1996) Cooperative action of hsp70, hsp90 and DnaJ proteins in protein renaturation. *Biochemistry,* **35,** 14,889–14,898.

12. Hohfeld, J., Minami, Y., and Hartl, F. U. (1995) Hip, a novel cochaperone involved in the eukaryotic hsc70/hsp90 reaction cycle. *Cell* **83,** 589–598.

13. Gross, M. and Hessefort, S. (1996) Purification and characterization of a 66–kDa protein from rabbit reticulocyte lysate which promotes the recycling of hsp70. *J. Biol. Chem.* **271,** 16,833–16,841.

14. Whitesell, L., Mimnaugh, E. G., De Costa, B., Myers, C., and Neckers, L. M. (1994) Inhibition of heat shock protein HSP90-pp60^{v-src} heteroprotein complex formation by benzoquinone ansamycins: Essential role for stress proteins in oncogenic transformation. *Proc. Natl. Acad. Sci. USA,* **91,** 8324–8328.

15. Matts, R. L. and Hurst, R. (1992) The relationship between protein synthesis and heat shock proteins levels in rabbit reticulocyte lysates. *J. Biol. Chem.* **267,** 18,168–18,174.

16. Matts, R. L., Schatz, J. R., Hurst, R., and Kagen, R. (1991) Toxic heavy metal ions activate the heme-regulated eukaryotic initiation factor-2α kinase by inhibiting the capacity of hemin-supplemented reticulocyte lysates to reduce disulfide bonds. *J. Biol. Chem.* **266,** 12,695–12,702.

17. Jackson, R. J. and Hunt, T. (1983) Preparation and use of nuclease-treated rabbit reticulocyte lysates for the translation of eukaryotic messenger RNA. *Methods Enzymol.* **96,** 50–74.

18. Merrick, W. C. (1983) Translation of exogenous mRNAs in reticulocyte lysates. *Methods Enzymol.* **101,** 606–615.

19. Alvares, K., Carrillo, A., Yuan, P. M., Kawano, H., Morimoto, R. I., and Reddy, J. K. (1990) Identification of cytosolic peroxisome proliferator binding protein as a member of the heat shock protein hsp70 family. *Proc. Natl. Acad. Sci. USA* **87,** 5293–5297.

20. Takenaka, I. M., Leung, S.-M., McAndrew, S. J., Brown, J. P., and Hightower, L. E. (1995) Hsc70–binding peptides selected from a phage display library that resemble organellar targeting sequences. *J. Biol. Chem.* **270,** 19,839–19,844.

21. Abravaya, K., Myers, M. P., Murphy, S. P., and Morimoto, R. I. (1992) The human heat shock protein hsp70 interacts with HSF the transcription factor that regulates heat shock gene expression. *Genes Dev.* **6,** 1153–1164.

22. Mosser, D. D., Duchaine, J., and Massie, B. (1993) The Dna-binding activity of the human HSF is regulated in vivo by hsp70. *Mol. Cell Biol.* **13,** 5427–5438.
23. Lee, B. S., Chen, J., Angelidis, C., Jurivich, D. A., and Morimoto, R. I. (1995) Pharmacological modulation of heat shock factor 1 by antiinflammatory drugs results in protection against stress-induced cellular damage. *Proc. Natl. Acad. Sci. USA.* **92,** 7207–7211.
24. Smith, D. F., Baggenstoss, B. A., Marion, T. N., and Rimerman, R. A. (1993) Two FKBP-related proteins are associated with progesterone receptor complexes. *J. Biol. Chem.* **268,** 18,365–18,371.
25. Smith, D. R., Whitesell, L., Nair, S. C., Chen, S., Prapapanich, V., and Rimerman, R. A. (1995) Progesterone receptor structure and function altered by geldanamycin, an hsp90–binding agent. *Mol. Cell. Biol.* **15,** 6804–6812.
26. Whitesell, L., Shifrin, S. D., Schwab, G., and Neckers, L. M. (1992) Benzoquinoid ansamycins possess selective tumoricidal activity unrelated to src kinase inhibition. *Cancer Res.* **52,** 1721–1728.
27. Matts, R. L., Hurst, R., and Xu, Z. (1993) Denatured proteins inhibit translation in hemin-supplemented rabbit reticulocyte lysate by inducing the activation of the heme-regulated eIf-2α kinase. *Biochemistry* **32,** 7323–7328.

IV

CELL-BASED ASSAYS

12

Genotoxic Sensors

Amy Cheng Vollmer

1. Introduction

Bacterial mutagenesis assays have been used as preliminary screens for the evaluation of chemicals because they are rapid, simple, and are correlated with carcinogeneity in humans *(1)*. The activation of bacterial DNA repair systems (recently reviewed; *2,3*) can be used as a measure of mutagenic and genotoxic effects of various chemical as well as physical treatments. Many of the gene products that act to repair DNA, however, are difficult to assay owing to the nature of their enzymatic activities and the particular substrates on which they act. Thus, investigators have used relatively inexpensive and rapid alternative approaches. Measuring the reversion of specific auxotrophic bacterial mutations is the strategy used in the Ames tests *(4,5)*. However, this type of method requires a substantial incubation and costly multistep manual procedures. The use of various transcriptional fusions also allows for the detection of agents that interact with DNA without a long incubation period. Many tests use *lacZ* as the reporter for measuring transcriptional activation of promoters that are induced by DNA damage *(6–9)*. In addition, *lacZ*-prophage induction assays *(10,11)* measures additional effects of the SOS response.

Bioluminescence allows for sensitive and rapid measurement of bacterial *lux* gene expression without the need for incubation on agar plates. The Mutatox® assay detects the restoration of a mutant non-bioluminescent bacterium to its normal light-emitting state after exposure to possible mutagens *(12)*. More recently, two systems involving luminescent lysogenic *Escherichia coli* prophage have been developed. One *(13)* utilizes firefly luciferase as the reporter, and the other *(14)* incorporates bacterial luciferase via a suicide plasmid carrying mini-Tn*5luxAB*. Although these measurements do not

From: *Methods in Molecular Biology, Vol. 102: Bioluminescence Methods and Protocols*
Edited by: R. A. LaRossa © Humana Press Inc., Totowa, NJ

require the interruption of cell growth, they still require exogenous addition of decanal or luciferin substrate to the samples.

LaRossa and colleagues *(15,16)* report the construction and initial characterization of an alternative panel of easily assayed bioluminescent transcriptional fusions useful for genotoxicity studies. Promoters for three *E. coli* genes, *recA, uvrA*, and *alkA,* have each been fused to the promoterless *Vibrio fischeri luxCDABE* operon present within the broad host range, multicopy plasmid pUCD615 *(17)*. *E. coli* strains containing these fusions allow visualization of the transcriptional responses induced by DNA damage, without the need to perform enzyme assays or to add luciferase substrates exogenously. The full *lux* operon encodes not only the catalytic luciferase (LuxAB), but also the enzymes (LuxC, LuxD, LuxE) required to shunt fatty acyl metabolites from central metabolism and to convert them to the endogenous aldehyde substrate for luciferase. These DNA repair promoter::*luxCDABE* fusions thus reflect the presence of genotoxic doses of stressors as an increase in the production of light. At the same time, the presence of lethal concentrations of toxicants may also be monitored by their inhibitory effect on luminescence denoted by decreases in the low to moderate constitutive level of expression from these fusions.

Although a total of three such promoter::lux fusions have been constructed (strain DPD 2794, *recA::lux*; DPD 2818 *uvrA::lux,* and DPD 2844 *alkA::lux*), this chapter describes the materials and techniques using the *recA::luxCDABE* fusion to provide real-time reporting of genotoxicity in a dose-dependent manner, without the need to add substrate exogenously. Instrumentation for quantitation of bioluminescence can be as sophisticated as microtiter format luminometers or as commonplace as liquid scintillation counters.

2. Materials

Materials described in this chapter are those that measure luminescence using a multititer plate format luminometer. Chapter 2 contains materials for measuring luminescence in a liquid scintillation counter.

1. Culture media: LB agar or broth containing 50 µg/mL of kanamycin sulfate (LB^+Kan_{50}).
2. Bacterial strain DPD 2794 carrying the *recA::lux* fusion on a kanamycin resistant plasmid, in host parent strain *E. coli* RFM443 (F^-, *galK2, lac74, rpsL200;* **ref. *18***).
3. Microtiter plates, opaque black or white, with adhesive covers (*see* **Note 1**).
4. Microplate luminometer (for a table of commercial luminometers, *see* Chapter 2 of this vol.).
5. Analytical grade mitomycin C (Sigma, St. Louis, MO) or a UV crosslinking apparatus (e.g., Stratalinker®, Stratagene, La Jolla, CA) may be used as positive controls for inducing bioluminescence in DPD 2794.

3. Methods

3.1. Measurements in Liquid Cultures

Methods described in this chapter are those to measure luminescence using a multititer plate format luminometer. Chapter 2 contains methods for measuring luminescence in a liquid scintillation counter.

1. Incubate cultures of the appropriate strain overnight in LB + Kan_{50} with aeration at 37°C.
2. Make a 1:100 dilution into LB or LB + Kan_{25} (*see* **Note 2**).
3. Incubate the culture with aeration at 37°C until early to midexponential phase (about 30 Klett units).
4. Place chemical to be tested into sample wells. Dilutions of the chemical to be tested may be made in sequence, from the first row (A) to the seventh (G); the last row (H) may be reserved for a control without addition of test chemical or treatment. Range-finding experiments may begin with 5- to 10-fold dilutions. More focused dose–response experiments may use two-fold dilutions (*see* **Note 3**).
5. Positive controls for the particular sensor strain should also be included in the plate. Mitomycin C in the range of 0.01–2 µg/mL may be used for DPD 2794 (*see* **Notes 4** and **5**).
6. Place 50-µL aliquots of the exponential culture into sample wells.
7. Cover the wells with a transparent, adhesive seal to prevent dehydration or volatilization of potential genotoxicants (*see* **Note 6**).
8. Place the sample in the luminometer detection chamber, and run a program that will read the samples at regular intervals, storing the data in a convenient spread sheet format per manufacturer's instructions (*see* **Note 7**) .
9. The data may be displayed as a kinetic curve plotting relative light units (RLU) vs time, or may be analyzed to reflect response ratio ($RLU_{induced\ sample}/RLU_{untreated\ control}$) as a function of concentration or dose at a particular time-point *(19)*. Response ratios greater than one indicate induction of transcription of the *lux* genes and, therefore, activation of the DNA repair response. Response ratios equal to one reflect no transcriptional activation relative to the untreated sample. Response ratios of less than one indicate loss of the treated sample's metabolic capacity, and reflect a toxic dose or concentration.
10. Cell viability may be determined by standard serial dilution and plating on LB + Kan_{50}.

3.2. Measurement of Plate Cultures by Disk Assay

1. Incubate liquid cultures of the appropriate strain overnight in LB + Kan_{50} with aeration at 37°C.
2. Plate 50–100 µL on LB + Kan_{50} agar by spreading.
3. On a sterile filter paper disk (3–10 mm diameter), pipet 1–10 µL of of stock solution of the test chemical. Positive controls, as stated above, should be included.
4. Carefully place disks on the agar surface using sterile forceps. Press gently to remove air bubbles trapped between the disk and the agar. Several small disks

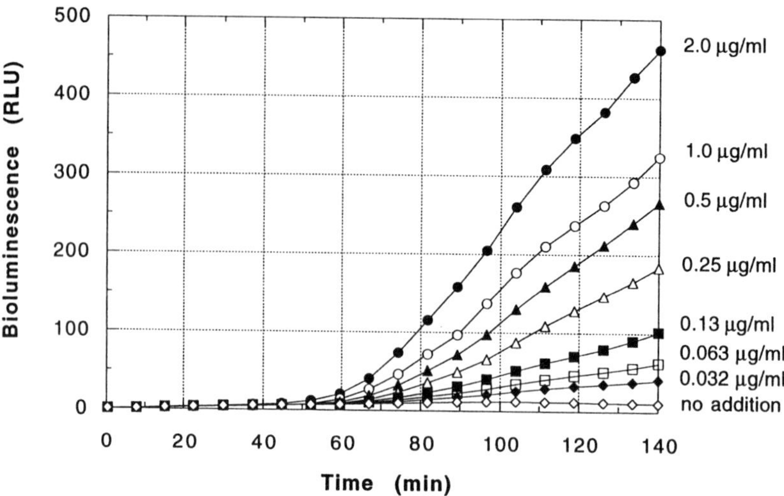

Fig. 1. Kinetic plot of the induction of DPD2794 (*recA::luxCDABE*) by mitomycin C. Each curve represents an average of two duplicates read in microtiter plate format.

may be placed on a single plate, so that they are spaced evenly apart and are at least 10 mm from the edge of the plate.

5. Plates may be incubated overnight at 37°C.
6. Zones of growth inhibition may be apparent. X-ray film may be placed on top of the plates in a darkroom. Exposure time may range from 10 s to hours. Film should be developed manually or by using an automated film processor according to manufacturer's instructions.

3.3. Results

Luminescence values are presented as RLU (per the particular instrument's output). The kinetic profile of DPD 2794 response had a 40–60 min lag followed by an increase in bioluminescence (**Fig. 1**). The response was dose-dependent in the range of 0–2 µg/mL of mitomycin C; mitomycin C concentrations higher than 2 µg/mL led to a decrease in luminescence to below the control levels (not shown). This was accompanied by a loss of viability, as judged by colony formation capacity (from 3×10^9 CFU/mL without mitomycin C to 4×10^7 CFU/mL at 4 µg/mL of mitomycin C). This set of bioluminescent sensors is unique in that there is the inherent ability to record both a "lights off" response as well as a "lights on" response from a single strain. Results of plate genotoxicity assays are displayed in figures contained in Chapter 2 of this volume, where more specific methods on the types of film and exposure settings are denoted. Sensitivity of this assay to mitomycin C (detection limit = 0.009 µg/mL) is about equal to or 2 logs more sensitive than the detection limits reported by

Maillard et al. *(14)* and Lee et al. *(13)*, respectively, for other biolumniescent assays. In prophage induction experiments, it is more difficult to distinguish between lethal and sublethal levels of treatment, since inevitable consequence of prophage induction is host cell lysis. The Mutatox® assay of Microbics has the advantage of being portable into field sites; however, as a reversion-based "lights on" assay, it requires incubation of about 1 d before results are obtained. The unique advantages of the system described here are:

1. That no exogenous substrate is needed, decreasing the manual or automated steps required;
2. Rapidity; and
3. The inherent duality for a "lights off" response indicative of metabolic disruption and a "lights on" (induction) response at lower concentrations that is correlated with genotoxicity.

4. Notes

1. Opaque microtiter plates should be used. Black plates exhibit less "crosstalk" owing to reflection of light from one well into neighboring wells. These are especially useful for strains with high background transcription and large induction potential, such as the DPD 2794 (*recA::lux*) strain. By reflection, white plates enhance light fluence at the detector from strains that produce lower amounts of light, such as the DPD 2818 (*uvrA::lux*). Either color plate may be used with DPD 2844 (*alkA::lux*), which displays an intermediate level of light production during exponential growth.
2. Kanamycin is required to maintain the plasmid during overnight growth. However, for growth into exponential phase during the day of the assay, kanamycin is not necessary and can be completely left out or supplemented to a final concentration of 25 µg/mL. Furthermore, a final concentration of 50 µg/mL of kanamycin has been shown to be slightly inhibitory in terms of the expression of the *lux* genes (Belkin, personal communication.)
3. It is recommended that duplicate samples be run in neighboring columns, so that in a 96-well format, columns 1 and 2 contain the same test chemical and sensor strain; similarly, columns 3 and 4 would be paired, and so forth.
4. Induction of liquid cultures may also be performed by exposing the cells to UV (254 nm) at doses from 5–2000 J/m^2 in an uncovered plastic dish using a Stratalinker®2400 (Stratagene) irradiation unit, prior to adding the cells to the 96-well plate.
5. Exposure of cells already in the microtiter plate (or exposure of an empty microtiter plate) to UV is not recommended. UV will produce chemiluminescence from pigment or plastic components in either black or white plates. If plates are to be resterilized by UV, allow 6–8 h between irradiation and use of the plates for bioluminescence measurements (Packard TopCount technical service information).
6. For prolonged incubations (>3 h), O_2 may become limited in the wells of a tightly covered plate.

7. Temperature control is important. The samples should be incubated at temperatures between 20 and 26°C. The *V. fischeri lux* gene products are thermally unstable at higher temperatures.

References

1. McCann, J., Choi, E., Yamasaki, E., and Ames, B. (1975) Detection of carcinogens as mutagens in the Salmonella microsome test: assay of 300 chemicals. *Proc. Natl. Acad. Sci. USA* **72,** 5135–5139.
2. Rupp, W. D. (1996) DNA repair mechanisms, in, *Escherichia coli and Salmonella typhimurium: Cellular and Molecular Biology* (Neidhardt, F. C., Curtiss, R., Ingraham, J. L., Linn, E. C. C., Low, K. B., Magasanik, B., Reznikoff, W. S., Riley, M., Schaechter, M., and Umbarger, H. E., eds.), American Society for Microbiology, Washington, DC, pp. 2277–2294.
3. Walker, G. C. (1996) The SOS response of *Escherichia coli,* in, *Escherichia coli and Salmonella typhimurium: Cellular and Molecular Biology* (Neidhardt, F. C., Curtiss, R., Ingraham, J. L., Linn, E. C. C., Low, K. B., Magasanik, B., Reznikoff, W. S., Riley, M., Schaechter, M., and Umbarger, H. E., eds.), American Society for Microbiology, Washington, DC, pp. 1400–1416.
4. Ames, B. N., Lee, F. D., and Durston, W. E. 91973) An improved bacterial test system for the detection and classification of mutagens and carcinogens. *Proc. Natl. Acad. Sci. USA* **70,** 2281–2285.
5. Maron, D. M. and Ames, B. N. (1983) Revised methods for the *Salmonella* mutagenicity test. *Mutation Res.* **113,** 173–215.
6. Oda, Y., Nakamura, S., Oki, I., Kato, T., and Shinagawa, H. (1985) Evaluation of the new system (*umu* test) for the detection of environmental mutagens and carcinogens. *Mutat. Res.* **147,** 219–229.
7. Nunoshiba, T. and Nishioka, H. 1991. "*Rec-lac* test" for detecting SOS-inducing activity of environmental genotoxic substances. *Mutat. Res.* **254,** 71–77.
8. Quillardet, P., Huisman, O., D'Ari, R., and Hofnung, M. (1982) SOS chromotest, a direct assay of induction of an SOS function in *Escherichia coli* K12 to measure genotoxicity. *Proc. Natl. Acad. Sci. USA* **79,** 5971–5975.
9. Orser, C. S., Foong, F. C. F., Capaldi, S. R., Nalezny, J., MacKay, W., Benjamin, M. and Farr, S. B. (1995) Use of Prokaryotic Stress Promoters as Indicators of the Mechanisms of Chemical Toxicity. *In Vitro Toxicol.* **8,** 71–85.
10. Elesperu, R. K., and White, R. J. (1983) Biochemical prophage induction assay: a rapid test for antitumor agents that interact with DNA. *Cancer Res.* **43,** 2819–2830.
11. Rojanapo, W., Nagao, M., Kawachi, T., and Sugimura, T. (1981) Prophage λ induction test (inductest) of antitumor antibiotics. *Mutat. Res.* **88,** 325–335.
12. Ulitzur S., Weiser, I., and Yannai, S. (1980) A new, sensitive and simple bioluminescence test for mutagenic compounds. *Mutation Res.* **74,** 113–124.
13. Lee, S., Suzuki, M., Kumagai, M., Ikeda, H., Tamiya, E., and Karube, I. (1992) Bioluminescence detection system of mutagen using firefly luciferase genes introduced in *Escherichia coli* lysogenic strain. *Anal. Chem.* **64,** 1755–1759.

14. Maillard, K. I., Benedik, M. J., and Willson, R. C. (1996) Rapid detection of mutagens by induction of luciferase-bearing prophage in *Escherichia coli. Environ. Sci. Technol.* **30,** 2478–2483.
15. Belkin, S., Vollmer, A. C., Van Dyk, T. K., Smulski, D. R., Reed, T. R., and LaRossa, R. A. (1995) Oxidative and DNA damaging agents induce luminescence in *E. coli* harboring *lux* fusions to stress promoters in *Bioluminescence and Chemiluminescence: Fundamentals and Applied Aspects,* (Campbell, A. K., Kricka, L. J., and Stanley, P. E., eds.), John Wiley, Chichester, pp. 509–512.
16. Vollmer, A. C., Belkin, S., Smulski, D. R., Van Dyk, T., and LaRossa, R. A. (1997) Detection of DNA damage by use of *Escherichia coli* carrying *recA'::lux, uvrA'::lux* or *alkA'::lux* reporter plasmids, *Appl. Environ, Microbiol.* **63,** 2566–2571.
17. Rogowsky, P. M., Close, T. J., Chimera, J. A., Shaw, J. J., and Kado, C. I. (1987) Regulation of the *vir* genes of *Agrobacterium tumefaciens* plasmid pTiC58. *J. Bacteriol.* **169,** 5101–5112.
18. Drolet, M., Phoenix, P., Menael, R., Massé, E., Liu, L. F., and Crouch, R. J. (1995) Overexpression of RNase H partially complements the growth defect of an *Escherichia coli ΔtopA* mutant: R-loop formation is a major problem in the absence of DNA topoisomerase I. *Proc. Natl. Acad. Sci. USA* **92,** 3627–3530.
19. Van Dyk, T. K., Majarian, W. R., Konstantinov, K. B., Young, R. M., Dhurjati, P. S., and LaRossa, R. A. (1994) Rapid and sensitive pollutant detection by induction of heat shock gene-bioluminescence gene fusions. *Appl. Environ. Microbiol.* **60,** 1414–1420.

13

Stress Detection Using Bioluminescent Reporters of the Heat-Shock Response

Tina K. Van Dyk

1. Introduction

The heat-shock response is the coordinated induction of a set of proteins in response to a variety of cellular stresses, including elevated temperature *(1)*. Several of the induced proteins function to reactivate or degrade denatured proteins, which are the signal initiating the response. The heat-shock response, therefore, functions to maintain the proteinacious component of the cell in an active form. Thus, by inducing heat-shock protein synthesis, the cell is better able to survive the stress condition. The regulation of the heat shock response typically occurs at the transcriptional level. In *Escherichia coli,* the heat-shock sigma factor, σ^{32}, drives transcription of about 20 genes *(2)*. These include *dnaK,* encoding the molecular chaperone Hsp70, *grpE,* encoding a protein that interacts with Hsp70 in the protein folding pathway, and *lon,* a protease that degrades unfolded proteins. Detection of the heat-shock response has been selected as a useful indicator of biological stress *(3–5)* because this response is universally found in biological systems and is induced by sublethal levels of a wide variety of cellular insults.

E. coli cells containing *E. coli* heat-shock gene promoters fused to *Vibrio fischeri luxCDABE* genes are convenient whole-cell biosensors for monitoring transcriptional induction of the heat-shock response *(6,7)*. This five-gene *lux* reporter system allows nondestructive, real-time analysis of heat shock response induction, which is quantitated by increased bioluminescence. Furthermore, use of this *lux* reporter system yields cells programmed to produce all components needed for bioluminescence, thereby obviating the costs associated with substrate acquisition and addition to the cultures. In order to function, the proteins encoded by the *lux* reporter system must be provided with

From: *Methods in Molecular Biology, Vol. 102: Bioluminescence Methods and Protocols*
Edited by: R. A. LaRossa © Humana Press Inc., Totowa, NJ

Table 1
Whole-Cell Biosensor Strains for Detection of the Heat-Shock Response

Strain	Plasmid	Host[a]	Reference
TV1061	pGrpELux5	RFM443 *(tolC⁺)*	**6**
TV1076	pGrpELux5	DE112 *(tolC⁻)*	**6**
WM1202	pRY002 *(dnaK-lux)*	RFM443 *(tolC⁺)*	**6**
WM1302	pRY002 *(dnaK-lux)*	DE112 *(tolC⁻)*	**6**
DPD1006	pLonLux2	RFM443 *(tolC⁺)*	**9**
DPD1008	pLonLux2	DE112 *(tolC⁻)*	Unpublished

[a]The genotype of *E. coli* RFM443 is *galK2 Δlac74 rpsL200* **(11)**. The genotype of *E. coli* DE112 is *tolC*::mini-Tn*10 galK2 Δlac74 rpsL200* (6).

energy (ATP) and reducing power (NADH$_2$ and FAD) by the cell. Furthermore, the temperature maximum for *V. fischeri lux*-encoded proteins is 30°C. Thus, actively growing cells at 26°C are the reagent used in tests for induction of the heat-shock response. Alternatively, freeze-dried and reconstituted cell cultures may be used for such tests (**8**, *see also* Chapter 10). Applications of these bioluminescent gene fusion strains have included characterizing the range of chemicals inducing this stress response *(7,9)*, and monitoring quality of influent and effluent streams of wastewater treatment facilities *(10)*.

2. Materials

2.1. E. coli *Strains and Growth*

1. *E. coli* strains that contain plasmid-borne transcriptional fusions of several *E. coli* heat-shock promoters to *V. fischeri luxCDABE* are listed in **Table 1**. As well as containing the genetic fusion of interest, each of the plasmids confers resistance to kanamycin and ampicillin. These plasmids are in pairs of host strains that are identical, except for the presence or absence of a mutation, *tolC*, caused by insertion of a miniTn10, which confers resistance to tetracycline. Inactivation of *tolC* increases the susceptibility of *E. coli* to hydrophobic chemicals *(12,13; see* **Note 1**).
2. LB medium *(14)* (per liter): 10 g Bacto-tryptone, 5 g Bacto-yeast extract, 10 g salt. Adjust to pH 7.0 with 5 *N* NaOH. For agar plates add 15 g/L of Bacto-agar. Sterilize by autoclaving (*see* **Note 2**).
3. Kanamycin sulfate stock solution: 1.0 g/100 mL water. Filter-sterilize and store in aliquots at –20°C. When required, add 250 µL of the 1% stock solution to 100 mL of LB medium to give 25 µg/mL final concentration.
4. Refrigerated shaking incubator set at 26°C, or a shaker platform placed in a refrigerated incubator set at 26°C (*see* **Note 3**).
5. Klett-Summerson colorimeter or, alternatively, a spectrophotometer to measure optical density.

2.2. Microplate Preparation

1. Sterile, white, flat-bottom microplates, such as Microlite™ (Dynatech, Chantilly, VA).
2. Multichannel pipeter.
3. Sterile tips for the multichannel pipeter.
4. Clear acetate plate sealers for microplates, such as from Dynex.

2.3. Luminometery

Use a microplate luminometer capable of controlling temperature at 26°C, such as the ML3000 microplate luminometer (Dynatech), modified to control temperatures below 30°C (*see* **Note 3**).

3. Methods

3.1. Cell Growth

1. Grow *E. coli* strain TV1061, or one of the other heat-shock–*lux* fusion strains, in LB plus kanamycin (25 µg/mL) for about 20 h at 26°C (*see* **Notes 4** and **5**) .
2. Dilute the overnight culture into 10.0 mL of LB medium without kanamycin (*see* **Note 6**) in flasks at 26°C. Typically, a 1:50 dilution is used.
3. Incubate, shaking at 250 rpm, at 26°C for approx 3 h. Measure and record Klett units at various times after dilution. Use cells in early log phase (Klett reading should be between 20 and 40; this corresponds approx to OD_{600} of 0.1–0.2). These cells must be immediately used for the stress-induction experiment (*see* **Note 7**).

3.2. Preparation of the Microplate

1. A dilution series of the compound to be tested is prepared in the wells of a microplate resulting in a volume of 50 µL/well. Typically, a stock solution of a compound in LB at 2X the highest concentration to be tested is put into the wells of row A and is further diluted as follows.

   To form a 1:2 dilution series: Start with 100 µL of the stock solution in the wells of row A. Put 50 µL of LB medium into all the other wells. Remove 50 µL of the solution in each well of row A, transfer to the corresponding wells of row B, and mix by pipeting up and down. Then transfer 50 µL from the wells of row B into those of row C and mix. Continue similarly down the plate until the row G wells. From these, take out 50 µL and dispose, leaving the row H wells with 50 µL of LB without chemical.

   To form a 1:3 dilution series, start with 75 µL of the stock solution in the wells of row A. Put 50 µL of LB medium without addition into all the other wells. Remove 25 µL of the solution in row A and transfer to row B, mix, and so forth, as for the 1:2 dilution series. The wells of row H should contain only LB medium, as for the 1:2 dilution series.

2. Add 50 µL of the early log phase cells to the prepared microtiter plate giving 100-µL final volume in each well, cover the plate (optional; *see* **Note 8**), and immediately place the microtiter plate in the luminometer at 26° C.

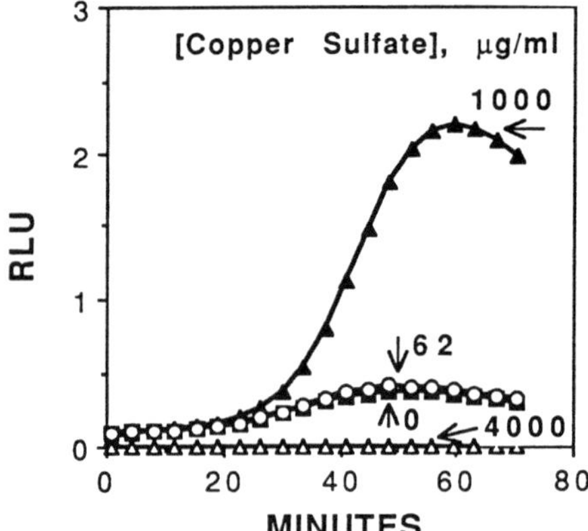

Fig. 1. Copper sulfate effects on a heat-shock bioluminescent gene fusion strain. *E. coli* strain TV1061, containing a fusion of the *grpE* promoter to *luxCDABE,* was tested with various concentrations of $CuSO_4 \cdot 5H_2O$ in LB medium, as described in this chapter. For clarity, the kinetic curves for only three concentrations of $CuSO_4 \cdot 5H_2O$ are shown in this figure.

3.3. Luminometery

1. Quantitate bioluminescence in a microplate luminometer with temperature controlled at 26°C using the cycle mode and the following settings (for a Dynatech ML3000 luminometer): Gain—Medium; Data—All; Cycles—20; Pause—300 s; Auto gain—On; Mixing—On (*see* **Note 9**).

2. When the cycle run is completed, save the data, and covert it to a form for graphing. For data collected on the ML3000, the following steps are used: the data are converted to Excel™ (Microsoft) on the ML3000, a specially written macro in Excel is used to convert the data to a format for following the kinetics of bioluminescence changes in each well, the data (usually the average of duplicates) is transferred to a graphing program, and relative light units (RLU) vs time are plotted for various concentrations of compound tested and the no addition control.

3.4. Interpretation of Data and Calculations

1. Interpretation of kinetic profiles: Several responses of these bioluminescent fusion strains to chemicals are expected, depending on the concentration of the chemical and whether it induces a heat-shock response. **Figure 1** illustrates a typical kinetic profile. In **Fig. 1**, three different effects of copper sulfate are seen. At the lowest concentration shown, 62 µg/mL, the bioluminescence kinetics are indistinguishable from the control, indicating that this concentration has little to no effect on cellular physiology. At the highest concentration shown, 4000 µg/mL, there is an immediate and essentially complete loss of bioluminescence as compared with the control. This "lights off" response indicates toxic activity of cop-

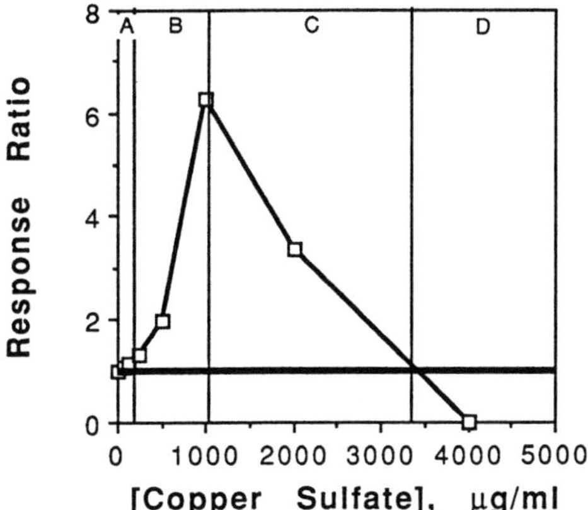

Fig. 2. Response ratios of strain TV1061 treated with $CuSO_4 \cdot 5H_2O$ in LB medium. The response ratios were calculated after 60 min of treatment with the chemical. The horizontal line at a response ratio of 1.0 represents the no response level. The vertical lines denote various regions of response *(see text)*.

per sulfate, most likely owing to prevention of cellular metabolism required for production of energy or reducing power *(15)*. The intermediate concentration, 1000 µg/mL, shows the typical profile for induction of the heat-shock response. There is a lag time of about 20 min followed by a rapid, but transient increase in the rate of light production. This "lights on" response is interpreted as a report of the increased transcriptional activity at the heat-shock gene's promoter, with the lag time presumably representing the time required for transcription and transla-tion of the five-gene *lux* operon at 26°C.

Chemicals that do not induce a heat-shock response show no effect at low concentrations and a "lights off" response at high concentration, but lack a "lights on" response at intermediate concentrations (*see* **Note 10**). Occasionally, certain chemicals, particularly solvents, elicit an increase in bioluminescence as com-pared with the untreated control; however, this increase occurs immediately without the typical lag time. The mechanism of such early increases in bioluminescence is unknown and may confound interpretation if the kinetic profile does not include early time-points.

2. Response ratios and interpretation: A convenient way to quantitate the degree of response, both of "lights on" and "lights off," is by dividing the light production in a treated sample by that of the control. Thus:

Response ratio = RLU (treated sample)/RLU (control, untreated sample)   (1)

It is often useful to calculate response ratios at a specific time following addition of cells to the chemical. For example, the response ratios from the experiment shown in **Fig. 1** were calculated at 60 min after addition and are shown plotted vs copper sulfate concentration in **Fig. 2.** As with the kinetic curves, the response of the fusion strain to copper sulfate depends on the concentration tested. These responses are:

a. Ambivalence at lowest concentrations;

b. Increased bioluminescence ("lights on") at intermediate concentrations; and

d. Decreased bioluminescence ("lights off") at highest concentrations.

Note also there is a region (c) that is between the inducing (b) and toxic (d) regions where the degree of bioluminescence produced is likely a balance between reporter induction and destruction of bioluminescent capacity.

3. Specific induction units (SIU): An alternative way of quantitating the "lights on" response uses the difference between the light production (RLU) in the treated samples from that in the untreated controls, normalized to the number of cells in each well. Thus, SIU with units of $RLU/10^7$ cells are calculated *(9)*:

$$SIU = [RLU(\text{treated sample}) - RLU (\text{untreated, control sample})]/10^7 \text{ cells},$$
$$\text{where } 10^7 \text{ cells} = (\text{mL culture in test}) (\text{Klett units}) (5.6 \times 10^6 \text{ cells/}$$
$$\text{mL/Klett unit})/10^7 \qquad (2)$$

Use of SIU to describe induction is important when assessing the additivity induction responses elicited by chemical combinations.

4. Notes

1. Use of host strains with altered susceptibility to chemicals, such as those with a *tolC* mutation, can have a dramatic effect on the concentration range yielding an inducing response *(6)*. This is one way to shift the response profile to detect lower concentrations.

2. Any medium that supports good growth of *E. coli* may be used. The concentration of chemical required to give a response may change substantially with alterations in growth medium. This is another convenient way to shift the concentrations yielding responses to a desired range.

3. If a refrigerated incubator or luminometer is not available, growth of cells and quantitation of bioluminescence may be done at room temperature.

4. Use of a working stock of the biosensor strain(s) for inoculation of the overnight culture is convenient and results in more consistent growth. This working stock is made by streaking a single colony to an LB kanamycin (25 µg/mL) plate. Following overnight incubation at 37°C, 2.5 mL of sterile 0.8% NaCl are added, and the cells are resuspended using a sterile spreader. Equal volumes of these resuspended cells and sterile glycerol are mixed in a sterile glass vial and stored at –20°C. The 50% final glycerol stocks do not solidify; 20 µL are used to inoculate 5.0 mL medium. Such stocks are useful for at least 6 mo.

5. It is a good practice to verify the phenotype of the cells in the culture used. To do so, streak to MacConkey, LB supplemented with 20 µg/mL of tetracycline, LB supplemented with 25 µg/mL of kanamycin, LB supplemented with 150 µg/mL of ampicillin, and LB plates. Incubate at 37°C overnight. Score the plates the following day. The expected results are:

| | Growth on: | | | | |
Host strain	MacConkey	LBTet	LBKan	LBAmp	LB
tolC$^+$ hosts	+	−	+	+	+
tolC$^-$ hosts	−	+	+	+	+

6. Leaving kanamycin out of the cultures grown up to log phase has little, if any, effect on plasmid stability, but typically results in a larger bioluminescent response.

7. Cells must be actively growing in early logarithmic phase rather than late logarithmic or early stationary phase for best results *(16)*. If using more than one strain in a microplate, it can be helpful to start several cultures of each, either at different initial dilutions of the overnight culture or at different times, so that the cultures in the appropriate growth stage will be available at the same time.

8. Use of a plate sealer is not required since evaporation is minimal during the time frame of these tests (<80 min). The measured responses are essentially the same with or without a cover. However, use of a sealer is recommended if volatile chemicals are tested.

9. Medium gain is usually a good choice; however, high gain (with the auto gain reset on) is recommended for biosensor strains with very low basal bioluminescence, such as *E. coli* WM1202 and WM1302.

10. Some chemicals that do not give an inducing response may do so in the presence of another inducer. An example is cadmium chloride, which when tested alone yields only a "lights off" or no response, but when tested in the presence of ethanol yields a "lights on" response greater than that given by the ethanol *(9)*. Deliberate use of this synergistic induction phenomena allows detection of responses from certain classes of chemicals, resulting in broader utility for this method.

References

1. Morimoto, R. I., Tissières, A., and Georgopoulos, C. (1994) Progress and perspectives on the biology of heat shock proteins and molecular chaperones, in *The Biology of Heat Shock Proteins and Molecular Chaperones* (Morimoto, R. I., Tissières, A., and Georgopoulos, C., eds.), Cold Spring Harbor Laboratory, Cold Spring Harbor, NY, pp. 1–30.

2. Gross, C. A. (1996) Function and regulation of heat shock proteins, in *Escherichia coli and Salmonella: Cellular and Molecular Biology* (Neidhardt, F. C., eds.), ASM, Washington, DC, pp. 1382–1399.

3. de Pomerai, D. (1996) Heat-shock proteins as biomarkers of pollution. *Hum. Exp. Toxicol.* **15**, 279–285.

4. Goering, P. L. (1995) Stress proteins. Molecular biomarkers of chemical exposure and toxicity. *Environ. Sci. Res.* **50**, 217–227.

5. Hightower, L. E. (1992) A brief perspective on the heat-shock response and stress proteins. *Marine Environ. Res.* **35**, 79–83.

6. Van Dyk, T. K., Majarian, W. R., Konstantinov, K. B., Young, R. M., Dhurjati, P. S., and LaRossa, R. A. (1994) Rapid and sensitive pollutant detection by induction of

heat shock gene-bioluminescence gene fusions. *Appl. Environ. Microbiol.* **60,** 1414–1420.

7. Van Dyk, T. K., Smulski, D. R., Reed, T. R., Belkin, S., Vollmer, A. C., and LaRossa, R. A. (1995) Responses to toxicants of an *Escherichia coli* strain carrying a *uspA'::lux* genetic fusion and an *E. coli* strain carrying a *grpE'::lux* genetic fusion are similar. *Appl. Environ. Microbiol.* **61,** 4124–4127.

8. Van Dyk, T. K., and Wagner, L. W. (1996) Lyophilized bioluminescent bacterial reagent for the detection of toxicants patent WO 96/16187.

9. Van Dyk, T. K., Reed, T. R., Vollmer, A. C., and LaRossa, R. A. (1995) Synergistic induction of the heat shock response in *Escherichia coli* by simultaneous treatment with chemical inducers. *J. Bacteriol.* **177,** 6001–6004.

10. Belkin, S., Van Dyk, T. K., Vollmer, A. C., Smulski, D. R., and LaRossa, R. A. (1996) Monitoring subtoxic environmental hazards by stress-responsive luminous bacteria. *Environ. Toxicol. Water Qual.* **11,** 179–185.

11. Drolet, M., Phoenix, P., Menzel, R., Massé, E., Liu, L. F., and Crouch, R. J. (1995) Overexpression of RNase H partially complements the growth defect of an *Escherichia coli* Δ*topA* mutant: R-loop formation is a major problem in the absence of DNA topoisomerase I. *Proc. Natl. Acad. Sci. USA* **92,** 3526–3530.

12. Schnaitman, C. (1991) Improved strains for target-based chemical screening. *ASM News* **57,** 612.

13. Fralick, J. A. (1996) Evidence that TolC is required for functioning of the Mar/AcrAB efflux pump of *Escherichia coli. J. Bacteriol.* **178,** 5803–5805.

14. Miller, J. H. (1972) *Experiments in Molecular Genetics.* Cold Spring Harbor, NY, Cold Spring Harbor Laboratory, 1972.

15. Chatterjee, J. and Meighen, E. A. (1995) Biotechnological applications of bacterial bioluminescence *(lux)* genes. *Photochem. and Photobiol.* **62,** 641–650.

16. Rupani, S. P., Gu, M. B., Konstantinov, K. B., Dhurjati, P. S., Van Dyk, T. K., and LaRossa, R. A. (1996) Characterization of the stress response of a bioluminescent biological sensor in batch and continuous cultures. *Biotechnol. Prog.* **12,** 387–392.

14

Real-Time Reporter of Protein Synthesis Inhibition

Matti Korpela, Marko Virta, and Matti Karp

1. Introduction

In studies concerning drug–receptor interaction, the fundamental need is the knowledge of the events that take place when a drug and a receptor combine. The current knowledge of microbial cell structure at the molecular level is restricted so that very often unsatisfyingly indirect methods must be used. The methods of analysis may be too severely narrowed down. There is a big risk in drawing conclusions without any real evidence of immediate responses to drugs by living bacteria.

This chapter presents a bioluminescent assay that describes a built-in amplification system for the screening and study of chemical substances with an inhibitory effect on protein synthesis. This real-time in vivo approach for protein synthesis inhibition using living bacteria fulfills a major need for the understanding of drug–receptor action. The method also works well with lyophilized bacteria, and the time needed for the assay is less than an hour. Thus, the assay is suitable for rapid and extremely sensitive screening of lead chemicals (antimicrobial drug candidates) from combinatorial libraries.

The assay described here is based on the measurement of real-time in vivo light production of recombinant *Escherichia coli* bacteria expressing luciferase genes. In the described assay, we use vectors with efficient regulation of protein (i.e., luciferase) synthesis for studying drugs affecting protein synthesis. The assay utilizes the very strong bacteriophage λ leftward promoter (P_L), which efficiently controls luciferase synthesis. Using this promoter, luciferase synthesis is repressed when bacteria are grown at suboptimal temperatures (<36°C). Protein synthesis can be efficiently switched on (i.e., induced) by a brief heat shock (42°C), which inactivates a mutant λ repressor protein. The incubation of a drug with bacterial cells is performed prior to the induction of

From: *Methods in Molecular Biology, Vol. 102: Bioluminescence Methods and Protocols*
Edited by: R. A. LaRossa © Humana Press Inc., Totowa, NJ

λ P_L-directed protein synthesis. After a heat shock, the luciferase synthesis is measured with a luminometer, and real-time results from protein synthesis will be collected. The difference in the results when compared to noninhibited control samples reveals the influence of the drug candidate on protein synthesis *in situ*. The real-time approach in studying protein synthesis inhibition enhances possibilities to simulate the drug action in a living target.

The bacterial luciferase enzyme is a dimeric protein with two different subunits encoded by the corresponding *luxA* and *luxB* genes *(1,2)*. All known bacterial luciferases catalyze a reaction that involves the oxidation of long-chain aldehyde and $FMNH_2$, and the reaction produces blue-green light according to the following *(3)*:

$$FMNH_2 + O_2 + RCHO \xrightarrow{\text{Luciferase}} FMN + RCOOH + H_2O + LIGHT \text{ (490 nm)} \qquad (1)$$

Owing to its fatty nature, the aldehyde substrate can easily pass the cell membrane and serve as a substrate for the enzymatic reaction resulting in luminescence. The other substrates are available in sufficient quantities inside the bacterial cell (*see* **Note 2**).

E. coli containing the cloned luciferase genes does not need high (> 1%) salt concentrations to be maintained, whereas the original donor strains of marine origin *(Vibrio, Photobacterium)* do. The need for a high salt concentration can cause problems when studying certain activities of compounds on naturally luminescent bacterial strains. The second advantage of using heterologous expression systems like *E. coli* is that their genetics and physiology are normally much better known than those of the donor strains, which facilitates the interpretation of results.

Hundreds of specific *E. coli* mutations are known, which makes possible optimized study of compound activity on specific reactions. It is possible to choose bacterial strains with selected resistance factors, membrane functions, and transport mutations. Furthermore, it is rather simple to also transfer other characteristics into bacterial cells by genetic engineering techniques. This broadens the applicability of microorganisms in bioassays.

The above-mentioned real-time in vivo protein synthesis assay accompanied by molecular-level methods will provide an efficient test panel for studying protein synthesis inhibitors (*see* **Notes 3** and **4**).

2. Materials

2.1. Bacteria and Plasmids

1. The bacterial strain used was *E. coli* K-12 strain M72 [Sm^r *lacZ*[Am] Δ*bio-uvrB* Δ*trpE42* [λ*N7*{Am}-*N53*{Am} *cI857* ΔHI]) *(4)*. This strain carries a chromosomal insertion of the *cI857* repressor gene that is essential for the temperature-sensitive regulation of luciferase synthesis.

2. The plasmid used was pCSS110 containing *Vibrio harveyi luxA* and *luxB* genes under the control of phage λ heat-inducible leftward p_L promoter (*see* **Note 1**).

2.2. Chemicals

1. L-broth: 10 g Bacto-tryptone, 5 g yeast extract (Difco, Detroit, MI), 5 g NaCl, H_2O add 1 L, pH 7.0 with NaOH, autoclave for 20 min at 121°C.
2. L-agar plates: 10 g Bacto-tryptone, 5 g yeast extract, (Difco), 5 g NaCl, H_2O add 1 L, pH 7.0. Autoclave and cool the agar media to 50°C in the water bath before adding ampicillin (final concentration is 100 µg/mL).
3. Ampicillin stock: weigh 1.0 g of ampicillin, Sigma (St. Louis, MO) cat. no. A-9518, and dissolve in 10 mL of distilled water. Filter-sterilize through 0.2-µm filter, and aliquot in 1-mL portions. Store at –20°C. In all steps, one should use a 1000-fold dilution of this stock solution, ie., final working concentration is thus 100 µg/mL.
4. Aldehyde: 0.01% (v/v) n-decyl aldehyde, Sigma cat. no. D-7384, dissolved in ethanol or 0.001% (v/v) sonicated in water.

2.3. Apparatus

1. A single-tube luminometer (such as BioOrbit 1250 manual luminometer or Turner Design model TD-20/20) or a tube luminometer with higher sample capacity (such as Bio-Orbit 1251 with a carousel for 25 tubes), and a microplate luminometer (such as Luminoskan, Labsystems Oy) with an in-built shaker and temperature control (*see* **Notes 12–14**).
2. Ellerman tubes or similar ones for the light emission measurements depending on what kind of tube luminometer is used. White microplates/strips or black ones for microplate luminometer.
3. A shaker with temperature-control and a temperature-controlled incubator.
4. A water bath with a temperature control.
5. Pipets with variable pipeting volumes.
6. A platinum wire inoculator or sterile inoculators made of plastic.
7. Bunsen burner (lamp).

3. Methods

3.1. Cultivation of Bacteria

1. Transfer *E. coli* K-12 M72/pCSS110 onto a fresh L-agar plate containing ampicillin (100 µg/mL) using a sterile platinum wire from a –70°C stock (*see* **Note 4**) and spread it so that single colonies are obtained. At a maximum, cultivate for 12 h at 30°C. Make a new plate weekly.
2. From a fresh plate containing *E. coli* K-12 M72/pCSS110, take a single colony into 5 mL of L-broth and 5 µL ampicillin stock solution (final concentration of ampicillin is 100 µg/mL). Cultivate for no more than 12 h at 30°C in a shaker at 250 rpm (*see* **Note 6**).
3. Make a 1:100 dilution into fresh medium containing ampicillin and continue cultivation for a few hours until an OD_{600} value of 0.5 is reached (*see* **Note 7**). Use these bacteria in assays. Lyophilized cells (*see* **Note 8**) may also be used.

4. Make a permanent stock of cells by adding glycerol to 20% (v/v) and aliquoting the cells into 1-mL Eppendorf tubes. These cells frozen at –70°C can be used for several years as permament stocks. Streaking for single colonies (**step 1**) allows the starting material to be regenerated.

3.2. Light Emission Measurement

1. *E. coli* K-12 M72/pCSS110 cells are grown as described in **Subheading 3.1.** (*see* **Note 7**).
2. After this, a suitable dilution (*see* **Note 9**) is made in LB medium, and 500 µL of this are added to luminometer tubes.
3. The tubes are incubated at 30°C, and different amounts of protein synthesis inhibitors are added to tubes.
4. The tubes are kept at 30°C for 20 min, after which the temperature is shifted to 42°C for 10 min (*see* **Notes 5** and **10**).
5. Thereafter the tubes are returned to a 30°C water bath for 10 min.
6. Light emission is measured after the addition of *n*-decyl aldehyde to 0.001% (in ethanol) with the tube luminometer (*see* **Note 9**).
7. Light emission is measured with a manual Bio-Orbit 1250 luminometer (Turku, Finland) and recorded on an LKB-Bromma chart recorder. The magnitude of light emission in millivolts (mV) recorded in this case is inversely proportional to the inhibition of bacterial luciferase synthesis. The bigger values in mV reflect smaller concentrations of inhibitor in the sample. The reduction in mV of the highest concentration of an inhibitory agent relative to a blank sample, without any inhibitory agents, is normally 100–300-fold.
8. It is preferable to use luminescence values from the linear parts of the kinetic real-time luminescence curves when calculating the effects of the compounds (**Fig. 1A**). Percent light inhibition caused by varying inhibitor concentrations (S1–S4) is calculated by comparing inhibited samples with uninhibited control (inhibition, 0%) at the same point of time (**Fig. 1B**). These inhibitory concentrations are generally lower than those needed to inhibit growth (*see* **Note 1**).

4. Notes

1. Construction of a plasmid pCSS110: Plasmids pPLcAT10 *(5)* and pWH102 *(6)* were both cut with restriction enzymes *Hind*III and *Pvu*I. The smaller fragment from pPLcAT10 (1 kb) containing the leftward promoter of phage λ and a part of the gene coding for β-lactamase and the bigger fragment from pWH102 (6.0 kb) containing *lux* genes, an origin of replication, and the rest of the β-lactamase gene were excised from an agarose gel, purified, and ligated with T4–DNA ligase. After transformation into electrocompetent *E. coli* MC1061 (*cI*⁺) cells, correct transformants were identified by screening for dim light-producing colonies as described in **Subheading 3.**
2. Luminescent species can be roughly divided into two categories according to the luminescence mechanism: Bacterial bioluminescence and eukaryotic bioluminescence. The common feature for bacterial luminescence is similar enzymatic reactions and highly homologous genetic structures. The enzyme responsible for light

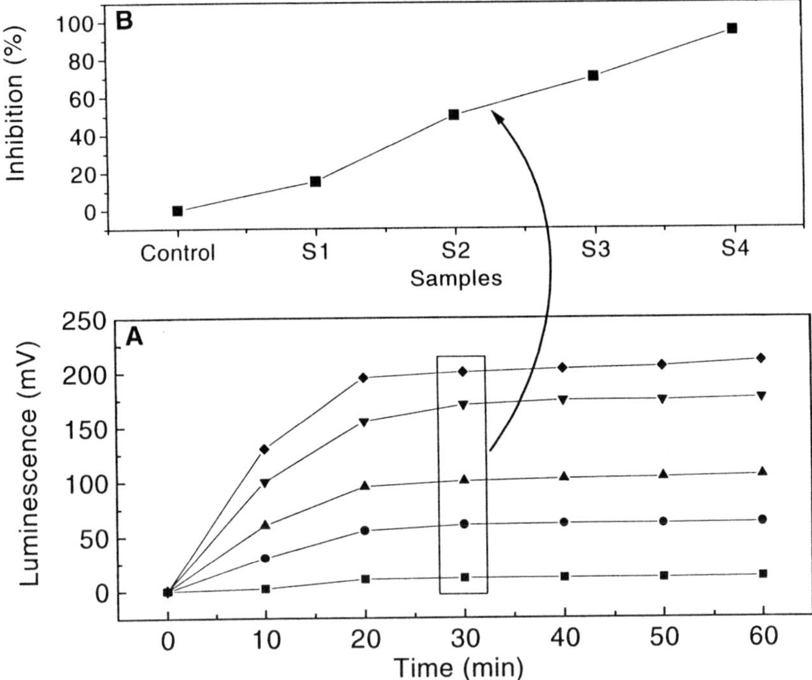

Fig. 1. Schematic real-time in vivo assay of a protein synthesis inhibitor (S). Different concentrations of protein synthesis inhibitor samples (S1–S4) and a control sample are incubated with *E. coli* K-12 M72/pCSS110 at 30°C for 20 min. Then the bacterial cells are heat-shocked for 10 min at 42°C, and thereafter, the samples are temperated at 30°C for 10 min. **(A)** Kinetic real-time luminescence curves. After addition of *n*-decyl aldehyde into the samples, a kinetic luminescence curve for each sample is measured. Luminescence is expressed as millivolts. **(B)** Response curve for inhibitor. Luminescence inhibition curve measured in 30-min time-point after the addition of substrate. The luminescence values of the inhibitor samples are compared to the value of the control sample, and inhibitions (%) are calculated. —■— Inhibitor (S4), —○— inhibitor (S3), —▲— inhibitor (S2), —▼— inhibitor (S1), —◆— control.

production is called bacterial luciferase, and it is a dimeric protein with two non-identical subunits and corresponding genes, *luxA* and *luxB (1,2)*. The pathway for the aldehyde biosynthesis (tetradecanal, R = 14C) is located on the same operon with two luciferase genes, but only the luciferase genes are essential for light production if aldehyde is exogenously added *(7)*.

3. Another type of luminescence mechanism is found in insects, e.g. in North American firefly (*Photinus pyralis*) and in luminous click beetle *Pyrophorus plagiophthalamus*. The luciferase reaction is based on energy transfer from ATP to the substrate, D-luciferin, yielding luminescence according to the following reaction *(8)*:

$$\text{ATP} + O_2 + \text{D-luciferin} \xrightarrow{\text{Luciferase}} \text{AMP} + PP_i + CO_2 + \text{oxyluciferin} \\ + \text{light (560 nm, } P.\ pyralis) \tag{2}$$

The click beetle luciferases are able to produce light of four different colors, the emission maximum ranging from 547–593 nm.

4. We have developed a set of different plasmids for luciferase expression in well-known host organisms, such as *E. coli* or *Bacillus subtilis (9–11)*. There exist clear differences between bacterial and eukaryotic luciferases with respect to light emission strength, kinetics, temperature stabilities, and so forth. Eukaryotic luciferases seem to be more sensitive indicators of inhibitor effects in almost all cases. This is probably based on the different connection of enzyme reaction to the central metabolite ATP, which reflects the intracellular state more directly than $FMNH_2$, the metabolic link exploited with bacterial luciferase when the aldehyde substrate is added to the incubations *(12,13)*.

5. Although *V. harveyi* luciferase is not thermostable, it is most useful for studies concerning protein synthesis inhibition, since any preformed luciferase is destroyed by the heat-induction treatment, thus amplifying the effects of protein synthesis inhibitors. If one wants to use a more thermostable luciferase, one could use plasmid pCSS118 containing *luxA* and *luxB* genes from the bacterium *Photorhabdus luminescens (13)*.

6. It is always essential to use fresh plates and cultivation methods, which keep the cells young and healthy either in solid or liquid cultivation, i.e., minimum period of growth in each stage in order to avoid mixed population of cells consisting of viable and dead cells.

7. It is possible to cultivate even further up to an OD_{600} value of 2.0, but then higher background light emission values will be obtained. This means lower induction factors.

8. For lyophilization of bacterial cells, a single colony is inoculated into 5 mL of LB-medium and grown overnight. Thereafter, the culture is diluted 1:100 with fresh LB-medium and grown at 30°C to an OD at 600 nm of 1.5. Bacterial cells are harvested by centrifugation (5200g, 10 min). The pellet is suspended in the same volume of protective medium containing 15% sucrose, and lyophilized as 0.5-mL aliquots by standard procedures. Lyophilized cells are rehydrated with 0.5 mL of H_2O and held for 30 min at 30°C. Rehydrated cells are diluted 1:20 in the L-broth to a final cell density of 5×10^6 mL^{-1}.

9. Before starting an assay, make sure that the bacterial cell dilution will give an appropriate level of luminescence signal. First, verify the level of the highest luminescence value in the assay by giving a heat shock without any inhibitory compounds (i.e., control sample). Also, measure the lowest luminescence value for the bacterial cell dilution without a heat shock. Be sure not to work below (or even just above) the background level or too high on the dynamic measuring range of the luminometer. Otherwise, a situation will occur where luminescence signals in a real assay possibly reach a level that exceeds the instrument's linear range or the signal is too low to give exact luminescence values in the assay. Normally the dynamic ranges of different luminometers are between 4 and 6

decades. Rule of thumb: the lowest usable signal has to be about one decade above the background level of the luminometer.

10. The incubation time before a heat shock depends on compounds that are being worked with. It is possible to start a heat shock after 5–10 min of incubation, because in this experimental setting, bacterial luciferase synthesis is not dependent on bacterial cell division cycle. A large amount of luciferase is synthesized during the 10 min heat-shock period at 42°C. During this heat-shock period, a small amount of bacterial luciferase initially present before induction will be denatured. Bacterial luciferase is extremely unstable in such a high temperature as 42°C. The temperature is changed to 30°C for efficient bacterial luciferase synthesis after the heat shock because of this reason. The highest light production is achieved within 30 min after substrate addition. Light production is rather stable for a few hours.

11. Detection limits of tested protein synthesis inhibitors with this assay concept are normally much lower than MIC values using the same bacterial strain. Thus, results from this assay format are not exactly directly comparable to conservative MIC values. The above-described assay is very sensitive based on the concept of the assay, i.e., the built-in protein synthesis amplification system. Light production is not dependent on the bacterial cell division cycle in the assay system described here. Using luciferase enzyme as the marker protein makes the assay more sensitive compared to spectrophotometric assays. It has been shown that light-measuring methods are from 100–1000 times more sensitive than corresponding spectrophotometric methods *(14)*.

12. Whatever luminometer is used, a very important aspect for data transferring and handling is an external computer connection. When purchasing a luminometer, be sure that the instrument can collect data in a continuous mode and transfer data to the application software. This is even more important when working with microplate luminometers because of the vast quantity of data that can be readily generated.

13. Working with tube luminometers, it is possible to use higher sample volumes, such as 0.5–2.0 mL, compared with microplate luminometers (0.1–0.25 mL). Using microplates, it is beneficial to use sonicated substrate (in water), because ethanol would affect the results in such a small working volumes.

14. Using a microplate luminometer, it is much more convenient to analyze replicates of different concentrations of affecting compounds. Timing is much more reproducible than working with tube luminometers. With a built-in shaker and temperature control (Peltier elements), it is possible to do the whole assay without a water bath and external shakers. Measurement efficiency is also superior compared to tube luminometers. It is possible to measure the whole microplate (96 samples) in <60 s. A robot-compatible microplate luminometer is a necessity, especially for large-scale screenings of lead compounds.

References

1. Cohn, D. H., Mileham, A. J., Simon, M. I., Nealson, K. H., Rausch, S. K., Bonam, D., and Baldwin, T. O. (1985) Nucleotide sequence of the *luxA* gene of *Vibrio*

harveyi and the complete amino acid sequence of the a subunit of bacterial luciferase. *J. Biol. Chem.* **260,** 6139–6146.

2. Johnston, T. C., Thompson, R. B., and Baldwin, T. O. (1986) Nucleotide sequence of the *luxB* gene of *Vibrio harveyi* and the complete amino acid sequence of the β subunit of bacterial luciferase. *J. Biol. Chem.* **261,** 4805–4811.

3. Hastings, J. W., Baldwin, T. O., and Nicoli, M. Z. (1978) Bacterial luciferase: Assay, purification and properties, in *Methods in Enzymology,* vol. 57 (DeLuca, M., ed.), pp. 135–152.

4. Bernard, H. U., Remaut, M. V., Hersfield, H. K., Das, D. R., Helinski, C., Yanofsky, C., and Franklin, N. (1979) Construction of plasmid cloning vehicles that promote gene expression from the bacteriophage λ p_L promoter. *Gene* **5,** 59–76.

5. Stanssens, P., Remaut, E., and Fiers, W. (1985) Alterations upstream from the Shine-Dalgarno region and their effect on bacterial gene expression. *Gene* **39,** 441–453.

6. Gupta, S. C., Reese, C. P., and Hastings, J. W. (1986) Mobilization of cloned luciferase genes into *Vibrio harveyi* luminescence mutants. *Arch. Microbiol.* **143,** 325–329.

7. Meighen, E. A., Riendeau, D., and Bognar, A. (1981) Bacterial bioluminescence: accessory enzymes, in *Bioluminescence and Chemiluminescence: Basic Chemistry and Analytical Applications* (DeLuca, M. and McElroy, W. D., eds.), Academic, New York, pp. 129–138.

8. McElroy, W. D. and DeLuca, M. (1985) Firefly luminescence, In *Chemi- and bioluminescence* (J. G. Burr, J. G., ed.), Marcel Dekker, New York., pp. 387–399.

9. Lampinen, J., Korpela, M., Saviranta, P., Kroneld, R., and Karp, M. (1990) Use of Escherichia coli cloned with genes encoding bacterial luciferase for evaluation of chemical toxicity. *Toxic Assess.* **5,** 337–350.

10. Lampinen, J., Virta, M., and Karp, M. (1995) Comparison of Gram positive and Gram negative bacterial strains cloned with different types of luciferase genes in bioluminescence cytotoxicity tests. *Environ. Toxicol. Water Qual.* **10,** 157–166.

11. Virta, M., Karp, M., and Vuorinen, P. (1994) Nitric oxide donor-mediated killing of bioluminescent *Escherichia coli. Antimicrob. Agents Chemother.* **38,** 2775–2779.

12. Koncz, C., Langridge, W. H., Olsson, O., Schell, J., and Szalay, A. A. (1990) Bacterial and firefly luciferase genes in transgenic plants: advantages and disadvantages of a reporter gene. *Dev. Genet.* **11,** 224–232.

13. Lampinen, J., Virta, M., and Karp, M. (1995) Use of controlled luciferase expression to monitor chemicals affecting protein synthesis. *Appl. Environ. Microbiol.* **61(8),** 2981–2989.

14. Lövgren, T., Peacock, R., Lavi, J., Karp, M., and Raunio, R. (1982) The bioluminescent assay of NADH and NADPH. *Int. Lab.* **12,** 58–61.

15

Luminescence-Based Cell Viability Testing

Ian A. Cree

1. Introduction

There are many reasons for testing cell viability. Simple tinctorial assays, such as trypan blue exclusion, have their place, but luminescence assays based on the detection of adenosine triphosphate (ATP) are particularly useful, since it is possible to detect the ATP present in fewer than 20 cells/mL using optimized reagents. These same reagents can also accurately determine the ATP from 2×10^7 cells/mL in an adjacent well or test tube with a completely linear relationship between cell number and light output, provided that the cells maintain a relatively invariant ATP content. Similar methods are used for bacteria and eukaryotic cells, but this chapter is restricted to the consideration of eukaryotic cells. Luminescence measurement of ATP levels uses the following reaction:

$$\text{ATP} + \text{D-Luciferin} + O_2 \xrightarrow[\text{Mg}^{2+}]{\text{Luciferase}} \text{AMP} + 2P + CO_2 + \text{light} \qquad (1)$$

This technology has found uses in a variety of circumstances. The first application was to provide simple measurement of biomass: "how many living cells are there in this sample?" *(1)*. Lundin and coworkers followed this by designing more complex assays that link other enzyme systems (pyruvate kinase and myokinase) to luciferase, allowing measurement of ATP, ADP, and AMP to produce an assessment of "energy charge" within the cells of interest *(1)*. At the same time, Kangas and coworkers *(2)* were the first to use the ATP cell viability assay to measure the effect of the extracellular environment on cell viability. This led several workers to produce cell viability assays designed to test the effect of drugs on cell viability *(3–6)*. Such assays were of particular interest to those seeking an assay suitable for testing the effect of chemotherapeutic drugs on cancer (*see* **Note 1**).

From: *Methods in Molecular Biology, Vol. 102: Bioluminescence Methods and Protocols*
Edited by: R. A. LaRossa © Humana Press Inc., Totowa, NJ

Human tumors of the same type show clinical heterogeneity of responsiveness to cytotoxic drugs. Any method that allowed the chemosensitivity of tumors to be predicted in individual patients would be welcome, since it would allow optimal treatment to be given to each patient. Many attempts have been made to do this, but no chemosensitivity test has yet achieved widespread clinical use *(7,8)*. Prediction of chemosensitivity on the basis of tumor growth rates, estimated by histological methods or by nucleotide incorporation has proven disappointing. It is unlikely that molecular methods will fare much better, since the response to cytotoxic agents is determined by a large number of different biochemical pathways. In vitro clonogenic assays have perhaps had more success, but the difficulty of obtaining clones from many solid tumors has limited their usefulness. Nonclonogenic assays, such as those based on tetrazolium salt reduction (the MTT assay) or ATP measurement, have the advantage of measuring tumor cell survival as well as growth, and can be successfully performed in a large proportion (>95%) of tumors. However, technical problems associated with growth of noncancerous cells from the tumor and an inability to test large numbers of drugs at different concentrations using small biopsies have as yet limited their usefulness. The MTT assay is considerably less sensitive than the tumor chemosensitivty assay (TCA) and cannot be used with small biopsies *(9)*.

The other recent application of this technology is the ATP-based lymphocyte transformation test. In this type of assay, the ability of cells to respond to mitogenic or antigenic stimulation is determined by the increase in ATP consequent on both blast transformation and cell division *(10)*. Similar assays can be used to study the cytokine-dependent cell growth, or growth inhibition with a large number of different cell types.

2. Materials

2.1. Cell Culture

Cell lines provide a simple and convenient supply of cells suitable for ATP assays of toxicity or growth enhancement. However, it must always be remembered that they represent a clone of cells adapted to culture rather than host tissue conditions, and that their response to any agent may be completely different relative to the tissue or tumor from which they were derived. Nevertheless, cell lines are a valuable scientific resource, and it is now possible to obtain lines from most tissues or tumors from commercial or academic sources. Adherence independent lines are mainly of lymphoid origin and are particularly useful for quality-assurance assays, since they grow simply using a base medium, such as RPMI 1640 or Dulbecco's Minimal Essential Medium (DMEM), supplemented with 10% fetal bovine serum (FBS) and antibiotics (penicillin and streptomycin). Some cell lines have special requirements, and the supplier should detail these. For experimental work, FBS is best avoided:

batch-to-batch variation is a problem, and defined FBS substitutes are gaining favor. Cells are usually grown in 750-mL plastic (polystyrene) flasks with angled necks, which encourage cell attachment and allow gas exchange *(11)*. For ATP assays, it is rarely necessary to have large numbers of cells, so roller flasks and more sophisticated methods of bulk cell culture are not required.

Cells can also be obtained by dissociating animal or human tissues *(6)*. Dissociation reagents may contain a variety of enzymes. Enzyme purity and concentration affect performance and cell viability.

1. Cell culture flasks (250/750 mL).
2. Culture medium, e.g., RPMI 1640, DMEM.
3. FBS or substitute with growth factors matching cell requirements. Aliquot and freeze at $-20°C$.
4. Antibiotic solution (penicillin-streptomycin): Some such solutions contain antimycotics which can be useful if there is a contamination problem. However, they can interfere with toxicity-type assays and are best avoided where possible.
5. Mycoplasma detection kit.
6. Trypsin (0.25%)–EDTA (1 mM) solution: This is required for adherent cell lines.
7. Dissociation enzyme solution.
8. Trypan blue solution (0.4%).
9. Universal (25-mL) polystyrene tubes to fit centrifuge.
10. Sterile disposable pipets.
11. Sterile 10-mL pipets to fit automated pipet bulb or electronic pipeter.
12. Set of air displacement pipets (20–1000 µL) with sterile racked tips.
13. Sterile disposable scalpels.
14. Sterile plastic Petri dishes.
15. Sterile Ficoll-Hypaque solution, such as Lymphoprep (Nycomed, Birmingham, UK).
16. Sterile polystyrene (adherent cells) or polypropylene (nonadherent cells) microplates (96-well).
17. Nonvolatile maximum inhibitor of cell growth (e.g., MI reagent, DCS, Hamburg, Germany).
18. Bench centrifuge with sealable buckets in case of spills: Refrigerated centrifuges are not necessary.
19. CO_2 incubator with 99% humidification (e.g., Napco, TIS Services, Bentworth, UK).
20. Modified Neubauer chamber.
21. Inverted microscope (e.g., Olympus CK2, London, UK).
22. Class 100-type laminar flow hood affording both operator and sample protection.

N.B.: Plasticware can be obtained from companies, such as Becton-Dickinson (Oxford, UK), Greiner (Dursley, UK), Costar (High Wycombe, UK), Sigma (Poole, UK), or Alpha Laboratories (Eastleigh, UK). Reagents can be purchased from Sigma or Gibco.

2.2. Cell Extraction

1. Extraction reagent to match luciferase-luciferin (both are generally available from the same company).
2. Universal (25-mL) polystyrene tubes to fit centrifuge.
3. Set of air displacement pipets (20–1000 µL) with sterile racked tips.
4. An eight-channel electronic automated pipet (e.g., Biohit, Alpha Laboratories) dispensing 25–250 µL.

2.3. ATP Measurement

1. Luminometer—microplate-type recommended with injector. The Berthold LB96P luminometer was designed with this type of assay in mind, although several others can do the same job.
2. Set of air displacement pipets (20–1000 µL) with sterile racked tips.
3. An eight-channel electronic automated pipet (e.g., Biohit) dispensing 25–250 µL.
4. White 96-well polystyrene microplates (Dynatech).
5. Luciferin-luciferase reagent—commercial agents are recommended, because of quality-control considerations, although it is possible to buy the basic reagents from suppliers, such as Boerhinger Mannheim (Lewes, UK) or Sigma.
6. ATP standard + dilution buffer.

3. Methods

A summary of the method is shown in **Fig. 1**. Cells from solid tumors, ascites, or cell culture are suitable, and it is relatively easy to alter the method to allow drug sequencing issues to be addressed.

3.1. Cell Handling and Culture (see Note 1)

1. Grow cell lines to confluence, and then passage every 7–14 d depending on growth rate to new flasks. Discard remainder, use for assays, or consider freezing aliquots if not immediately required.
2. To obtain cell suspensions from adherent cells, wash in HBSS, add 0.25% trypsin/ 1 m*M* EDTA in calcium-free HBSS or similar buffer, and incubate at 37°C for 10 min. Then shake to loosen cells. The appropriate volume is usually 5 mL for a 75-mL flask. Add RPMI + serum to neutralize trypsin, and pour off cell suspension into sterile Universal tube.
3. To obtain cells directly from tissues (normal/tumor), mince tissue with sterile scalpel in Petri dish, add with dissociation medium to Universal tube, and incubate at 37°C for 4–24 h according to manufacturers' instructions. Following dissociation, wash cells twice and count.
4. Wash cell suspension at least twice by centrifugation at 200*g* for 10 min to sediment cells, discard supernatant, resuspend in fresh medium for assay/subculture, and repeat centrifugation to remove all enzyme.
5. Assay medium should contain all growth factors required by cells in base medium, together with antibiotics. HEPES is also added if not already present to enhance the buffering capacity of the medium.

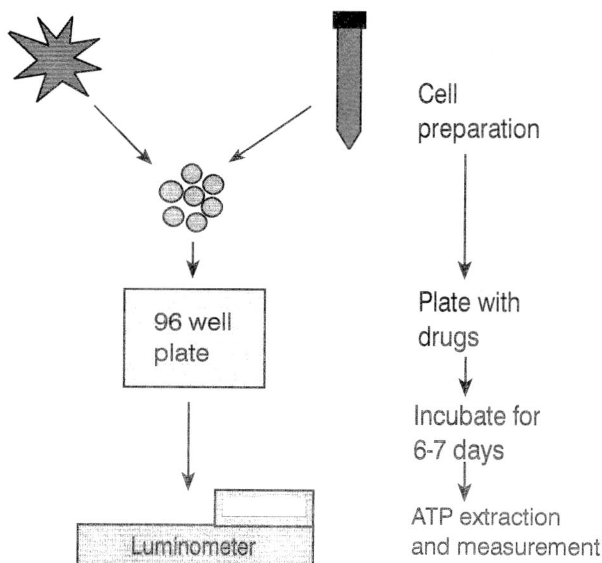

Fig. 1. Summary of ATP-TCA method. Cells are prepared by enzymatic dissociation of solid tumor, harvesting of cell cultures, or density centrifugation of blood/bone marrow samples. Following this, they are plated out with drugs according to the layout shown in **Fig. 2** and incubated at 37°C for 6–7 d. Extractant is added to each well, and the ATP content measured by addition of luciferin-luciferase reagent in a luminometer.

6. If there is <50% cell viability or large amounts of nonviable tissue debris, the cell suspension can be layered over a Ficoll-Hypaque solution, such as Lymphoprep (Nycomed) and centrifuged at 400*g* for 25 min. The washing **step 4** must be repeated following this. Lymphoprep can also be used to obtain mononuclear cells from venous blood for lymphocyte transformation testing.

7. Count cell suspension in modified Neubauer chamber, and adjust with sterile culture medium to 50,000 cells/mL (cell lines) or 200,000 cells/mL (tumor-derived cells).

8. Choose clear polystyrene or polypropylene 96-well plate, depending on cell adherence characteristics. For chemosensitivity assays using tumor-derived cells, round-bottom polypropylene plates (Costar) are recommended. Flat or round-bottom 96-well plates are suitable for other cells types, but obviously cells are easiest to monitor visually in flat-bottom plates.

9. Add 100 µL cell medium to wells of plate. This permits agents to be added and dilutions made within the plate. A suitable layout is shown in **Fig. 2**. Add a nonvolatile maximum inhibitor of cell growth (e.g., MI reagent, DCS, Hamburg, Germany) to at least six wells, and leave six wells as a no-agent control. Avoid agents, such as thiomersal, which might interfere with the luciferase. Detergents such as Tween-20 or Triton X are suitable at low concentration.

	1	2	3	4	5	6	7	8	9	10	11	12
A	No drug control (MO)						Maximum inhibitor (MI)					
B	Drug 1 - 200% TDC			Drug 2 - 200% TDC			Drug 3 - 200% TDC			Drug 4 - 200% TDC		
C	Drug 1 - 100% TDC			Drug 2 - 100% TDC			Drug 3 - 100% TDC			Drug 4 - 100% TDC		
D	Drug 1 - 50% TDC			Drug 2 - 50% TDC			Drug 3 - 50% TDC			Drug 4 - 50% TDC		
E	Drug 1 - 25% TDC			Drug 2 - 25% TDC			Drug 3 - 25% TDC			Drug 4 - 25% TDC		
F	Drug 1 - 12.5% TDC			Drug 2 - 12.5% TDC			Drug 3 - 12.5% TDC			Drug 4 - 12.5% TDC		
G	Drug 1 - 6.25% TDC			Drug 2 - 6.25% TDC			Drug 3 - 6.25% TDC			Drug 4 - 6.25% TDC		
H	Drug 1 - 3.13% TDC			Drug 2 - 3.13% TDC			Drug 3 - 3.13% TDC			Drug 4 - 3.13% TDC		

Fig. 2. Suggested microplate layout. Drugs are prepared at 8 × the 100% test drug concentration (TDC) and diluted within the plate prior to addition of cells to each well.

10. Add 100 µL of cell suspension to each well.
11. Place plate in a loosely covered plastic tub with wet towels in its base. Ensure that there are holes in the lid to permit gas exchange. Place tub with plate in 5% CO_2 at 37°C.
12. Incubation can be continued for up to 10–14 d if required: 72 h are usual for lymphocyte transformation tests, and 7 d for chemosensitivity assays.

3.2. Preparation of Cell Extracts

1. At end of the incubation period, extract cells by addition of ATP extractant. Most systems allow the addition of 50–75 µL to a 200-µL vol within the culture wells.
2. Mix thoroughly, changing pipet tips as necessary.
3. If ATP analysis is not to be done immediately, freeze at –20°C and store. We have stored samples for up to a month and still obtained satisfactory ATP counts after 1 mo using the extraction reagent available from DCS Innovative Diagnostik Systeme.

3.3. Measurement of ATP by Luciferase-Luciferin (see Notes 3–5)

1. Transfer aliquot (usually 50 µL) to while microplate or luminometer tubes as appropriate.
2. Make up ATP standard with 6–10 dilutions, incorporating range of ATP values expected.
3. Load tubes or plate into luminometer: ATP standards should be run before and after test samples to ensure stability of the luciferin-luciferase reagent over time.
4. Set luminometer to inject a similar quantity of the luciferin-luciferase reagent, and load reagent. Most machines are automated at this point.

3.4. Analysis of the Results

1. Check the raw counts to ensure that the luciferase maintained its activity during the analysis and that background readings were acceptable.
2. Analysis of the data is performed using a spreadsheet (e.g., Excel or Quattro Pro). It is helpful to enter data automatically if the luminometer has an RS232 interface and can be physically linked to a PC.

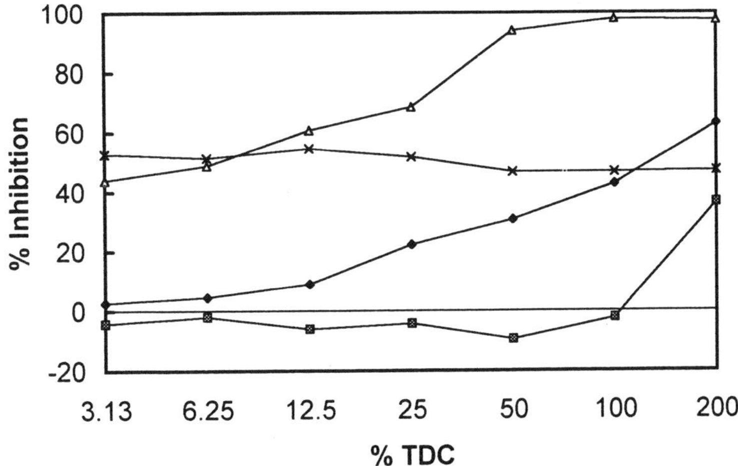

Fig. 3. Example results from a melanoma skin metastasis, showing sensitivity to actinomycin D, weak sensitivity to cisplatin, and resistance to 4–hydroperoxycyclophosphamide (4–HC). There is 50% inihibition across the range of concentrations of vindesine tested, a common finding with spindle-active agents. —●— Cisplatin, —■— 4HC, —△— actinomycin D, —X— vindesine.

3. The degree of inhibition of ATP is expressed as a percentage of the no drug/agent (MO) control, subtracting the maximum inhibitor values (MI) as% Inhibition = $1 - (Test - MI)/(MO-MI)$. Graphs of inhibition vs concentration (**Fig. 3**) are prepared using a simple spreadsheet *(6)*.
4. Further analysis involves calculation of indices, such as the IC90, IC50, minimum inhibitory concentration (MIC), and maximum nontoxic concentration (MNTC).
5. For tumor chemosensitivity work *(6)*, two summary indices are useful: the area under the concentration-inhibition curve (AUC) and a sensitivity index defined as the sum of the inhibition at each concentration (Index). These allow comparison of individual tumors (**Fig. 4**) *(6,12,13)*.

4. Notes

1. Most cell biology or immunology laboratories are suitable for this assay *(11)*. Care must be taken in two main areas: handling of drugs and handling of fresh (potentially infected) human tissue.
2. Cytotoxic drugs are inherently dangerous and often carcinogenic. They must be handled in accordance with local safety regulations. Powdered drugs are particularly hazardous and should be handled in stoppered containers within a hood with extraction to the outside. Hospital pharmacies are a good source of advice, as well as spare drugs. Many cytotoxic drugs can be successfully divided into aliquots and frozen *(14)*.

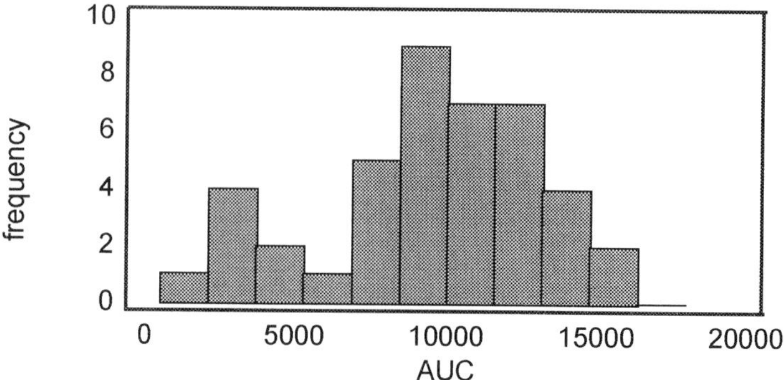

Fig. 4. Summary data from 42 primary breast adenocarcinomas for epirubicin, a topoisomerase II inhibitor.

3. Most luminometers have considerable dead space, and care must be taken to avoid dilution effects—wash thoroughly after use.
4. If backgrounds are high with luciferin-luciferase alone (blanks), flush system with a 0.01% solution of Tween-20 or other detergent (check with manufacturer first) to remove ATP containing algae or bacteria in system. Then wash thoroughly with water.
5. To prevent blocking, we routinely charge the luminometer with sterile distilled water before loading the luciferase, and wash afterward, with water, then with 70% alcohol, and then flush with air to prevent bacterial or algal growth.

References

1. Lundin, A., Hasenson, M., Persson, J., and Pousette A. (1986) Estimation of biomass in growing cell lines by adenosine triphosphate assay. *Methods Enzymol.* **133,** 27–42.
2. Kangas, L., Gronroos, M., and Nieminen, A. L. (1984) Bioluminescence of cellular ATP: a new method for evaluating cytotoxic agents in vitro. *Med. Biol.* **62,** 338–343.
3. Andreotti, P. E., Thornthwaite, J. T., and Morse, I. S. (1991) ATP tumor chemosensitivity assay, in *Bioluminescence and Chemiluminescence: Current Status,* (Stanley, P. E. and Kricka, L. J., eds.), John Wiley, Chichester, pp. 417–20.
4. Sevin, B. U., Perras, J. P., Averette, H. E., Donato, D. M., and Penalver, M. (1993) Chemosensitivity testing in ovarian cancer. *Cancer* **71,** 1613–1620.
5. Andreotti, P. E., Linder, D., Hartmann, D. M., Cree, I. A., Pazzagli, M., and Bruckner, H. W. (1994) TCA-100 tumor chemosensitivity assay: differences in sensitivity between cultured tumor cell lines and clinical studies. *J. Bioluminescence Chemiluminescence* **9,** 373–378.
6. Andreotti, P. E., Cree, I. A., Kurbacher, C. M., Hartmann, D. M., Linder, D., Harel, G., Gleiberman, I., Caruso, P. A., Ricks, S. H., Untch, M., Sartori, C., and Bruckner, H. W. (1995) Chemosensitivity testing of human tumors using a

microplate adenosine triphosphate luminescence assay: clinical correlation for cisplatin resistance of ovarian carcinoma. *Cancer Res.* **55,** 5276–5282.

7. Bellamy, W. T. (1992) Prediction of response to drug therapy of cancer. A review of *in vitro* assays. *Drugs* **44,** 690–708.

8. Bosanquet, A. G. and Bell, P. B. (1996) Novel ex vivo analysis of nonclassical, pleiotropic drug resistance and collateral sensitivity induced by therapy provides a rationale for treatment strategies in chronic lymphocytic leukemia. *Blood* **87,** 1962–1971.

9. Petty, R. D., Sutherland, L. A., Hunter, E. M., and Cree, I. A. (1995) Comparison of MTT and ATP-based assays for the measurement of viable cell number. *J. Bioluminescence Chemiluminescence* **10,** 29–34.

10. Crouch, S. P., Kozlowski, R., Slater, K. J., and Fletcher, J. (1993) The use of ATP bioluminescence as a measure of cell proliferation and cytotoxicity. *J. Immunol. Methods* **160,** 81–88.

11. Freshney, R. I. (1984) *Culture of Animal Cells. A Manual of Basic Technique.* 3rd ed., Wiley-Liss, New York.

12. Hunter, E. M., Sutherland, L. A., Cree, I. A., Dewar, J. A., Preece, P. E., Wood, R., A., Linder, D., and Andreotti, P. E. (1993) Heterogeneity of chemosensitivity in human breast carcinoma: use of an adenosine triphosphate (ATP) chemilumi-nescence assay. *Eur. J. Surg. Oncol.* **19,** 242–249.

13. Cree, I. A., Kurbacher, C. M., Untch, M., Sutherland, L. A., Hunter, E. M. M., Subedi, A. M. C., James, E. A., Dewar, J. A., Preece, P. E., Andreotti, P. E., and Bruckner, H. W. (1996) Correlation of the clinical response to chemotherapy in breast cancer with ex vivo chemosensitivity. *Anti-Cancer Drugs* **7,** 630–635.

14. Hunter, E. M., Sutherland, L. A., Cree, I. A., Subedi, A. M. C., Hartmann, D., Linder, D., and Andreotti, P. E. (1994) The influence of storage on cytotoxic drug activity in an ATP-based chemosensitivity assay. *Anti-Cancer Drugs* **5,** 171–176.

16

Phagocyte Chemiluminescence

Ian A. Cree

1. Introduction

Phagocytes form an essential defence against microbial infection and have an important role in debridement following tissue injury. In human subjects, there are essentially two classes of phagocyte: polymorphonuclear leukocytes (PMNL) and mononuclear phagocytes, both derived from myelomonocytic bone marrow cells. PMNL circulate in the bloodstream, and are subdivided into neutrophils, eosinophils, and basophils. Mononuclear phagocytes circulate in the blood as monocytes as a heterogeneous population, a proportion of which becomes tissue macrophages (e.g., Kupffer cells in liver, microglia in brain). While blood-borne PMNL and monocytes are easily accessible for study, many tissue macrophages are difficult to obtain for in vitro chemiluminescence. Although methods of obtaining relatively pure populations of tissue macrophages exist, all tend to activate the cells to a variable degree, which impedes interpretation of any results obtained. Most in vitro macrophage studies are therefore conducted with monocyte-derived cells, although alveolar (lung) and peritoneal macrophages can be studied by direct sampling with minimal preparation.

Both PMNL and monocytes have a large and diverse arsenal of antimicrobial weapons. However, the importance of oxygen radical production during the respiratory burst is shown by the occurrence of chronic granulomatous disease in those who possess mutant respiratory enzyme chain proteins *(1)*. Natural phagocyte chemiluminescence (CL) was first noted by Allen et al. *(2)* and is dependent on the reaction of superoxide with surrounding molecules to produce photons. Since many reactions do not produce photons, natural CL is weak. The addition of specific enhancing agents, luminol and lucigenin *(3,4)*, produces much greater CL, which has been used to assess phagocyte function

From: *Methods in Molecular Biology, Vol. 102: Bioluminescence Methods and Protocols*
Edited by: R. A. LaRossa © Humana Press Inc., Totowa, NJ

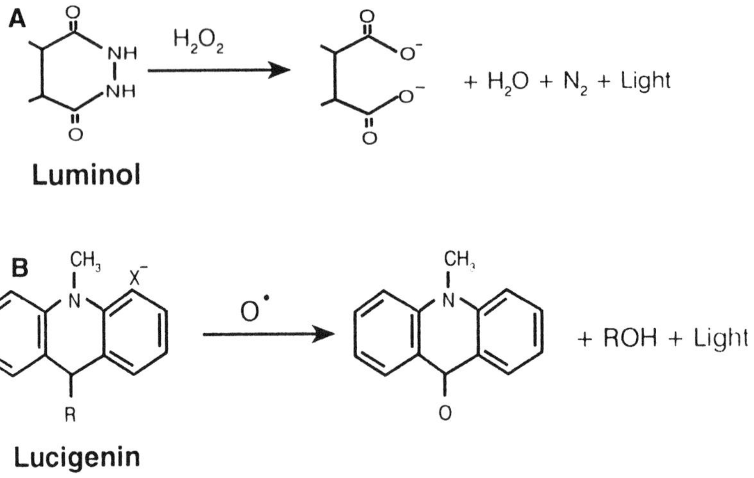

Fig. 1. The reactions of **(A)** luminol and **(B)** lucigenin with peroxide and super-oxide, respectively, to produce light.

in a large number of basic and clinical research settings. The reaction of luminol with superoxide is catalyzed by myeloperoxidase *(4)* (**Fig. 1A**), whereas lucigenin appears to directly react with superoxide *(4)* (**Fig. 1B**). Other differences include the ability of the chemicals to enter the cell: luminol does this much more readily than lucigenin *(4)*, although it should be remembered that both will be taken into phagosomes during fusion in experiments with particulate stimuli. Furthermore, approx 15% of phagosomes never entirely fuse, permitting entry of lucigenin to the site of superoxide synthesis even without sub-total phagocytic activation.

The ex vivo phagocyte CL method given here allows the assay of both PMNL and monocyte CL consecutively from four patients with two enhancers and four stimuli *(5)*. The total assay time is around 6 h, reduced to 4 h if only PMNL are assayed. The method can be easily adapted to study in vitro drug effects *(6)*.

2. Materials

As noted above, phagocytes are usually obtained from blood. It is usually possible to obtain 4 million PMNL and 1 million mononuclear cells (MNC) from 1 mL of peripheral venous blood. Since lymphocytes do not produce appreciable CL in comparison with phagocytes, they can be approximated to monocytes, which comprise about 15% of the cells in most samples *(5)*. The precise number of monocytes can be quickly estimated by a phagocytic assay using opsonized zymosan. Other methods, such as esterase staining, FACS analysis, or immuno-histochemistry are too slow to be useful in the assay. There are some myelomonocytic cell lines able to produce a respiratory burst. These may have specific uses, particularly in assessing the effects of drugs, although in practice, the easy availability of blood from transfusion centres and volunteers makes this largely unnecessary. Monocyte-macrophages require cell culture of monocytes for several weeks with frequent changes of medium and removal of dead cells.

It is best to prepare all except the stock solutions of PMA (freeze in aliquots) and Zymosan on the day of their use.

2.1. Cell Separation

1. Universal 25-mL plastic bottles with screw tops and conical plastic bottoms are ideal for both cell separation and blood collection.
2. Sodium heparin 1000 U/mL (Leo Laboratories, Princes Risborough, UK).
3. Sterile plastic 1- or 2-mL disposable bulb pipets (graduated and individually wrapped).
4. Mono-Poly Resolving Medium (M-PRM, 16–980–49, Flow Labs, High Wycombe, Bucks, UK) or similar Ficoll-Hypaque solution permitting resolution of both PMNL and monocytes.
5. Sterile distilled water.
6. 10X Hank's balanced salt solution (HBSS) (cat. no. 042-04065H, Gibco, Paisley, Scotland).
7. Automatic pipets: range 50–1000 µL.
8. Trypan blue solution (0.4%).
9. Modfied Neubauer hemocytometer or access to a Coulter counter.
10. Bench centrifuge: refrigeration is not necessary, but covered buckets are.
11. Disposable rubber gloves.

2.2. Stimulants

1. Zymosan (Sigma [Poole, UK], adjust to 1×10^9 particles/mL stock).
2. Pooled serum (aliquots).
3. Automatic pipets: range 50–1000 µL.
4. Plastic tubes 1.5–3.0 mL vol.

2.3. Enhancers

1. HBSS.
2. Automatic pipets: range 50–1000 µL.
3. Plastic tubes: 1.5–3.0 mL vol.

2.4. Chemiluminescence Assay

1. Automatic pipets: range 50–1000 µL.
2. Eight-channel automatic pipet: range 25–250 µL.
3. Incubator: a simple plate incubator is fine.
4. White 96-well microplates (e.g., Dynatech [Billingshurst, UK], Berthold [Wildbad, Germany], Luminoskan [Basingstoke, UK]).
5. Microplate luminometer (e.g., Dynatech). Injectors are unnecessary for this assay.

2.5. Preparation of Buffers, Enhancers, and Stimulants

1. HBSS: add 90 mL distilled water to 10 mL of 10X concentration HBSS with 0.035 g sodium bicarbonate. Check pH.
2. HBSS/bovine serum albumin (BSA): add 135 mL distilled water to 15 mL 10X concentration HBSS solution with 0.525 g sodium bicarbonate and 0.15 g BSA (Sigma, Poole Dorset, UK, A7906). Check pH.
3. Lucigenin: add 0.0063 g lucigenin (M8010 Sigma) + 50 mL HBSS/BSA.
4. Luminol: Make up stock luminol (A8511, Sigma) as $10^{-2}M$ solution in dimethyl sulfoxide (DMSO, Sigma, D8779). Dilute 1:100 for use in HBSS/BSA on day of assay.
5. Make up PMA from 1 mg/mL stock in DMSO (Sigma, D8779). Dilute aliquot to 1 µg/mL in HBSS/BSA.
6. Serum opsonized zymosan (SOZ): dilute stock (1×10^9/mL) 1 in 100 to obtain 1×10^7/mL particles in HBSS/BSA, and add an equal volume of 40% autologous human serum. Incubate at 37°C for at least 20 min, and then wash in HBSS/BSA by centrifugation if experiment requires this.
7. Zymosan: nonopsonized zymosan is made up by diluting stock (*see* **Subheading 2.5., item 3**) 1 in 200 with HBSS/BSA to obtain a 5×10^6 particle/mL suspension.

3. Methods

A summary of the method used is shown in **Fig. 2.** It is best to prepare all reagents except the stock solutions of PMA (freeze in aliquots) and zymosan on the same day as the assay.

3.1. Cell Separation

1. Take 15 mL venous peripheral blood from the antecubital fossa using a 19- or 21-gage needle according to the size of the attached syringe. Above 20 mL a 19-gage needle is preferred (*see* **Note 1**).
2. Transfer to a tube containing either EDTA or sodium heparin (25 U/mL). There are some reports that Lithium adversely affects CL responses.
3. Move samples to a laminar flow hood permitting both operator and sample protection. *Caution:* Wear gloves.
4. Pipet 11 mL of M-PRM into labeled 25-mL universal tubes (*see* **Note 2**).
5. Carefully layer (*see* **Note 3**) up to 13 mL undiluted blood on top of M-PRM, and replace the tube tops (**Fig. 3**). The remaining 2 mL can be used for hematology or FACS analysis.
6. Centrifuge tubes in sealed buckets at 400*g* for 30 min. The blood separates into layers as shown in **Fig. 3** (*see* **Note 4**).
7. Label tubes for MNC and PMNL for each patient.
8. Pipet off the cell layers using a disposable 1-mL bulb pipet.
9. Make up to 20 mL vol with HBSS, and centrifuge at 300*g* for 10 min.
10. Remove supernatant and resuspend white cell pellet in HBSS, making up to 20-mL vol (*see* **Note 5**).

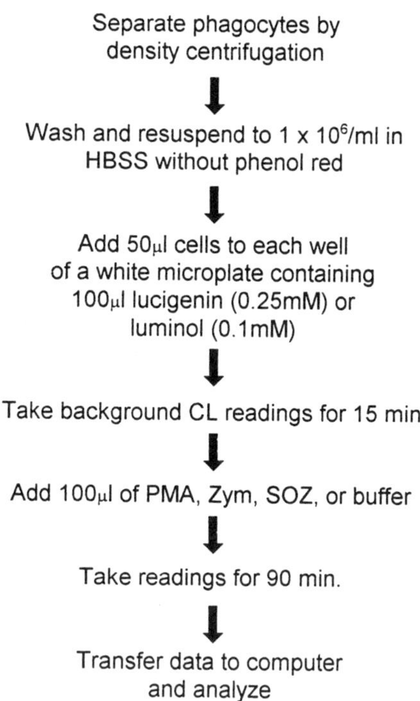

Fig. 2. Summary of method.

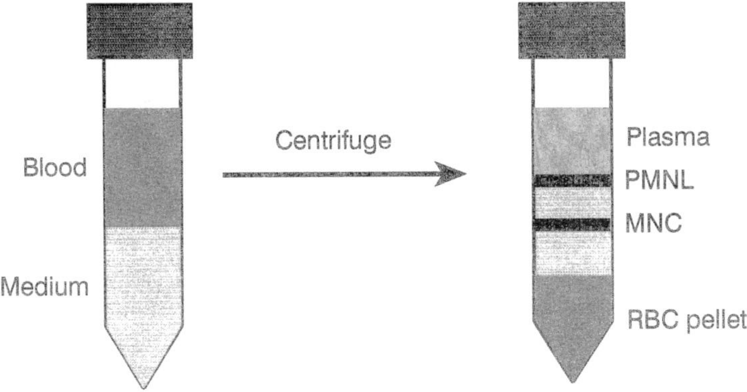

Fig. 3. Blood separation using a dual-phase separation medium, such as MRPM or Polyprep. Anticoagulated blood is layered on top of the medium and centrifuged to separate polymorphonuclear leukocytes (PMNL) and mononuclear cells (MNC, monocytes, and lymphocytes) into two bands that can be pipeted off.

	PMA			SOZ			ZYM			HBSS				
	1	2	3	4	5	6	7	8	9	10	11	12		
A													Lucigenin	Subject 1
B														Subject 2
C														Subject 3
D														Subject 4
E													Luminol	Subject 1
F														Subject 2
G														Subject 3
H														Subject 4

Fig. 4. A sample plate layout. All cell/stimulant combinations are tested in triplicate to improve reproducibility, and either PMNL or MNC from up to four subjects can be tested in one plate with both luminol and lucigenin.

11. Centifuge again for 10 min at 300*g*.
12. Resuspend pellet in 1 mL HBSS/BSA.
13. To assess cell number using a hemocytometer, dilute a 50-μL vol 1:1 with 0.4% trypan blue, and count as directed. Alternatively use a Coulter counter. The trypan blue allows cell viability to be assessed as well as cell number, but is less accurate. Count the number of erythrocytes present in each fraction, too: this should be <10% of PMNL and <1% of MNC. There should be <1% PMNL in the MNC fraction.
14. Monocytes can be counted by adding 50 μL MNC to 50 μL of opsonized zymosan and incubating for 15 min at 37°C. Place a drop of the cells on a slide after this time and view directly, counting the proportion of cells with ingested zymosan particles. This approximates well to the number of monocytes estimated by esterase staining or immunocytochemistry, which may be done later to check results.
15. Dilute cells to desired concentration: both PMNL and MNC (15% of which are probably monocytes) are used at 0.5×10^6/mL in the following CL assay.

3.2. Chemiluminescence Assay

1. Switch on luminometer (*see* **Note 6**), and plate incubator to warm up to 37°C.
2. Add 50 μL cell suspension to each well as shown in the plate layout (**Fig. 4**), and then 50 μL of enhancer as shown (**Fig. 4**).
3. Leave for 30 min in plate incubator to equilibrate.
4. Place plate in luminometer, and commence four cycles of background readings (*see* **Note 7**). The luminometer should be programmed to read each well every 5 min allowing for the time taken to read the plate: this usually means a delay between cycles of 3.5–4.0 min.
5. If background CL <10% of expected peak CL, add 100 μL prewarmed (37°C) stimulants to wells as shown in **Fig. 4**.
6. Commence 21 cycles of CL readings.

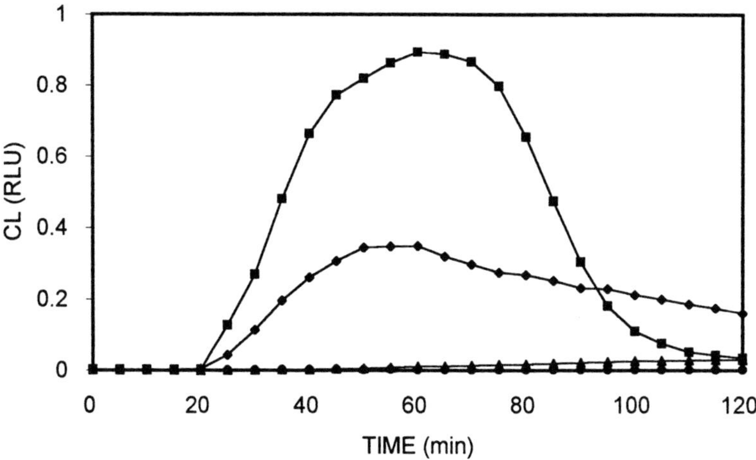

Fig. 5. An example of results obtained with normal human phagocytes. The PMA produces an intense and prolonged response, whereas the SOZ induces a prolonged, but less intense response. The zymosan produces a small, but measurable effect at this concentration, whereas there is no CL produced before or after stimulation with buffer alone. —●— HBSS, —■— PMA, —◆— SOZ, —▲— ZYM.

3.3. Analysis of the Results

1. The data from 25 cycles are automatically downloaded to an on-line computer and should also be printed out by a printer attached either to the computer or luminometer at the same time *(7)*.
2. The 2400 readings are analyzed in triplicate according to the plate layout shown in **Fig. 5**. Modern spreadsheets allow this process to be completely automated and graphs produced a few minutes after completion of the assay. The macro takes in the data and arranges it into columns and rows so that data for each set of three wells are separated, and merges the arranged data with a blank spreadsheet containing the formulas for mean and range, peak, slope, and total CL to be calculated for each data triplet. Graphs form part of this blank spreadsheet, which will rapidly recalculate the merged data and can then be saved to a new file name.
3. Graphs of CL against time (**Fig. 5**) are produced and viewed to ensure that the background CL was <10% of peak CL (PMA-stimulated) and that the control wells (HBSS only) remained <10% of peak CL throughout the assay.
4. The peak CL, maximum slope, and total CL are the main descriptors used to compare patients. In general, total CL has least variance <1% of mean).

4. Notes

1. When taking blood, ensure that the cells are not inadvertently activated by mechanical shear forces within the needle. A larger needle may frighten the donor, but they are usually sharper and cause no damage of any significance

(needles larger than 19-gage are routinely used by the blood transfusion service for 500-mL vol).

2. There are a number of good-quality cell separation media on the market, including Mono-Poly Resolving Medium (Flow Laboratories, High Wycombe, England) and Polyprep (Nycomed). There tends to be some batch-to-batch variation, and it is necessary to discard poorly separated specimens from time to time in most laboratories. In my experience, the donor often has or is incubating an infection at the time of venepuncture. The alternative is to use dense Ficoll layered on top of Ficoll-Hypaque mixture or Lymphoprep.

3. Layering blood requires practice. The initial 1 mL is pipeted with great care down the side of the tube with the plastic pipet tip just above the meniscus of the M-PRM. If a small bubble is then produced in the blood by allowing 10–20 µL of air to enter the pipet barrel and then expelling this to form the bubble, this disperses further blood over a larger area keeping the blood–medium interface intact.

4. It is best to make up all reagents fresh on the day of assay: this can be done while the blood is being separated.

5. Resuspension of cells in HBSS/BSA seems to help prevent cell clumping. Many authors put cells on ice during separation, but I have never found this to make any significant difference to their background CL readings prior to stimulation.

6. Choice of luminometer is often dictated by circumstances (i.e., funding). In general, all microplate luminometers are suitable. This work was performed with a Dynatech ML1000 luminometer with internal temperature control, the ability to cycle (making measurements of each well every 5 min), and dump data to an on-line microcomputer *(5,7,8)*. There is still debate over shaking or stirring. Both enhance phagocyte CL, especially with particulate stimuli, but mechanical activation may occur, and neither are really necessary.

7. Measurement of background CL before stimulants are added is necessary to ensure that the cells have not been activated during separation *(7)*. The most common cause of this is contamination of glassware, plasticware, or reagents by bacterial toxins. Cells from some seriously ill patients may be particularly easily activated owing to priming of phagocytes as part of septic processes *(9,10)*.

References

1. Segal, A. W. (1993) Structure of the NADPH-oxidase: membrane components. *Immunodeficiency* **4,** 167–179.
2. Allen, R. C., Sternholm, R. L., and Steele, R. H. (1972) Evidence for the generation of an electronic excitation state(s) in human polymorphonuclear leukocytes and its participation in bacterial activity. *Biochem. Biophys. Res. Commun.* **47,** 679–684.
3. Blair, A. L., Cree, I. A., Beck, J. S., and Hastings, M. J. G. (1988) Measurement of phagocyte chemiluminescence in a microtitre plate format. *J. Immunol. Methods* **112,** 163–168.
4. Lundqvist, H., Kricka, L. J., Stott, R. A., Thorpe, G. H., and Dahlgren, C. (1995) Influence of different luminols on the characteristics of the chemiluminescence reaction in human neutrophils. *J. Biolumin. Chemilumin.* **10,** 353–359.

5. McCafferty, A. C., Cree, I. A., and McMurdo, M. E. T. (1995) The influence of age and sex on phagocyte chemiluminescence. *J. Bioluminescence Chemiluminescence* **10,** 41–48.

6. Ramage, L., Blair, A. L., Cree, I. A., and Dhillon, D. P. (1993) Effect of salmeterol on polymorphonuclear leukocyte (PMNL) function in vitro. *J. Bioluminescence Chemiluminescence* **8,** 247–252.

7. Cree, I. A. (1991) Assays of human phagocyte function using microtitre plate luminometers, in *Bioluminescence and Chemiluminescence. Current Status.* (Stanley, P. E. and Kricka, L. J., eds.), John, Chichester, pp 261–264.

8. Ramage, L., Cree, I. A., and Dhillon, D. P. (1994) Comparison of salmeterol with placebo in mild asthma: effect on peripheral blood phagocyte function and cytokine levels. *Int. Arch. Allergy. Immunol.* **105,** 181–184.

9. Cree, I. A. (1993) Longitudinal studies of phagocyte chemiluminescence in patients with lung disease, in *Chemiluminescence and Bioluminescence. Status Report* (Szalay, A., Kricka, L. J., and Stanley, P. E., eds.), John, Chichester, 451–455.

10. Moussa, K., Michie, H. J., Cree, I. A., MacCafferty, A., Winter, J. H., and Brown, R. A. (1994) Phagocyte function and cytokine production in community-acquired pneumonia. *Thorax* **49,** 107–111.

17

Detection of Oxidants Using
lux Fusions to Oxidative Stress Promoters

Shimshon Belkin

1. Introduction

This chapter provides basic instructions in the use of genetically engineered *Escherichia coli* strains that luminesce in response to the presence of oxidants. These strains carry plasmid derivatives of pUCD615 *(1)* in which *Vibrio fischeri luxCDABE* is driven by selected promoters of genes responsive to oxidative stress. The construction of such plasmids has been reported before *(2–5)* and will not be detailed here.

Bacteria have developed several complex mechanisms, with a considerable degree of overlap, to allow them to cope with potential oxidative hazards *(6)*. Two *E. coli* global regulatory circuits that appear to be dedicated to the fight against deleterious oxygen species are controlled by *oxyR* and *soxRS*, responsible for the H_2O_2 and superoxide responses, respectively. Other global regulators involved are *rpoH (7)*, *rpoS (8)*, *soxQ (9)*, *fur, arcA, fnr (10)*, and possibly other circuits. In the plasmids, whose use is described here, the *lux* operon was fused to promoters of genes belonging to the first two regulatory circuits: *katG*, the catalase (HPI) gene under the control of *oxyR*, and *micF*, a member of the *soxRS* regulon. The nature of the selected promoters dictates the type of oxidants detected: since *oxyR* is primarily induced by hydrogen peroxide *(6)*, cells containing the *katG'::lux* fusion are expected to emit light in the presence of H_2O_2 and related compounds. Similarly, the *micF'::lux* fusion should report the presence of superoxide radicals, the active oxygen species the *soxRS* regulon is known to combat *(6)*. Similar fusions can also be readily constructed for other promoters, which may allow the detection of other oxidative species (*see 2–5* for general strategy).

From: *Methods in Molecular Biology, Vol. 102: Bioluminescence Methods and Protocols*
Edited by: R. A. LaRossa © Humana Press Inc., Totowa, NJ

2. Materials

1. Luminometer: Numerous instruments are available for the sensitive quantification of photon fluxes (*see 11* for a recent review); to a certain extent, details of the experimental procedure will be dictated by luminometer choice or availability. The methodology described in this chapter is structured to suit a microtiterplate luminometer, which allows sensitive quantification of the light emitted from all 96 wells of a standard size microtiter dish. Some simple modifications would be needed to adapt the technique to other light-measuring devices. Two microtiter plate instruments were used: a Dynatech (Chantilly, VA) ML3000 and an Anthos Labtech (Salzburg, Austria) Lucy 1. Both allow incubation of the plate at a controlled temperature with shaking, and reading of the luminescence emitted from each of the wells at predetermined intervals. Similar instruments are available from other manufacturers, and the use of the specific models mentioned above does not imply their superiority to others.

2. Microtiter plates: Standard-size 96-well plates are used. To prevent light transfer between wells, it is essential that opaque (either white or black) plates are selected. Such plates are available from various manufacturers, and the author did not find one brand to be preferable. The methods outlined below refer to the standard A–H and 1–12 notation for rows and columns, respectively.

3. Bacterial strains: A large number of plasmids are available which combine different *E. coli* stress promoters with *V. fischeri lux* genes. Only two of those are referred to in the text below.

   The two *E. coli* tester strains used were DPD2511 and DPD2515, containing the *katG'::lux* and the *micF'::lux* fusions, respectively, in host strain RFM443 *(12)*. The design and construction of the former have been previously described in detail *(4)*, and the use of the latter has already been reported in several publications *(13,14)*.

   Growth and maintenance of these strains are routinely carried out in LB medium *(15)* in the presence of kanamycin (25 mg/L). Although routine strain maintenance can be carried out at 37°C, growth at 26°C is recommended in preparation for experiments (*see* **Note 1**).

4. Experimental media: Two simple sterile media are used in the procedures below— single- and double-strength LB broth, without antibiotics.

5. Sample preparation: The procedure below differentiates between assaying the effects of known chemicals, from which concentrated stock solutions can be prepared, and aqueous samples of unknown composition.

   As a rule, if permitted by the compound's solubility, prepare a concentrated stock solution at a concentration 50- to 100-fold higher than the highest concentration you wish to test, in a pH 7.0 buffer. If such a concentrated solution cannot be prepared and a lower dilution will be required, prepare your stock daily in LB.

   For unknown aqueous samples, the only pretreatments required are neutralization to pH 7.0 and, if necessary, clarification by filtration or centrifugation.

6. Photometer or colorimeter: To obtain reproducibility, it is essential that cells from a constant growth phase or physiological state are routinely used. It is therefore

important that growth is monitored for several hours until the desired cell density is reached, by any device allowing the determination of optical density. An old-fashioned Klett-Sumerson colorimeter, coupled with side-armed growth flasks, is very practical.

3. Methods

3.1. Experimental Design

It is important that the plate is designed well in advance, preferably with the aid of a blank 8 × 12 table. The procedure below is designed for testing seven double dilutions of each sample down microtiter plate column (*see* **Notes 2** and **3**). Duplicate tests, preferably in adjacent columns, are routinely performed. In this format, six different compound/strain combinations can be tested in a single plate.

For each compound to be tested, the highest concentration will be in row A. Medium volumes and sample concentrations necessary to add to the wells in row A are calculated, so that the final volume will be 100 µL and the compound's concentration will be twofold higher than the highest concentration to be tested (for example, 4 µL of a 5000 mg/L (0.5%) H_2O_2 stock solution, into 96 µL LB, to yield a temporary concentration of 200 mg/L in 100 µL).

3.2. Plate Preparation—Known Chemicals

1. Prepare the plate with the various dilutions of the tested samples in advance, so that when the cells are ready, they can be immediately introduced into the plate and luminescence monitoring initiated. It is recommended to do this approx 30–60 min before the cells are expected to reach their desired density (*see* **Subheading 3.4.**).
2. Use a brand-new opaque white microtiter plate.
3. Clearly mark column pairs destined for each compound or strain.
4. Place 50 µL LB medium in all wells in rows B–H.
5. In the wells in row A, place the appropriate amounts of medium and sample to yield a concentration twofold higher than the highest concentration to be tested (*see* **Subheading 3.1.**) in 100 µL.
6. Generate a twofold dilution series "downward" along the columns, by progressively transferring 50 µL from well to well, lightly mixing at each step. Avoid splatter. Discard 50 µL after mixing the contents of row G. **Do not touch row H.**
7. The plate should now hold 50 µL in all wells, each column containing a dilution series of the tested compound. Row A holds a concentration twofold higher than the highest concentration to be tested, and row G twofold higher than the lowest. In the example presented in **Subheading 3.1.** above, these values will be 200 and 3.125 mg/L, respectively (final concentrations will be halved after addition of cells; *see* **Notes 4** and **5**). Row H will serve as the zero control.

3.3. Plate Preparation—Unknown Samples

1.–4. as in **Subheading 3.2.** *above.*
5. In the wells in row A, place 50 µL of a twofold concentrated LB medium.
6. Add 50 µL from the tested sample to wells in row A to yield a temporary concentration of 50% in single-strength LB.
7. Generate a twofold dilution series "downward" along the columns, by progressively transferring 50 µL from well to well, lightly mixing at each step. Avoid splatter. Discard 50 µL after mixing the contents of row G. **Do not touch row H.**
8. The plate should now hold 50 µL in all wells, each column containing a dilution series of the tested compound, from 50% in row A to 0.78% in row G. Row H, containing LB only, will serve as the zero control. Actual sample concentrations will be halved after addition of cells, to range between 25 and 0.39%.

3.4. Cell Preparation and Initiation of Experiment

1. Grow cells overnight, in LB medium, at 26°C, with shaking, in the presence of 25 mg/L kanamycin (or 50 mg/L ampicillin; resistance to both antibiotics is coded for by the plasmid containing the promoter::*lux* fusion).
2. Dilute the cells 100-fold into fresh LB without antibiotics (*see* **Note 7**), and reincubate at 26°C with shaking.
3. Follow growth of the culture for a few generations, until a predetermined cell concentration is reached. We have routinely used a cell density yielding 20–40 Klett units (filter 54), approximately corresponding to $2–4 \times 10^8$ cells/mL. Time from inoculation is approx 3–4 h (generation time of *E. coli* at 26°C is close to 1 h).
4. Approximately 30–60 min before the expected cell density is reached, prepare the plate with the tested samples according to **Subheadings 3.2.** or **3.3.** (*see* **Notes 8** and **9**).
5. Remove the culture from the shaker, add 50 µL to each of the wells in the freshly prepared plate, insert into the luminometer, and immediately start monitoring luminescence.

3.5. Data Collection and Analysis

1. The luminometer should be set up for readings at intervals of 5–15 min, at 26°C, with intermittent shaking. Collect luminescence data for 90 or 180 min for strains DPD2511 and DPD2515, respectively (*see* **Note 6**).
2. Different luminometers present data in different modes, not always immediately amenable to simple plotting of time-courses and dose–responses. It is therefore important to first transform the data generated by the luminometer into a table with the time-points in the first column and the luminescence reading for each well, column by column, in the following 96 columns. In this seemingly trivial suggestion often lies the solution for a reasonable viewing of the enormous amount of data that may be generated in a single run (*see* **Note 10**). A simple Excel (or equivalent) spreadsheet or macro can be designed for this purpose and

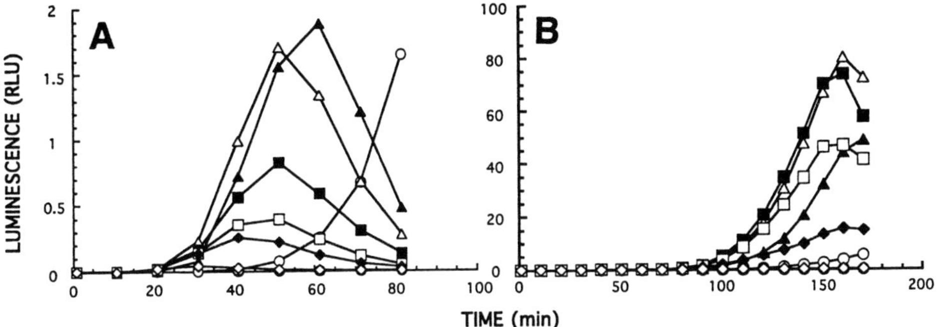

Fig. 1. Time-course curves of luminescence development in strain DPD2511 *(katG'::lux)* in response to H_2O_2 **(A)** and in strain DPD2515 *(micF'::lux)* in response to methyl viologen **(B)**. H_2O_2 (mg/L): —●— 100, —○— 50, —▲— 25, —△— 12.5, —■— 6.3, —□— 3.1, —◆— 1.6, —◇— 0. Methyl viologen (mg/mL): —●— 5.00, —○— 2.50, —▲— 1.25, —△— 0.63, —■— 0.31, —□— 0.16, —◆— 0.08, —◇— 0.00.

routinely used (*see* **Subheading 4.**). If samples were prepared in duplicate, as recommended above, the program should also calculate averages, yielding 48 data sets.

3. Plot the kinetic response of the cells for each sample concentration (*see* **Note 9**). **Figure 1**, for example, presents the luminescence induced by hydrogen peroxide and methyl viologen (a redox cycling agent that intracellularly generates super-oxide radicals). The data can be exhibited either as actual luminescence value (normally presented as the arbitrary light units of the specific instrument used, as in **Fig. 1**) or as the ratio of the luminescence of the induced samples to that of the uninduced control (response ratio).

4. Plot the responses as a function of sample concentration. In **Fig. 2**, as an example, the data from **Fig. 1** are used; each of the curves in that figure is represented by one point only—the response ratio at the peak; alternatively, luminescence at a selected time-point can be used.

5. From plots such as the ones in **Fig. 2**, the characteristic responses of the tester strains to the tested samples can be quantified. One such parameter is the maximal response ratio obtained (a ratio of 386, at 625 mg/L methyl viologen, for DPD2515; a ratio of 277, at 50 mg/L H_2O_2, for DPD2511). Another convenient parameter is the sample concentration causing a twofold increase in luminescence, recently referred to as EC_{200} *(15,16)*. The lower this value is, the more sensitive the strain, or the more oxidative the sample. As demonstrated in **Fig. 2**, it can be determined from the intersection of the "upward" slope of the dose–response curve or of its extrapolation, with the line representing a response ratio of 2. A more detailed discussion of the EC_{200} concept and its calculation can be found in refs. *(16)* and *(17)*.

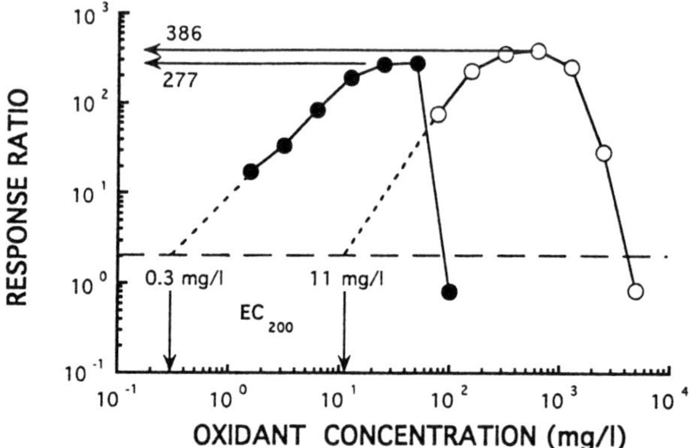

Fig. 2. Dose–response curves for the response of strain DPD2511 *(katG'::lux)* to H$_2$O$_2$ (●) and of strain DPD2515 *(micF'::lux)* to methyl viologen (○).

4. Notes

1. The assays described above are carried out at 26°C, a compromise between the optimal temperature for the host (37°C) and that for the luminescence apparatus (<20°C). Changes in the host strain or in the source of the luminescence genes can make this compromise unnecessary *(5)*.

2. In the preparation of the samples in the microtiter plates, as well as for the subsequent addition of the cells, the use of adjustable 12-channel multipipets and suitable reservoirs is highly recommended.

3. It is essential that the effect of each compound or sample is tested in a dilution series rather than in a single concentration for two main reasons:
   a. To characterize the dose-response;
   b. To cover a broad concentration spectrum, and thus identify the concentration above which a toxic effect may occur, masking or abolishing the induction.

4. It may sometimes be necessary to screen a broader concentration range, and plate preparation may thus deviate from the recommendations presented here. Different volumes can be used to achieve different dilution series, and any other dilution factor may be used.

5. Similarly, if fewer than seven concentrations are sufficient, the plate can be designed differently. For instance, if the testing of only 4 concentrations/sample is appropriate, two separate sets can be prepared in rows A–D and E–H, doubling the number of samples tested per plate. In this case, one four-well segment should be kept sample-free to serve as a common zero control.

6. For a significant increase in the number of samples simultaneously screened, the following approach may be adopted: prepare a large number of plates, and incubate them not in the luminometer, but rather in a 26°C incubator with slow shaking. In this case, all plates are in turn periodically taken out of the incubator, a single reading is taken in the luminometer, the data are immediately saved, and the plate is returned to the

incubator. Since a single reading may last up to 2 min, at least six plates can be sequentially monitored while maintaining a 15-min interval between readings. Data collation is somewhat more laborious (unless a special computer program is generated), but for the screening of numerous samples, the effort is certainly worthwhile.

7. The presence of some antibiotics drastically affects microbial bioluminescence; thus, although the presence of the drugs is essential for routine strain maintenance, it should be avoided during the actual assay.

8. Designing the appropriate controls to validate the oxidative nature of the tested samples is left to the imagination of the researcher, as dictated by the individual study. The author routinely uses catalase (600 U/mL) to confirm activation by peroxides. An altogether different type of control may be obtained by transforming, with the same plasmids, *E. coli* strains of different genetic backgrounds (for example, the *micF'::lux* plasmid into an SOD mutant; *18*).

9. In all microtiter plate luminometers tested by the author, there is a potential danger of a "spillover" of photons to adjacent wells. This may amount to approx 0.1% of the original luminescence, and may therefore be considered a problem if luminescence is at least a thousand-fold higher then the controls. In such high-luminescence instances, it is recommended that empty columns of wells separate between samples or strains.

10. An enormous amount of data can be generated even by a simple luminometer run: a plate read every 10 min for 3 h will yield nearly 2000 data points. To simplify handling of these data, it is recommended that a simple procedure is used to assimilate and reduce all the numbers to a conveniently handled format. Although many commercially available scientific data handling programs can carry this out, even a simple Excel (or equivalent) macro is sufficient. This macro should, ideally, carry out the following functions:
    a. Calculate averages of duplicates;
    b. Rearrange data, if necessary, in a plotable time-dependent matrice;
    c. Calculate response ratios for each time point;
    d. Select maximal luminescence values obtained for each sample concentration;
    e. Select maximal response ratios calculated for each sample concentration; and
    f. Plot, on demand, the desired time-course and dose–response figures.

Acknowledgments

The procedures outlined in this chapter were developed in the laboratory of, and in cooperation with, R. A. LaRossa of DuPont Central Research and Development, Wilmington, DE. His contribution, and that of his coworkers, T. K. Van Dyk, D. R. Smulski, and A. C. Vollmer (Swarthmore College, PA) were in many ways more significant than that of the author and are gratefully acknowledged.

References

1. Rogowsky, P. M., Close, T. J., Chimera, J. A., Shaw, J. J., and Kado, C. I. (1987) Regulation of the *vir* genes of *Agrobacterium tumefaciens* plasmid pTiC58. *J. Bacteriol.* **169**, 5101–5112.

2. Van Dyk, T. K., Majarian, W. R., Konstantinov, K. B., Young, R. M., Dhurjati, P. S., and LaRossa, R. A. (1994) Rapid and sensitive pollutant detection by induction of heat shock gene-bioluminescence gene fusions. *Appl. Environ. Microbiol.* **60,** 1414–1420.

3. Van Dyk, T. K., Smulski, D. R., Reed, T. R., Belkin, S., Vollmer, A. C., and LaRossa, R. A. (1995) Responses to toxicants of an *Escherichia coli* strain carrying a *uspA'::lux* genetic fusion and an *E. coli* strain carrying a *grpE'::lux* fusion are similar. *Appl. Environ. Microbiol.* **61,** 4124–4127.

4. Belkin, S., Smulski, D. R., Vollmer, A. C., Van Dyk, T. K., and LaRossa, R. A. (1996) Oxidative stress detection with *Escherichia coli* bearing a *katG'::lux* fusion. *Appl. Environ. Microbiol.* **62,** 2252–2256.

5. Van Dyk, T. K. and Rosson, R. H. Chapter 7, this vol.

6. Demple, B. 1991. Regulation of bacterial oxidative stress genes. *Ann. Rev. Genet.* **25,** 315–337.

7. Neidhardt, F. C. and VanBogelen, R. H. (1987) Heat shock response, in *Escherichia coli and Salmonella typhimurium: Cellular and Molecular Biology* (Neidhardt, F. C., Ingraham, J. L., Low, K. B., Magasanik, B., Schaechter, M., and Umbarger, H. E., eds.), ASM, Washington, DC, pp. 1334–1345.

8. Lange, R. and Hengge-Aronis, R. (1991) Identification of a central regulator of stationary-phase gene expression in *Escherichia coli. Mol. Microbiol.* **5,** 49–59.

9. Greenberg, J. T., Chou, J. H., Monach, P. A., and Demple, B. (1991) Activation of oxidative stress genes by mutations at the *soxQ/cfxB/marA* locus of *Escherichia coli. J. Bacteriol.* **173,** 4433–4439.

10. Compan, I. and Touati, D. (1993) Interaction of six global transcription regulators in expression of manganese superoxide dismutase in *Escherichia coli* K-12. *J. Bacteriol.* **175,** 1687–1696.

11. Stanley, P. E. (1996) Commercially available luminometers and imaging devices for low-light level measurements and kits and reagents utilizing bioluminescence or chemiluminescence: survey update 4. *J. Bioluminescence Chemiluminescence* **11,** 175–191.

12. Menzel, R. (1989) A microtiter plate-based system for the semiautomated growth and assay of bacterial cells for β-galactosidase activity. *Anal. Biochem.* **181,** 40–50.

13. Belkin, S., Vollmer, A. C., Van Dyk, T. K., Smulski, D. R., Reed, T. R., and LaRossa, R. A. (1994). Oxidative and DNA damaging agents induce luminescence in *E. coli* harboring *lux* fusions to stress promoters, in *Bioluminescence and Chemiluminescence: Fundamentals and Applied Aspects* (Campbell, A. K., Kricka, L. J., and Stanley, P. E., eds.), John Wiley, Chichester, pp. 509–512.

14. Belkin, S., Van Dyk, T. K., Vollmer, A. C., Smulski, D. R., and LaRossa, R. A. (1996). Monitoring sub-toxic environmental hazards by stress-responsive luminous bacteria. *Environ. Toxicol. Water Quality* **11,** 179–185.

15. Miller, J. H. (1972) Experiments in Molecular Genetics. Cold Spring Harbor Laboratory, Cold Spring Harbor, NY.

16. Belkin, S., Smulski, D. R., Dadon, S., Vollmer, A. C., Van Dyk, T. K., and LaRossa, R. A. A panel of stress-responsive luminous bacteria for the detection of specific classes of toxicants. *Water Res.,* in press.

17. Belkin, S. Stress responsive luminous bacteria for toxicity and genotoxicity monitoring, in *Microscale Aquatic Toxicology—Advances, Techniques and Practice* (Wells, P. G., Lee, K., and Blaise, C., eds.), CRC Lewis, Florida, in press.
18. Dukan, S., Dadon, S., Smulski, D. R., and Belkin, S. (1996). Hypochlorous acid activates the heat shock and *soxRS* systems of *Escherichia coli*. *Appl. Environ. Microbiol.* **62,** 4003–4008.

V

ENVIRONMENTAL APPLICATIONS

18

Luciferase-Based Measurement
of Water Contaminants

Michael A. Costanzo, Julie Guzzo, and Michael S. DuBow

1. Introduction

Aquatic environments are continually being subjected to a wide variety of contaminants from industrial, agricultural, and municipal sources. These contaminants include bacterial and parasitic pathogens, as well as toxic organic and inorganic chemical compounds.

As a potential carrier of pathogenic organisms, water can endanger human health and life. Many conventional tests are currently available to monitor microbial contamination. However, owing to the slow growth of many pathogenic microbes, conventional methodology does not allow for rapid identification of such contaminants, necessary for the prevention of disease and maintenance of the environment. Luciferase, an enzyme that emits light as a product of its reaction, is being used to develop biosensors that detect these pathogens more sensitively and expediently. One approach is the use of "reporter bacteriophages" containing the luciferase encoding *lux* (prokaryotic) or *luc* (eukaryotic) genes. This bacterial detection system was first described in 1987, when Ulitzur and Kuhn *(1)* cloned the bacterial luciferase *(lux)* genes into the bacteriophage λ genome. More recently, two groups have used this system to identify drug-resistant strains of *Mycobacterium tuberculosis* or *Listeria spp.* in contaminated food *(2,3)*. These species-specific, recombinant *lux/luc*-containing phages work by infecting specific bacterial pathogens present in food or human sputum, and on analysis via luminometry, the presence of viable pathogens is detectable through infected phage-based luciferase expression increases in light emission. This system is extremely sensitive, rapid, and relatively inexpensive. Although these pathogens are not water-borne, it may be possible to use this method for the detection of water-borne pathogens in

From: *Methods in Molecular Biology, Vol. 102: Bioluminescence Methods and Protocols*
Edited by: R. A. LaRossa © Humana Press Inc., Totowa, NJ

aquatic environments, especially after a rapid filtration step to collect and concentrate bacteria.

Water pollution from toxic chemicals has also received a great deal of attention owing to increased industrialization and a concomitant higher demand for chemicals *(4)*. Consequently, many countries are now facing serious ecological and toxicological problems resulting from the release of toxic effluents into aquatic environments *(5)*. These chemical contaminants include both organic and inorganic compounds. Once released into aquatic systems, the effects of these chemicals may be observed through several seasonal cycles, after which they can become biologically unavailable through incorporation into organic and inorganic particles in sediments *(6)*. The concentrations of chemical contaminants in aquatic environments can be assessed via numerous detection methods. These include atomic absorption spectroscopy (AAS) and chromatographic analysis tools, such as high-performance liquid chromatography (HPLC) and gas chromatography (GC), which measure specific chemical components in water samples *(7)*. Fiber-optic research has also expanded in this area *(8–10)*. For example, one assay has been developed to measure benzo(a)pyrene concentrations based on the interaction of benzo(a)pyrene with antibody-coated optical fibers. Antibody-bound benzo(a)pyrene is detected by measuring laser-induced fluorescence *(11,12)*. Continued research in this area will undoubtedly provide biosensors for the detection of a wide array of other chemical compounds. However, the analytical methods described above still retain several disadvantages. First, they can be costly and labor-intensive. Second, they are unable to determine toxicity to living organisms. For example, although a toxic chemical is present in a sample, it may not be present in a form that is harmful to living cells. Finally, analytical methods may not fully account for possible synergistic or antagonistic effects on living organisms by chemicals present in complex mixtures *(13)*.

As a result of these shortcomings, a trend has developed toward the use of living organisms to screen for toxic substances, since biologically relevant concentrations can be detected. These tests have been adopted using mostly fish and invertebrates as test organisms *(14)*. Although effective, these assays can also be extremely labor-intensive and expensive. Therefore, scientists have now turned to short-term toxicity assays employing microorganisms as model test organisms. These toxicity tests have been based on many criteria, such as bioluminescence, motility, growth viability, ATP production, oxygen uptake, nitrification, and heat production *(15)*. Assays based on the inhibition of light emission in bioluminescent bacteria include Microtox™, Biotox™, and Lumistox™ *(13,16–18)*. Other assays are based on growth inhibition of algal cells *(19)*. Commercially available toxicity tests include MetPad™, PolyTox, ECHA Biocide Monitor, and Toxi-Chromotest. MetPad™ is a bioassay kit designed specifically for the detection of heavy metal toxicity. This test is based on enzyme inhibition in a mutant strain of *Escherichia coli* caused by bioavailable heavy metals in aqueous samples *(15)*. A variation on this assay is the MetPlate™ assay, which is performed in microtiter plates *(20)*. The PolyTox kit utilizes a mixture of bacterial cultures isolated from wastewater. The assay is based on the reduction of respiratory activity of rehydrated bacterial cultures in the presence of contaminants *(21)*. The ECHA Biocide Monitor is a dipstick test for monitoring toxicity. It incorporates a test microorganism, present on the dipstick, and an oxidoreduction dye, tetrazolium salt, which is used to indicate the growth of the test bacterium. Therefore, toxicity detection is based on the inability of the bacteria to grow in the presence of contaminants *(22)*. Finally, the Toxi-Chromotest involves the inhibition of β-galactosidase biosynthesis by toxic metals *(15)*.

Many *lux*-based bacterial systems for the detection of metals and organics have been recently developed. One of these is based on the fact that expression of certain heat-shock genes is triggered by general stress responses, such as exposure to environmental toxins *(23)*. The *lux* operon from *Vibrio fischeri* was used to prepare a transcriptional fusion to two heat-shock promoters creating biosensors that emit light on exposure to various contaminants *(23)*. A series of gene fusion bioassays have also been developed to assess the mutagenic potential of certain chemicals. The most commonly used mutagenicity assay is the *Salmonella*/microsome assay, otherwise known as the Ames Test *(24)*. This assay employs histidine auxotrophs of *Salmonella typhimurium,* which revert to prototrophy and form colonies on media deficient in histidine. Mutagenicity is therefore detected by increases in the frequency of revertant colonies. Another example is the λ Inductest, which involves the use of bacterial prophages. Lysogenic *E. coli*, containing a DNA-damage-inducible λ prophage, is grown in the presence of a particular aqueous sample. If DNA damage occurs, the "SOS response" for DNA repair is triggered, and the prophage lytic cycle is induced. The appearance of increased plaque-forming units, therefore, indicates the presence of mutagens in the water sample *(25)*. A similar technique, which combines bacteriophage-induction and luciferase gene fusions, has also been developed *(26)*. In this assay, bacteriophages were constructed to contain a promoterless *luxAB* reporter cassette fused randomly to different locations in the phage genome. The recombinant phages were then used to lysogenize specific bacterial species, and one lysogen was chosen that exhibited increased luminescence in the presence of mutagens. This bacterial lysogen can therefore be grown in the presence of an aqueous sample. If the sample contains compounds that are mutagenic, induction of the prophage lytic cycle will be triggered, the luciferase genes will be expressed, and light will be emitted, thus providing for sensitive detection of mutagenic compounds if used with appropriate controls for cytotoxicity *(26)*. Another bacterial genotoxicity assay, the Mutatox™ test, employs dark mutants of *Vibrio fisheri* and determines the ability of chemicals to restore luminescence through the induction of mutations *(27,28)*.

The tests described above measure overall toxicity or genotoxicity to living organisms on exposure to chemical contaminants. Our laboratory, along with others *(29–38)*, has taken a different approach. We have created bacterial biosensors that detect biologically relevant concentrations of specific classes of numerous chemical compounds. These biosensors are based on luminescent reporter gene fusion technology. In order to identify genes in the *E. coli* genome whose expression changes in the presence of aqueous contaminants, a modified Tn*5* transposon, containing the *luxAB* genes (*see* **Note 1**) from *Vibrio harveyi* (*see* **Note 2**) and a tetracycline resistance gene, was transposed into the *E. coli* chromosome *(30)*. Three thousand *E. coli* clones that contained the Tn*5–luxAB* cassette inserted in single, random chromosomal locations were selected on the basis of tetracycline resistance. These clones were then screened in the absence and presence of toxic chemicals. Clones that exhibited changes in luminescence in the presence of water-soluble contaminants were selected. The light emitted from these biosensors in the presence of water contaminants can be qualitatively or quantitatively observed (*see* **Note 3**). Similar methods have been used by other groups in the development of biosensors that detect bioavailable concentrations of mercury and naphthalene *(29,33–35,37)*. These biosensors are ideal, since they not only test for specific compounds, but are also extremely sensitive, cost-effective, and the visible light emitted is easily quantifiable. Furthermore, continuous-flow systems have now been constructed incorporating these biosensors to continually measure luminescence in contaminated water samples *(7,35)*. The assays outlined here will complement traditional analytical techniques as well as short-term toxicity assays presently used to monitor environmental contaminants. Furthermore, these are simple, rapid, reproducible, sensitive, and inexpensive assays for the evaluation of water quality. Bacteria can be easily stored and rapidly grown, and toxic effects can be measured on several generations over a relatively short period of time. One problem that has still not been overcome by microbial bioassays is that no single microbioassay can detect all categories of environmental toxicants with equal sensitivity *(5)*. Therefore, a battery of test approaches, incorporating several microbial assays, has been recommended *(39)*.

2. Materials

1. *E. coli strains: E. coli* biosensor strains, LF20110 (Guzzo et al., manuscript in preparation), LF20111 *(31)*, LF20112 *(36)*, LF20113 *(32)*, LF20116 *(32)*, DMSO *(40)*, and TBT1-3 *(40)*, which detect iron and aluminum, heavy metal ions, nickel,

selenium (two clones), dimethyl sulfoxide, and tributyltin, respectively, were derived from *E. coli* strain DH1 containing the Amp^R, ColE1 RNA1-overproducing plasmid pTF421 *(41)* and the Tet^R transcriptional *luxAB* fusion cassette *(30)*. *E. coli* biosensor strain LF20012 *(38)*, which detects toxic arsenic and antimony oxyanions, was derived from *E. coli* strain 40 containing the Tet^R transcriptional *luxAB* fusion cassette in the chromosomal *ars* operon *(arsB::luxAB) (38)*.

2. Chemicals: Test chemicals used for identification of contaminant-responsive clones were obtained from several sources. Sodium arsenate was purchased from Fisher Scientific (Montréal, Canada) and sodium arsenite was obtained from American Chemicals (Montréal, Canada). Aluminum chloride, nickel sulfate, and sodium selenite were obtained from Anachemia Canada (Montréal, Canada). Ferric chloride was purchased from BDH (St. Laurent, Canada), and tributyltin chloride was obtained from Aldrich Chemical (Milwaukee, WI). Finally, sodium selenate and dimethyl sulfoxide were acquired from Sigma (St. Louis, MO). All chemical solutions were prepared using deionized H_2O as the solvent, followed by filter sterilization.

3. Luria-Bertani (LB) broth: 1% (w/v) Bacto-tryptone, 1% (w/v) NaCl, 0.5% (w/v) Bacto-yeast extract.

4. 2X LB Broth: 2% (w/v) Bacto-tryptone, 2% (w/v) NaCl, 1% (w/v) Bacto-yeast extract.

5. LB agar: 1% (w/v) Bacto-tryptone, 1% (w/v) NaCl, 0.5% (w/v) Bacto-yeast extract, 1.5% (w/v) agar.

6. 2X LB agar: 2% (w/v) Bacto-tryptone, 2% (w/v) NaCl, 1% (w/v) Bacto-yeast extract, 3% (w/v) agar.

7. Film: Agfa Curix RP-1 or Kodak XAR-5 X-ray film.

3. Methods

3.1. Bioassay—Solid Media

1. Grow 1-cm^2 patches of cells on a Petri dish containing 25 mL of LB agar (*see* **Note 4**) containing antibiotics appropriate to the particular strain. The pH of the LB solution should be adjusted accordingly (*see* **Note 6**). Add 0.01 mg/mL tetracycline and, if required, 0.4 mg/mL ampicillin. Grow overnight at 37°C.

2. Prepare 2X LB agar (*see* **Note 5**) and autoclave. Add an equal volume of appropriate dilutions of the putative contaminated water sample, and adjust the pH accordingly (*see* **Note 6**), along with the appropriate antibiotics (i.e., 0.01 mg/mL tetracycline and, if required, 0.4 mg/mL ampicillin). Once solidified, invert plates and allow to dry overnight at 32°C.

3. Replica plate (*see* **Note 7**) mastered colonies onto media containing LB agar alone plus the appropriate antibiotics, as well as LB agar containing various amounts of the contaminated water sample, along with the appropriate antibiotics. The order of replica plating should be as follows: LB agar plates alone followed by LB agar containing water contaminants. Grow at various times at 37°C (*see* **Note 8**).

4. Add aldehyde (Aldrich) to the lids of the Petri dishes (*see* **Notes 1** and **9**). Seal Petri dishes with parafilm.

5. Invert plates and expose to X-ray film at 23°C (*see* **Note 10**).

6. X-ray films are developed at various times, depending on the nature of the particular biosensor (*see* **Note 11**).
7. Develop the X-ray film, preferably using an X-ray film processor to ensure reproducibility. Luminescent clones will appear as dark spots on the film, with the intensity of the dark spot corresponding to the amount of light emitted.
8. Quantify light emission by scanning the X-ray film on a flatbed scanner linked to a phosphorimager SF apparatus (Molecular Dynamics, Sunnyvale, CA) and quantify the amount of light emitted from each patch of cells (*see* **Note 3**) using the ImageQuant software program (Molecular Dynamics). **Figure 1** schematically describes the luminescence assay performed on solid media.

3.2. Bioassay—Liquid Media

1. Grow a culture of a contaminant-responsive *E. coli* clone in 10 mL of LB broth (*see* **Note 4**) containing the appropriate antibiotics, and adjust pH of the solution accordingly (*see* **Note 6**). Add 0.01 mg/mL tetracycline and, if required, 0.4 mg/mL ampicillin. Grow overnight at 37°C.
2. Prepare a solution of 2X LB broth (*see* **Note 5**), adjust the pH accordingly (*see* **Note 6**), and autoclave. Dilute the overnight *E. coli* culture 20-fold into the autoclaved solution of 2X LB broth along with the appropriate antibiotics (i.e., 0.01 mg/mL tetracycline and, if required, 0.4 mg/mL ampicillin). Divide the bacterial culture into three flasks, and grow for 1 h to ensure logarithmic growth of the bacteria.
3. To one flask, add 1 vol contaminated water (*see* **Note 6**). To the second, "negative control" flask, add 1 vol of sterile deionized H_2O. To the final flask, add one volume of sterile deionized H_2O containing a known optimum concentration of the particular chemical being assayed. Incubate for 1 h at 37°C (*see* **Note 8**).
4. Measure the absorbance at 600 nm (A_{600}) of the three cultures to ensure that they are all the same (*see* **Note 12**).
5. Dilute aldehyde (Aldrich) in deionized H_2O (*see* **Notes 1** and **9**), and attach it to the injection system of a luminometer (if available) such that the aldehyde is added as a final concentration of 28 μM to the 1-mL sample. If the luminometer does not contain an injection system, aldehyde must be added manually before reading the sample.
6. Aliquot 1-mL samples into three luminometer cuvets, and quantify light emission via luminometry (*see* **Notes 11, 13,** and **14**). **Fig. 2** schematically describes the luminescence assay performed in liquid media.

4. Notes

1. Our *E. coli* biosensor library was constructed using only a portion *(luxAB)* of the luciferase operon *(luxABCDE)* from the marine bacterium, *V. harveyi (30)*. In doing so, the operon is reduced to 2 kb of DNA, as opposed to the 7.5-kb full-sized operon. Since our library contains only the *luxAB* portion of the operon, an aldehyde substrate must be added exogenously for the luciferase enzyme to catalyze the light-emitting reaction (*see* **Note 9**).

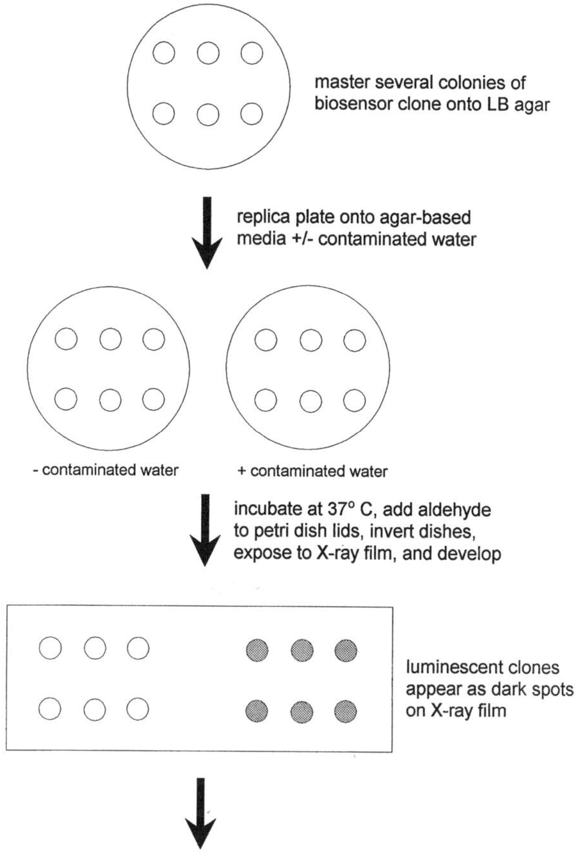

Fig. 1. Schematic diagram depicting a luminescence bioassay performed on solid media. Master 1-cm^2 patches of the biosensor clone onto agar-based media. Replica plate onto media with and without contaminated water. Grow cells, add decyl or dodecyl aldehyde to the covers of the Petri dishes, and expose X-ray films for a specified time. Develop X-ray film and quantify light emission via phosphorimaging.

2. A wide variety of luminescent reporter genes exist and have been used in a number of applications. We have employed the *V. harveyi luxAB* genes for the construction of our biosensor library. However, bacterial luciferases can be obtained from sources other than *V. harveyi*. For example, luciferase enzymes have also been cloned from species, such as *Vibrio fischeri, Photobacterium phosphoreum, Xenorhabdus luminescens,* and *Kryptophanaron alfredi (42,43)*. These bacterial luciferases differ in their temperature stability. For example, *V. fischeri* and *P. phosphoreum* luciferases are more heat-labile than other luciferases *(42)*. The

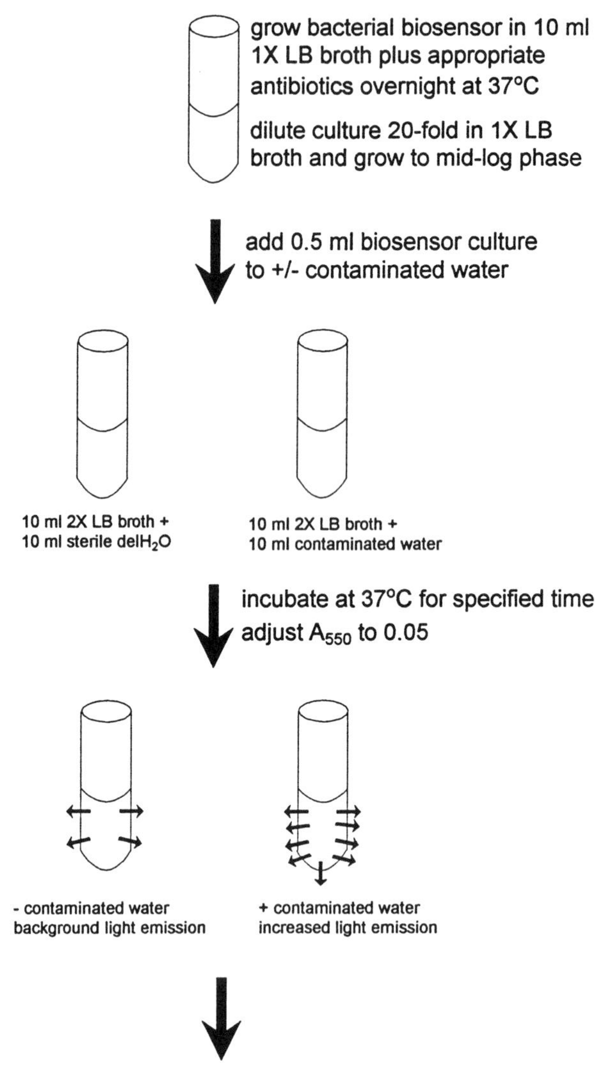

Fig. 2. Schematic diagram depicting a luminescence bioassay performed in liquid media. Grow the bacterial biosensor in nutrient broth overnight. Dilute, grow, add the biosensor culture to samples with and without contaminated water, and incubate for a specified time. Aliquot samples into luminometer cuvets, and measure luminescence using decyl or dodecyl aldehyde.

creation of *luxAB* gene fusions has also been reported in yeast and plant cells *(44)*. Eukaryotic luciferase enzymes are also available. One example is the firefly luciferase, luciferin *(luc) (45–49)*. The *luc* gene from the American firefly, *Photinus pyralis,* was first cloned by deWet et al. *(50,51)*. Subsequently, the luciferase genes from several other firefly species *(Luciola mingrelica, Luciola lateralis)* and click beetle species *(Pyrophorus plagiophthalamus)* were also cloned *(52,53)*.

3. Results obtained from luminescence assays may be both qualitatively and quantitatively measured. Qualitative measurements are possible using high-speed instant photographic film, X-ray film, or simple visual observation. These results can be obtained when luminescence assays are performed on solid LB agar plates. In general, this type of assay should be used for initial screening of the library to isolate responsive clones. Although primarily used to obtain qualitative results, luminescence assays performed on solid media, using X-ray film, can also be quantified via densitometry (*see* **Subheading 3.1.**). Since the luciferase-catalyzed reaction results in the production of light at a wavelength of 490 nm, quantitative results can also be directly obtained when assays are performed in liquid media and photons of light are counted via a luminometer (*see* **Subheading 3.2.**) (*see* **Note 15**). Furthermore, liquid luminescence assays, unlike assays performed on solid media, provide additional information regarding the kinetics of light induction on exposure to water contaminants. For example, time-course experiments can be performed to determine the precise time at which the *luxAB* reporter gene is maximally expressed (or repressed) on exposure to a particular water contaminant *(38)*.

4. Cultures of bacterial biosensors are stored in 25% (v/v) glycerol at –70°C. It is important to restreak cultures for single colonies and to retest these clones, using known concentrations of contaminant, when cultures have been in storage for >2 mo *(54)*. This is to ensure that cultures are pure, since mutant cells may reduce observable light emission.

5. Media used in the luminescence assays may vary. The most common types of media used in the luminescence assay are LB broth and LB agar. However, depending on the nature of the contaminant in question, LB broth and LB agar may not be ideal media, since they are undefined. LB is composed of yeast extract and digested proteins *(55)*. Therefore, many of the components comprising LB may act as ligands capable of interacting with contaminants present in the water sample potentially affecting the results of the assay *(56)*. Furthermore, LB broth and LB agar form a yellowish solution that may lead to decreased detection of luminescence. If any of these problems are encountered, use of a minimal media, such as TCMG broth *(57)*, or defined media, such as M9 broth *(58)*, for liquid luminescence assays and TCMG/noble agar or M9/noble agar for solid luminescence assays is suggested. Finally, deionized or double-deionized water should always be used in the preparation of media to limit the number of excess ions present in the final solution.

6. It has been shown that the pH of water strongly affects heavy metal solubility and speciation *(59)*. Normally, at acidic pH, metals exist as free cations in solution. However, increases in pH can lead to precipitation of metals as insoluble hydroxides or oxides *(60)*. We have previously shown that low pH (pH 5.5) leads to

increases in the activity of aluminum, resulting in higher luminescence *(61)*. Thus, for optimal results, the pH of the media should be adjusted to 5.5 when growing the aluminum-responsive biosensor (LF2011). When preparing media for growth of other *E. coli* biosensors, the pH may be adjusted by adding 2 *N* NaOH to 1X LB solutions and 4 *N* NaOH to 2X LB solutions.

It has been previously demonstrated that the oxidation-reduction potential (Eh) of aquatic environments also has an impact on toxicity assessment of water samples *(62)*. Metals, when combined with the reducing agent hydrogen sulfide, may form nontoxic metal sulfides *(62)*. The ionic composition of test waters has been shown to affect the speciation and toxicity of heavy metals. For example, hydroxy, chloride, phosphate, carbonate, sulfide, and bicarbonate ions can decrease metal toxicity via precipitation reactions *(63)*. Metals can also form complexes with inorganic anions *(64)*.

Finally, salinity and hardness can also affect metal toxicity. Ajmal and Khan *(65)* previously demonstrated that the toxicity of cadmium to microorganisms decreased with increasing sample hardness. For example, water hardness influences the ability of Cd(II) to bind inorganic anions, causing variations in cadmium toxicity *(66,67)*. This has also been demonstrated for copper toxicity in water samples with elevated calcium hardness *(68)*. The factors that affect heavy metal toxicity in water are extensively reviewed by Kong et al. *(62)*.

7. Replica plating allows for transfer of approximately equal amounts of cells from one plate to another and is therefore a quick means of screening all clones that comprise the biosensor library. Replica plating should be performed on thin plates containing 12.5 mL of LB agar, as opposed to 25 mL of media normally found in 100-mm Petri plates, in order to decrease light dispersion through the agar.

8. Blouin et al. *(69)* demonstrated that when aldehyde is supplied exogenously and dissolved O_2 is in excess, reduced $FMNH_2$ is the only remaining substrate directly required for the luminescence reaction. However, $FMNH_2$ has a relatively short life-time owing to its susceptibility to auto-oxidation. Oxidoreductase enzymes are therefore recruited to maintain $FMNH_2$ in sufficient supply for the luminescence reaction to occur *(69)*. Furthermore, luciferase substrate complexes are thought to promote the formation of long-lived intermediates. Blouin et al. *(69)* suggested the formation of a luciferase–FMNHOOH complex before addition of decanal aldehyde. The presence of decanal aldehyde, along with the luciferase–FMNHOOH complex, drives the reaction to completion, resulting in light emission. Therefore, Blouin et al. *(69)* proposed that if insufficient incubation time is allowed for the complex to form before addition of decanal, light emission will not be optimized.

9. The *luxAB* genes encode the heterodimeric bacterial enzyme, luciferase, which catalyzes a light-emitting reaction in the presence of an aldehyde substrate. It is believed that the natural aldehyde for the luminescence reaction is tetradecanal, since tetradecanal yields the highest luminescent responses catalyzed by bacterial luciferases in vitro *(70–72)*. However, tetradecanal is much less soluble, and gives a lower response in vivo than shorter chained aldehydes on addition to

bacteria expressing the *luxAB* genes. The lower response is presumably owing to the greater difficulty with which longer-chained aldehydes cross cell membranes *(73)*. Therefore, there seems to be a preference for using the shorter, decanal aldehyde over the longer-chained dodecanal aldehyde. However, for the purposes of assays we describe here, dodecanal aldehyde is preferred when luminescent clones exhibit extremely high light production. This avoids light measurements exceeding the maximum relative light units (RLUs) (*see* **Note 14**) measurable by many luminometers.

It has been shown that the luciferase enzyme has two aldehyde binding sites, a catalytic site and an inhibitory site *(69)*. Aldehyde should therefore be supplied in sufficient excess to ensure that it is nonlimiting *(74,75)*. However, at higher concentrations, decanal aldehyde becomes strongly inhibiting, and it is therefore important to use an appropriate concentration to ensure that the concentration remains optimal throughout the course of the assay *(69)*. With respect to biosensor applications, Blouin et al. *(69)* recommend measuring the luminescence intensity at a concentration of 28 μM decanal aldehyde.

10. Media preparation and luminescence assays on solid media should always be performed on a flat surface. If X-ray film is simply placed over Petri dishes, without applying an even amount of pressure on top, resulting luminescence in the form of dark spots on the X-ray film may appear to have very little contrast with the background (i.e., the spots have a "fuzzy" appearance on the film). In order to achieve defined contrast between spots and the background, we suggest snuggly affixing plates inside an X-ray film box with several pieces of cardboard. The Petri dishes are inverted prior to being placed inside the box, allowing light emission to be measured through the bottom of the plates. The X-ray film can then be placed (i.e., "slid into the box") on top of the plates such that the weight of the X-ray film is distributed evenly over the top all of the Petri dishes. This method not only ensures that the Petri dishes are contacting the X-ray film with equal pressure, but also secures the film such that it cannot be accidentally moved.

11. We have described two luminescence assays that can be used to detect the presence of water contaminants. However, many variations of these two assays exist. For example, it is also possible to measure light emitted from cells growing on solid media using a luminometer, as opposed to X-ray film. To do so, LB agar containing the contaminated water sample is first prepared and aliquoted into luminometer cuvets. A fixed amount of the liquid culture of the *E. coli* biosensor ($A_{600} = 0.05$) is then applied to the surface of the media. After incubation at 37°C, luminescence can be measured in a luminometer. Another variation involves incubation of the *E. coli* biosensor in liquid media for a specified time, in the presence of the contaminated water sample. Samples of this liquid suspension can then be aliquoted to the wells of a microtiter plate. X-ray film can then be placed on top of the microtiter plate and exposed for a given period of time. Luminescent samples will appear as dark spots on the film, with the intensity of the dark spot corresponding to the degree of light emission. Moreover, luminometers designed to measure samples in microtiter plates are now commercially available.

12. Cell density measurements become critical when performing luminescence assays in liquid media. Low RLU (*see* **Note 14**) readings may be encountered if the absorbance at 600 nm (A_{600}) of the bacterial sample is too high (i.e., $A_{600} > 0.1$), because high cell densities may cause obstruction of light emission from neighboring cells to the photomultiplier tube of the luminometer. Optical densities should also be adjusted so that they remain constant between control and test samples, thus ensuring that any changes (i.e., increases or decreases) in light emission reflect gene expression, rather than fluctuations in the numbers of bacterial cells present in the sample.

13. At least three samples are measured via luminometry to allow the calculation of mean values and standard deviations from mean values. These values are important in determining whether increases in luminescence are significant and are therefore essential in data analysis.

14. Luminometers detect and count emitted photons using a photomultiplier tube that converts photons into electrical pulses. The photon pulses counted are directly proportional to the light emitted by the reaction and are displayed as RLUs. RLUs are calculated as (photons counted/10) × calibration factor, where the calibration factor is used to standardize all units to RLUs when measuring a standard reference source (Operations Manual; Optocomp I Luminometer, MGM Instruments, Hamden, CT).

15. A variety of luminometers are currently available. Some models hold single samples in disposable cuvets, whereas others read microtiter plates and can therefore measure up to 96 samples at once *(76)*. Many luminometers are coupled with injection systems that can be used for the rapid and precise addition of aldehyde or other reagents. Furthermore, computers can be linked to many of luminometers, thus permitting rapid documentation and analysis of data. Finally, luminometers small enough to be used in the field have also been developed.

Acknowledgments

We thank C. Diorio and J. Cai for many helpful discussions. This work was supported by a grant (97043) from the Center for the Alternatives to Animal Testing (USA).

References

1. Ulitzur, S. and Kuhn, J. (1987) Introduction of *lux* genes into bacteria: a new approach for specific determination of bacteria and their antibiotic susceptibility, in *Bioluminescence and Chemiluminescence New Perspectives* (Schlomerich, J., Andreesen, R., Kapp, A., Ernest, M., and Woods, W. G., eds.), Bristol-Wiley, New York, pp. 463–472.

2. Jacobs, W. R., Jr., Barletta, R. G., Udani, R., Chan, J., Kalkut, G., Sosne, G., Kieser, T., Sarkis, G. J., Hatfull, G. F., and Bloom, B. R. (1993) Rapid assessment of drug susceptibilities of *Mycobacterium tuberculosis* by means of luciferase reporter phages. *Science* **260,** 819–822.

3. Loessner, M. J., Rees, C. E. D., Stewart, G. S. A. B., and Scherer, S. (1996) Construction of luciferase reporter bacteriophage A511::*luxAB* for rapid and sensitive detection of viable *Listeria* cells. *Appl. Environ. Microbiol.* **62,** 1133–1140.

4. Cofins, J. C., Perez-Garcia, A., Romero, P., and deVincente, A. (1993) A comparison of microbial bioassays for the detection of metal toxicity. *Arch. Environ. Contam. Toxicol.* **25,** 250–254.

5. Bitton, G. and Dutka, B. J. (1986) Introduction and review of microbial and biochemical toxicity screening procedures, in *Toxicity Testing Using Microorganisms,* vol. I (Bitton, G. and Dutka, B. J., eds.), CRC, Boca Raton, FL, pp. 1–9.

6. Peterson, H. G., Nyholm, N., Nelson, M., Powell, R., Huang, P. M., and Scroggins, R. (1996) Development of aquatic plant bioassays for rapid screening and interpretive risk assessments of metal mining liquid waste water. *Water Sci. Technol.* **33,** 155–161.

7. Gu, M. B., Dhurjati, P. S., VanDyk, T. K., and LaRossa, R. A. (1996) A miniature bioreactor. *Biotechnol. Prog.* **12,** 393–397.

8. Arnold, M. A. (1990) Fiber-optic biosensors. *J. Biotechnol.* **15,** 219–228.

9. Coulet, P. R. and Blum, L. J. (1992) Bioluminescence/chemiluminescence based sensors. *Trends Anal. Chem.* **11,** 57–61.

10. Blum, L. J., Gautier, S. M., and Coulet, P. R. (1993) Design of biolumiescence-based fiber optic sensors for flow-injection analysis. *J. Biotechnol.* **31,** 357–368.

11. Vo-Dinh, T., Tromberg, B. J., Griffin, G. D., Ambrose, K. R., Sepaniak, M. J., and Gardenhire, E. M. (1987) Antibody-based fiber optics biosensor for the carcinogen benzo(a)pyrene. *Appl. Spectroscopy* **41,** 735–738.

12. Tromberg, B. J., Sepaniak, M. J., Alarie, J. P., Vo-Dinh, T., and Santella, R. M. (1988) Development of antibody-based fiber-optic sensors for detection of a benzo(a)pyrene metabolite. *Anal. Chem.* **60,** 1901–1908.

13. Kahru, A., Kurvet, M., and Kulm, I. (1996) Toxicity of phenolic wastewater to luminescent bacteria *Photobacterium phosphoreum* and activated sludges. *Water Sci. Technol.* **33,** 139–146.

14. Peltier, W. H. and Weber, C. I. (1985) *Methods for Measuring the Acute Toxicity of Effluents to Freshwater and Marine Organisms,* 3rd ed., U. S. E. P. A., Cincinnati, OH, Report EPA-600/4–85/013.

15. Bitton, G. and Koopman, B. (1992) Bacterial and enzymatic bioassays for toxicity testing in the environment. *Rev. Environ. Contam. Toxicol.* **125,** 1–24.

16. Bulich, A. A. and Isenberg, D. L. (1981) Use of the luminescent bacterial system for the rapid assessment of aquatic toxicity. *ISA Trans.* **20,** 29–33.

17. Bulich, A. A. (1982) A practical and reliable method for monitoring the toxicity of aquatic samples. *Process Biochem.* **17,** 45–47.

18. Kahru, A., Tomson, K., Pall, T., and Kulm, I. (1996) Study of toxicity of pesticides using luminescent bacteria *Photobacterium phosphoreum. Water Sci. Technol.* **33,** 147–154.

19. Abdel-Hamid, M. I. (1996) Development and applications of a simple procedure for toxicity testing using immobilized algae. *Water Sci. Technol.* **55,** 129–138.

20. Bitton, G., Jung, K., and Koopman, B. (1994) Evaluation of a microplate assay specific for heavy metal toxicity. *Arch. Environ. Contam. Toxicol.* **27,** 25–28.

21. Elnabarawy, M. T., Robideau, R. R., and Beach, S. A. (1988) Comparison of three rapid toxicity test procedures: Microtox, Polytox, and activated sludge respiration inhibition. *Toxic. Assess.* **3,** 361–370.

22. Dutka, B. J. and Gorrie, J. F. (1989) Assessment of toxicant activity in sediments by the ECHA biocide monitor. *Environ. Pollut.* **57,** 1–7.

23. Van Dyk, T. K., Majarian, W. R., Konstintinov, K. B, Young, R. M., Dhurjati, P. S., and LaRossa, R. A. (1994) Rapid and sensitive pollutant detection by induction of heat shock gene-bioluminescence gene fusions. *Appl. Environ. Microbiol.* **60,** 1414–1420.

24. Maron, D. M. and Ames, B. N. (1983) Revised methods for the *Salmonella* mutagenicity test. *Mutat. Res.* **113,** 173–215.

25. Moreau, P., Bailone, A., and Devoret, R. (1976) Prophage lambda induction of *Escherichia coli* K12 *envA uvrB:* a highly sensitive test for potential carcinogens. *Proc. Natl. Acad. Sci. USA* **73,** 3700–3704.

26. Maillard, K. I., Benedik, M. J., and Willson, R. C. (1996) Rapid detection of mutagens by induction of luciferase-bearing prophage in *Escherichia coli. Environ. Sci. Technol.* **30,** 2479–2483.

27. Ulitzur, S. (1986) Determination of antibiotic activities with the aid of luminous bacteria. *Methods Enzymol.* **133,** 275–284.

28. Sun, T. S. C. and Stahr, H. M. (1993) Evaluation and application of a bioluminescent bacterial genotoxicity test. *J. AOAC Int.* **76,** 893–898.

29. King, J. M. H., DiGrazia, P. M., Applegate, B., Burlage, R., Sanseverino, J., Dunbar, P., Larimer, F., and Sayler, G. S. (1990) Rapid, sensitive bioluminescent reporter technology for naphthalene exposure and biodegradation. *Science* **249,** 778–780.

30. Guzzo, A. and DuBow, M. S. (1991) Construction of stable, single-copy luciferase gene fusions in *Escherichia coli. Arch. Microbiol.* **156,** 444–448.

31. Guzzo, A. and DuBow, M. S. (1991) Transcription of the *Escherichia coli fliC* gene is regulated by metal ions. *Appl. Environ. Microbiol.* **57,** 2255–2259.

32. Guzzo A. and DuBow, M. S. (1993) Selenium-induced gene expression to create luminescent biosensors and to elucidate genetically-programmed responses to selenium, in *Heavy Metals in the Environment,* vol. 1, 9th International Conference (Allan, R. J. and Nriagy, J. O., eds.), CEP Consultants, Edinburgh, UK, pp. 407–410.

33. Selifonova, O., Burlage, R., and Barkay, T. (1993) Bioluminescent sensors for detection of bioavailable Hg(II) in the environment. *Appl. Environ. Microbiol.* **59,** 3083–3090.

34. Corbisier, P., Ji, G., Nuyts, G., Mergeay, M., and Silver, S. (1993) *luxAB* gene fusions with the arsenic and cadmium resistance operons of *Staphlococcus aureus* plasmid pI258. *FEMS Microbiol. Lett.* **110,** 231–38.

35. Heitzer, A., Malachowsky, K., Thonnard, J. E., Bienkowski, P. R., White, D. C., and Sayler, G. S. (1994) Optical biosensor for environmental on-line monitoring of naphthalene and salicylate bioavailability with an immobilized bioluminescent catabolic reporter bacterium. *Appl. Env. Microbiol.* **60,** 1487–1494.

36. Guzzo, A. and DuBow, M. S. (1994) A *luxAB* transcriptional fusion to the cryptic *cel*F gene of *Escherichia coli* displays increased luminescence in the presence of nickel. *Mol. Gen. Genet.* **242,** 455–460.

37. Virta, M., Lampinen, J., and Karp, M. (1995) A luminescent-based mercury biosensor. *Anal. Chem.* **67,** 667–669.
38. Cai, J. and DuBow, M. S. (1996) Expression of the *Escherichia coli* chromosomal *ars* operon. *Can. J. Microbiol.* **42,** 662–671.
39. Blaise, C. (1991) Microbiotests in aquatic ecotoxicology: characteristics, utility, and prospects. *Environ. Toxicol. Water Qual.* **8,** 145–155.
40. Briscoe, S. F., Diorio, C., and DuBow, M. S. (1996) Luminescent biosensors for the detection of tributyltin and dimethyl sulfoxide and the elucidation of their mechanisms of toxicity, in *Environmental Biotechnology: Principles and Applications* (Moo-Young, M., Anderson W. A., and Chakrabarty, A. M., eds.), Kluwer Academic, Netherlands, pp. 645–655.
41. Fitzwater, T., Tamm, J., and Polisky, B. (1984) RNA1 is sufficient to mediate plasmid ColE1 incompatibility *in vivo. J. Mol. Biol.* **175,** 5–13.
42. Sakharov, G. N., Ismailov, A. D., and Danilov, V. S. (1988) Temperature dependances of the reaction of bacterial luciferase from *Beneckea harveyi* and *Photobacterium fischeri. Biochemistry (USSR)* **53,** 770–776.
43. Bronstein, I., Fortin, J., Stanley, P. E., Stewart, G. S. A. B., and Kricka, L. J. (1994) Chemiluminescent and bioluminescent reporter gene assays. *Anal. Biochem.* **219,** 169–181.
44. Kirchner, G., Roberts, J. L., Gustafson, G. D., and Ingolia, T. D. (1989) Active bacterial luciferase from a fused gene: expression of a *Vibrio harveyi luxAB* translational fusion in bacteria, yeast and plant cells. *Gene* **81,** 349–354.
45. Tatsumi, H., Masuda, T., and Nakano, E. (1988) Synthesis of enzymatically active firefly luciferase in yeast. *Agric. Biol. Chem.* **52,** 1123–1128.
46. Gould, S. J. and Subramani, S. (1988) Firefly luciferase as a tool in molecular and cell biology. *Anal. Biochem.* **175,** 5–13.
47. Kricka, L. J. (1988) Clinical and biochemical applications of luciferases and luciferins. *Anal. Biochem.* **175,** 14–21.
48. Alam, J. and Cook, J. L. (1990) Reporter genes: application to the study of mammalian gene expression. *Anal. Biochem.* **188,** 245–254.
49. Klimowski, L., Rayms-Keller, A., Olson, K. E., Yang, R. S. H., Tessari, J., Carlson, J., and Beaty, B. (1996) Inducibility of a molecular bioreporter system by heavy metals. *Environ. Toxicol. Chem.* **15,** 85–91.
50. DeWet, J. R., Wood, K. V., Helinski, D. R., and DeLuca, M. (1985) Cloning of firefly luciferase cDNA and the expression of active luciferase in *Escherichia coli. Proc. Natl. Acad. Sci. USA* **82,** 7870–7873.
51. DeWet, J. R., Wood, K. V., DeLuca, M., Helinski, D. R., and Subramani, S. (1987) Firefly luciferase gene: structure and expression in mammalian cells. *Mol. Cell. Biol.* **7,** 725–737.
52. Wood, K. V., Lam, Y. A., Seliger, H. H., and McElroy, W. D. (1989) Complementary DNA coding click beetle luciferases can elicit bioluminescence of different colors. *Science* **244,** 700–702.
53. Tatsumi, H., Kajiyama, N., and Nakano, E. (1992) Molecular cloning and expression in *Escherichia coli* of a cDNA clone encoding luciferase of a firefly, *Luciola lateralis. Biochim. Biophys. Acta* **1131,** 161–165.

54. Rainina, E. I., Efremenco, E. N., Varfolomeyev, S. D., Simonian, A. C., and Wild, J. R. (1996) The development of a new biosensor based on recombinant *E. coli* for the direct detection of organophosphorus neurotoxins. *Biosens. Bioelectron.* **11,** 991–1000.

55. Sambrook, J., Fritsch, E. F., and Maniatis, T. (1989) *Molecular Cloning: a Laboratory Manual,* 2nd ed., Cold Spring Harbor Laboratory Press, Cold Spring Harbor, NY.

56. Harris, W. R., Berthon, G., Day, J. P., Exley, C., Flaten, T. P., Forbes W. F., Kiss, T., Orvig, C., and Zatta, P. F. (1996) Speciation of aluminum in biological systems. *J. Toxicol. Environ. Health* **48,** 543–68.

57. Baker, T. A., Howe, M. M., and Gross, C. A. (1983) Mu dX, a derivative of Mu d1 (*lac* Ap^R) which makes stable *lacZ* fusions at high temperatures. *J. Bacteriol.* **156,** 970–974.

58. Miller, T. G. and Mackay, W. C. (1980) The effects of hardness, alkalinity and pH of test water on the toxicity of copper to rainbow trout *(Salmo gairdneri). Water Res.* **14,** 129–133.

59. Schubauer-Berigan, M. K., Dierkes, J. R., Monson, P. D., and Ankley, G. T. (1993) pH-dependent toxicity of Cd, Cu, Ni, Pb and Zn to *Ceriodaphnia dubia, Pimephales promelas, Hyalella azeteca,* and *Lumbriculus variegatus. Environ. Toxicol. Chem.* **12,** 1261–1266.

60. Gadd, G. M. and Griffiths, A. J. (1978) Microorganisms and heavy metal toxicity. *Microbiol. Ecol.* **4,** 303–307.

61. Guzzo J., Guzzo, A., and DuBow, M. S. (1991) Characterization of the effects of aluminum on luciferase biosensors for the detection of ecotoxicity. *Toxicol. Lett.* **64/65,** 687–693.

62. Kong, I. C., Bitton, G., Koopman, B., and Jung, K. H. (1995). Heavy metal toxicity testing in environmental samples. *Rev. Environ. Contam. Toxicol.* **142,** 119–147.

63. Sengul, F. and Turkman, A. (1989) Chromium treatment of wastewaters by chemical methods, in *Metal Speciation in the Environment,* NATO ASI series G, vol. 23 (Broekaert, J. A. C., Gucer, S., and Adams, F., eds.), Springer-Verlag, Berlin, pp. 613–624.

64. Raspor, B. (1991) Metal and metal compounds in water, in *Metals and Their Compounds in the Environment: Occurrence, Analysis, and Biological Relevance* (Merian, E., ed.), Weinheim, NY, pp. 233–256.

65. Ajmal, M. and Khan, A. U. (1984) Effect of water hardness on the toxicity of cadmium to microorganisms. *Water Res.* **12,** 1487–1491.

66. Calamari, D., Marchetti, R., and Vailati, G. (1980) Influence of water hardness on cadmium toxicity to *Salmo gairdneri. Water Res.* **14,** 1421–1426.

67. Hung, Y-W. (1982) Effects of temperature and chelating agents on cadmium uptake in the american oyster. *Bull. Environ. Contam. Toxicol.* **28,** 546–551.

68. Miller, J. H. (1992) *A Short Course in Bacterial Genetics.* Cold Spring Harbor Laboratory, Cold Spring Harbor, NY.

69. Blouin, K., Walker, S. G., Smit, J., and Turner, R. F. B. (1996) Characterization of in vivo reporter systems for gene expression and biosensor applications based on *luxAB* luciferase genes. *Appl. Environ. Microbiol.* **62,** 2013–2021.

70. Ulitzur, S. and Hastings, J. W. (1979) Evidence for tetradecanal as the natural aldehyde in bacterial bioluminescence. *Proc. Natl. Acad. Sci. USA* **76,** 265–267.
71. Meighen, E. A., Slessor, K. N., and Grant, G. G. (1982) Development of a bioluminescence assay for aldehyde pheromones of insects. I. sensitivity and specificity. *J. Chem. Ecol.* **8,** 911–921.
72. Meighen, E. A. and Grant, G. G. (1985) Bioluminescence analysis of long chain aldehydes: detection of insect pheromones, in *Bioluminescence and Chemiluminescence: Instruments and Applications,* vol. 2 (VanDyke, K., ed.), CRC, Boca Raton, FL, pp. 253–268.
73. Meighen, E. A. (1991) Molecular biology of bacterial bioluminesence. *Microbiol. Rev.* **55,** 123–142.
74. Blisset, S. J. and Stewart, G. S. A. B. (1989) *In vitro* bioluminescence determination of apparent Km's for aldehyde in recombinant bacteria expressing *luxAB*. *Lett. Appl. Microbiol.* **9,** 149–152.
75. Stewart, G. S. A. B. and Williams, P. (1992) *Lux* genes and the applications of bacterial bioluminescence. *J. Gen. Microbiol.* **138,** 1289–1300.
76. Blaise, C., Forghani, R., Guzzo, J., and DuBow, M. S. (1994) A bacterial toxicity assay performed with microplates, microluminometry and Microtox® Reagent. *Biotechniques* **16,** 932–937.

19

Bioluminescence-Based Metal Detectors

Marko Virta, Sisko Tauriainen, and Matti Karp

1. Introduction

1.1. General

This chapter consists of two methodologically rather divergent topics: the construction of a metal sensor bacterial strain and the measurement of bioavailable metal using such a strain. The construction part is written assuming that readers are familiar with basic recombinant DNA techniques, such as isolation and purification of DNA, the use of restriction enzymes, and ligation. However, if that is not the case, plasmids and bacterial strains for metal bioavailability measurements are available from the authors' laboratory.

1.2. Bacterial Biosensor

The use of living cells as biosensors offers several advantages over enzyme-based or other biosensors *(1)*. Analytical systems that require a sequence of biochemical reactions are greatly simplified by using cells, because all the reactions are conveniently packaged inside the cells and efficiently carried out. Furthermore, enzymes are in an optimal environment within the cell. Several methods based on microbial biosensors have been introduced for total toxicity measurements. Various parameters have been used as indicators of cells' responses to toxic compounds, for example, O_2 consumption (cell respiratory) *(2)*, luminescence from naturally luminescent bacteria *(3)*, bacteria expressing genes for luminescence *(4)*, inhibition of protein synthesis *(5)*, and nitrification measurement *(6)*. A novel approach for a microbial biosensor is to connect a strictly regulated promoter to a sensitive reporter gene **(Fig. 1)**. Very interesting promoters for environmental analysis are found in bacteria that survive in environments contaminated by, for example, heavy metals or organic compounds. The ability of the bacteria to survive in a contaminated environment is

From: *Methods in Molecular Biology, Vol. 102: Bioluminescence Methods and Protocols*
Edited by: R. A. LaRossa © Humana Press Inc., Totowa, NJ

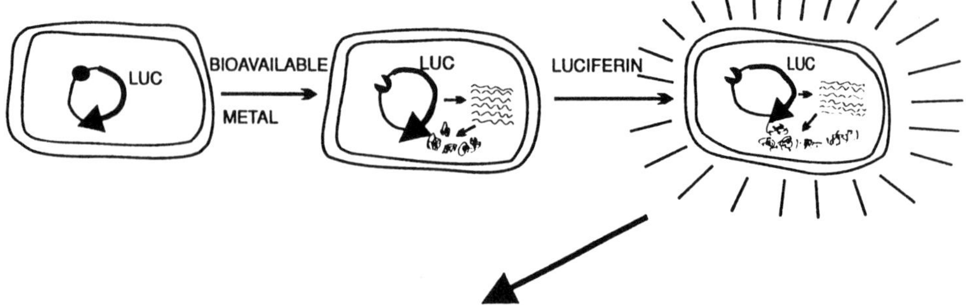

Fig. 1. The principle of bacterial biosensor for the detection of bioavailable metals.

usually based on a genetically encoded resistance system, which is very precisely regulated. Some biosensors using the promoter-reporter gene concept have been developed, for example, for the detection of mercury *(1,7)* and xenobiotic compounds *(8)*. This chapter describes the isolation of regulative unit from a metal resistant bacterial strain, construction of the sensor plasmid **(Fig. 2)**, and the measurement protocol for bioavailable metal using arsenate as an example.

1.3. Luciferases

Luciferases are a class of enzymes that produce light in their catalysis. Insect luciferases (e.g., from American firefly, *Photinus pyralis* or Jamaican click beetle, *Phyrophorous plagiophthalamus*) catalyze the reaction shown in **Fig. 3.**

The firefly luciferase gene (*luc*FF) was cloned *(9)* a decade ago and it has since become a widely used reporter gene in prokaryotic as well as in eukaryotic systems, because it provides sensitive and simple detection of the gene regulation. The cells that express recombinant insect luciferase produce light only if the luciferase substrate, D-luciferin, is added.

Another commonly used luciferase is bacterial luciferase from marine bacteria (e.g., *Vibrio harveyi*), which allows the construction of a self-luminescent bacterial biosensor. However, strains carrying the firefly luciferase *(P. pyralis)* or other insect luciferase genes offer several advantages over bacterial luciferase: the enzyme is more heat-stable, it is not subject to substrate inhibition *(10)*, and its quantum yield is approx 90% and therefore, considerably higher than that of bacterial luciferase (5–10%, depending of the strain used), which in practice means that lower protein levels will be detected.

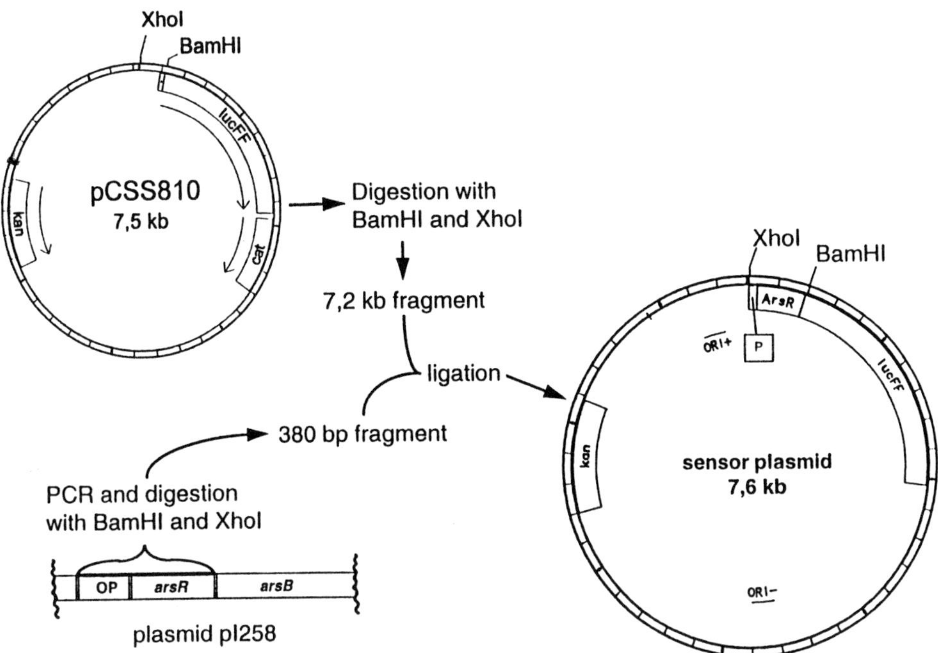

Fig. 2. The construction of a metal sensor plasmid and structure of plasmid pCSS810. Abbreviations used: *kan* = gene encoding kanamycin resistance, *cat* = gene encoding chloramphenicol resistance, *P* = operator/promoter, *lucFF* = gene encoding firefly luciferase, ORI^+ = origin of replication in Gram-positive hosts, ORI^- = origin of replication in Gram-negative hosts, *arsR* = gene encoding transcriptional regulatory protein.

1.4. Metal Resistance Genes

Regulatory elements from bacteria resistant to a heavy metal can provide a sensitive and selective receptor that will work in a physiological concentration range. Genetic determinants for these resistances are usually found on plasmids and transposons of soil bacteria, which facilitates their analysis and manipulation by molecular genetic techniques. Numerous genetic determinants encoding metal resistances have been characterized, including mercury *(11)*, arsenite *(12)*, cadmium *(13)*, zinc *(14)*, cobalt *(14)*, and copper *(15)*. These systems are composed of a regulatory protein that serves as a switch modulating gene expression from a responsive promoter. Plasmid construction is simplified when the gene encoding the regulatory protein and the responsive promoter are nearby or adjacent, Those cloned and characterized heavy metal resistance genes can be used for construction of biosensors specific for each of

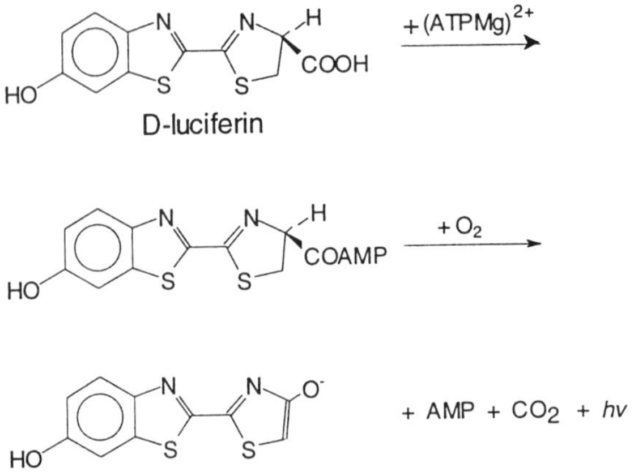

Fig. 3. Reactions catalyzed by insect luciferases.

these metals, which allows quite a rapid construction of a biosensor. Isolation of the regulatory unit by PCR allows the exact construction of the desired vector for the specific detection of the metal.

1.5. Metal Bioavailability

Bioavailability is the most important factor determining the toxicity of metals in biological systems *(16)*. However, the measurement of the bioavailability of metals is quite difficult or even impossible with traditional analytical methods.

2. Materials

2.1. Bacterial Strains and Plasmids

1. Metal sensor plasmid and strain: Plasmids with conveniently interchangeable regulatory units are available from the authors' laboratory, for example, plasmid pCSS810 *(17)*. The pCSS810 is a shuttle vector containing replication units originating from Gram-positive and Gram-negative strains. Naturally, the principle described here can also be used with other suitable plasmids. The best choice for a host strain carrying the sensor plasmid is the original strain that was used as a source for the regulatory unit, and for that purpose, it must usually be cured of original plasmids. However, it is usually sufficient to use a characterized strain of the same bacterial species. Naturally, a method to introduce plasmids into that strain must be known.
2. Toxicity test strain: A strain that constitutively expresses the *luc* gene is needed for the determination of metal-independent toxicity of a sample. The parental plasmid pCSS810 is usually suitable for this use, since the *luc* expression is controlled by the *lac* operator.

2.2. Cultivation

1. LB-broth *(18)*: 10 g tryptone, 5 g yeast extract, 5 g NaCl/L, pH 7.0.
2. LA-agar: LB medium supplemented with 16 g agar/L.
3. M9 media *(18)* supplemented with 0.1% hydrolyzed casein: 20 mL 5X M9 salts, 0.2 mL 1 M MgSO$_4$, 1 mL 20% glucose, 10 μL 1 M CaCl$_2$, 2 mL 5% hydrolyzed casein. Make up to 100 mL with H$_2$O.
4. 5× M9 salts (per liter): 6 g Na$_2$HPO$_4$, 3 g KH$_2$PO$_4$, 5 g NaCl, 1 g NH$_4$Cl. Adjust to pH 7.4 with NaOH or HCl.

2.3. PCR and DNA Manipulation

1. DNA-polymerase used in PCR should have a 3' to 5' proofreading activity in order to minimize the errors in the replication of DNA during the PCR. Different suppliers sell their own enzymes, for example, New England Biolabs (Beverly, MA) sells Vent®-polymerase and Stratagene (La Jolla, CA) sells *Pfu*-polymerase. The proofreading activity of an enzyme can lead to the shortening of primers from the 3'-end (*see* **Note 1**). Basic reaction conditions are, e.g., 2 mM MgSO$_4$, 200 μM of each dNTP, 0.4 μM of both primers, the template and 1 U/100 μL reaction volume of DNA polymerase in addition to a reaction buffer supplied with enzyme. The optimal Mg concentration can vary (*see* **Note 2**).
2. Appropriate thermal cycler for PCR (e.g., Perkin-Elmer, Foster City, CA).

2.4. Primer Design for PCR

1. Primers can be designed by a computer program (e.g., Wisconsin Sequence Analysis Package™, Genetics Computer Group, Madison, WI) or manually according to their C + G content using the following simplified formula to calculate the melting temperature:

$$T_m = 4°C \times (C + G) + 2°C \times (A + T) \tag{1}$$

where C, G, A, and T are the number of respective bases.

   Complementary primers have to be long enough to ensure unique binding to target DNA. Generally, the complementary part should be at least 18 bases long. Appropriate restriction sites should be added to the 5'-ends of the complementary primers. It should be noted that most restriction enzymes (including *Xho*I and *Bam*HI as well) need few extra bases to each side of their recognition sequences in order to correctly digest DNA. One pair of primers with their template are shown in **Fig. 4** as an example.

2.5. Luminescence Measurements

1. Luciferase substrate for measuring luciferase activity in vivo *(20)*: 1 mM D-luciferin (either from Bio-Orbit, Turku, Finland or Sigma, St. Louis, MO), 100 mM Na-citrate, pH 5.0. Store at –20°C and protect from light.
2. Luminometer: A wide range of luminometers are commercially available from a simple manual one-tube luminometer to a computer-controlled 96-well plate instrument.

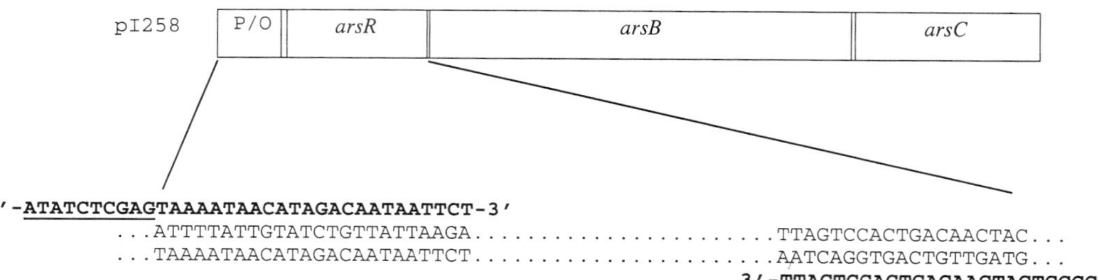

Fig. 4. The structure of the *ars* operon from plasmid pI258 *(19)* and sequence of PCR primers (shown bold). The nonhybridizing bases of the primers are underlined.

3. Methods

3.1. PCR

Following cycles are designed for the primers shown in **Fig. 4**.

    94°C 50 s   (denaturation)
    52°C 60 s   (annealing)
    72°C 60 s   (extension)

Twenty-five of above cycles should give sufficient product.

In addition to the actual amplification reaction, prepare two control reactions, one without template and one without primers. Use hot start, i.e., first mix all reaction components except DNA polymerase, then incubate the mixture for 5 min at 98°C, lower temperature to 80°C, add polymerase, and start cycles.

Analyze all reactions by electrophoresis through 1% agarose gels with known mol-wt markers. Control reactions should not result in any amplification product. If the size of the product is that expected, purify the product, and digest it (*see* **Subheading 3.2.2.**).

3.2. Other DNA Manipulations

Standard methods for DNA manipulations are described elsewhere *(18)*.

3.2.1. Purification

In addition to the DNA purification procedures described by Sambrook et al. *(18)*, commercial kits are available from many suppliers, e.g., from Qiagen (Qiagen GmbH, Hilden, Germany).

3.2.2. Digestion

Digest both pCSS810 and the PCR product with *Bam*HI and *Xho*I enzymes, purify (*Bam*HI cannot be heat-inactivated), and treat digested pCSS810 with alkaline phosphatase.

3.2.3. Ligation

Ligate digested and phosphatase-treated pCSS810 with digested PCR product. Purify ligation mix, and dissolve in water for electroporation.

3.3. Transformation

Transformation of *Escherichia coli* cells is described elsewhere as well as preparation of competent cells *(21)*. Any strain intended for recombinant DNA work should be suitable, e. g., MC1061 *(22)*. Transformation of the actual sensor strain depends on the strain used. The example used here, *Staphylococcus aureus,* can be transformed by electroporation with high efficiency *(23)*.

3.4. Selection of Correct Plasmids

Plasmid pCSS810 includes resistance genes for kanamycin *(kan)* and for chloramphenicol *(cat)*. However, it should be noted that the latter is located immediately downstream from luciferase gene and is under the control of the same promoter as the *luc* gene. Therefore, if the expression of *luc* is controlled with a promoter that is repressed in normal growth conditions, the expression of *cat* gene may be too weak to produce resistance to chloramphenicol, which makes the use of kanamycin as a selective agent more favorable. Suitable media for the selection of transformants are LA plates supplemented with 30 μg/mL of kanamycin sulfate.

The colonies that grow on LA-Km-plates can be checked for light production by picking colonies to 0.5 mL of LB-Km with toothpicks. After 6 h of culturing in a shaker at 37°C, luminescence can be measured as described in **Subheading 3.7.**, and luminescent clones can be selected for restriction analysis.

After the confirmation of the anticipated plasmid structure by restriction analysis and DNA sequencing, the sensor plasmid can be transformed to actual sensor strain, here *S. aureus* RN4220 *(24)*.

3.5. Bacterial Cultivation for Measurements

All the following procedures are done with three different plasmids.

1. Grow cells in LB with appropriate antibiotic (constructions based on pCSS810: 30 μg/mL of kanamycin) to logarithmic phase in shaker at 37°C (*see* **Note 3**).
2. Harvest the cells by centrifugation 4000*g* for 10 min (*see* **Note 4**).
3. Wash the cells once with M9 media supplemented with 0.1% hydrolyzed casein, and suspend them in the same media.
4. Immediately before induction reactions dilute cell suspension with the same casein supplemented M9 to obtain about 10^6 cells/mL (*see* **Note 4**).

3.6. Induction Reactions

Induction reactions should be performed with two different bacterial strains: the metal sensor strain and the general toxicity strain.

1. Pipet 50 μL of standard solution of a metal (or an unknown sample) to a luminometer tube (*see* **Note 5**).
2. Add 50 μL of bacterial dilution. Mix well, preferably with vortex.
3. Incubate for 60 min at 37°C.

3.7. Luminescence Measurements

1. Add 100 μL of luciferase substrate to induction reaction tubes.
2. Incubate for 5 min at room temperature.
3. Measure luminescence. Use a single measurement at room temperature. By comparing luminescent values among only samples of a single experimental series, variations owing to temperature fluctuation can be negated.

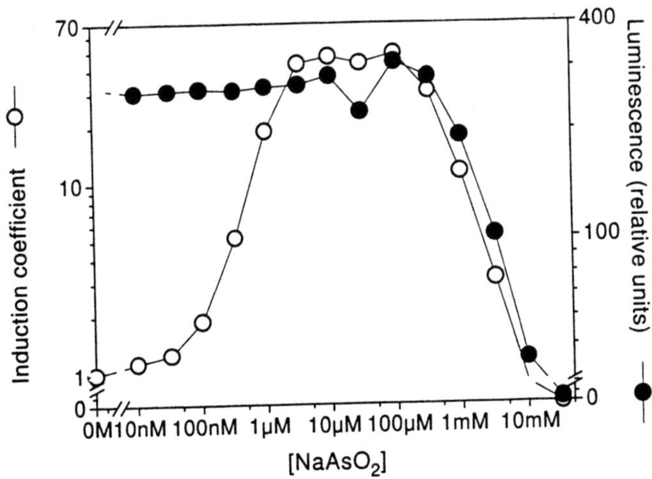

Fig. 5. The response of a biosensor plasmid constructed by using primers described in **Fig. 4** (open circles) and toxicity testing plasmid pCSS810 (closed circles). The host strain for both plasmids was *S. aureus* RN4220. Luminescence was determined by a single measurement with a Labsystems (Helsinki, Finland) Luminoskan luminometer.

3.8. Interpretation of the Results

1. Calculate induction coefficient from luminescence values by the following formula:

$$I = L_i/L_b \tag{2}$$

   where L_i = luminescence value from standard solution or sample and L_b = luminescence value from blank solution.
2. Draw a standard curve from the induction coefficients of standard solutions. The induction coefficient of the blank is 1.
3. Determine the concentrations of the samples from the standard curve.
4. Use the toxicity test strain for assessing the possible inhibitory effect of the sample.

An example of the standard curve obtained using strains containing either the plasmid constructed in **Fig. 4** or the toxicity indicating fusion is shown in **Fig. 5**.

4. Notes

1. The 3' to 5' proofreading activity of DNA polymerase can lead to degradation of the PCR primers. This can usually be minimized by using an appropriate amount of DNA polymerase (i.e., not in excess) and a high concentration of dNTPs (even 400 μM can be necessary). The hot-start procedure minimizes the contact time between polymerase and primers, and thus also diminishes the degradation problem.

2. The optimal Mg^{2+} concentration in the PCR reaction is usually in the range of 2–6 mM for proofreading polymerases. If 2 mM does not give satisfactory results, the concentration should be increased first to 4 mM and possibly to 6 mM.
3. The bacteria can be stored for a couple of hours in LB on ice. It is not recommended to store cells in M9 medium.
4. The cell number can be adjusted according to the sensitivity of luminometer used. The luminescence from the blank sample should be sufficiently above the instrument background to yield a sufficient dynamic range for measurement.
5. If an appropriate luminometer is available, luminometer tubes can be replaced with 96–well plates. Reaction volumes can be maintained as described.

References

1. Tescione, L., and Belfort, G. (1993) Construction and evaluation of a metal ion biosensor. *Biotechnol. Bioeng.* **42,** 945–952.
2. Kong, Z., van Rollegheim, P. A., and Verstraete, W. (1993) An activated sludge-based biosensor for rapid IC50 estimation and on-line toxicity monitoring. *Biosens. Bioelectron.* **8,** 49–58.
3. Ribo, J. M. and Kaiser, K. L. E. (1987) Photobacterium phosphoreum, toxicity bioassay. *Toxicol. Assess.* **2,** 305–323.
4. Lampinen, J., Korpela, M., Saviranta, P., Kroneld, R., and Karp, M. (1990) Use of *Escherichia Coli* Cloned with genes Encoding Bacterial Luciferase for Evaluation of Chemical Toxicity. *Toxicol. Assess.* **5,** 337–50.
5. Lampinen, J., Virta, M., and Karp, M. (1995) Use of controlled luciferase expression for monitoring of chemicals affecting protein synthesis. *Appl. Environ. Microbiol.* **61,** 2981–2989.
6. Holland, G. J., and Green, A. (1975) Development of a gross pollution detector: laboratory studies. *Water Treatment Exam.* **4,** 81–99.
7. Virta, M., Lampinen, J., and Karp, M. (1995) A luminescence-based mercury biosensor. *Anal. Chem.* **67,** 667–669.
8. King, J. M. H., DiGrazia, P. M., Applegate, B., Burlage, R., Sanseverino, J., Dunbar, P., Larimer, F., and Sayer, G. S. (1990) Rapid, sensitive bioluminescent reporter technology for naphtalene exposure and biodegradation. *Science* **249,** 778–781.
9. deWet, J. R., Wood, K. V., DeLuca, M., Helinski, D. R., and Subramani, S. (1987) The firefly luciferase gene: structure and expression in mammalian cells. *Mol. Cell. Biol.* **7,** 725–737.
10. Li, Z. and Meighen, E. A. (1994) The turnover of bacterial luciferase is limited by a slow decomposition of the ternary enzyme-product complex of luciferase, FMN, and fatty acid. *J. Biol. Chem.* **269,** 6640–6644.
11. Summers, A. O. (1986) Organization, expression, and evolution of genes for mercury resistance. *Ann. Rev. Microbiol.* **40,** 607–634.
12. Silver, S. K., Budd, K. M., Leahy, W. V., Shawn, D., Hammond, R. P., Novick, G. R., Wilsky, M. H., Malamy, H. M., and Rosenberg, H. (1981) Indicible plasmid-determined resistance to arsenate, arsenite and antimony (III) in *Escherichia coli* and *Staphylococcus aureus. J. Bacteriol.* **146,** 983–996.

13. Nucifora, G., Chu, L., Misra, T. K., and Silver, S. (1989) Cadmium resistance from *Staphylococcus aureus* plasmid pI258 *cadA* gene results from cadmium-efflux ATPase. *Proc. Natl. Acad. Sci. USA* **86,** 3544–3548.

14. Nies, D. H. and Silver, S. (1989) Plasmid-determined indicible efflux is responsible for resistance to cadmium, zinc, and cobalt in *Alcaligenes eutrophus. J. Bacteriol.* **171,** 896–900.

15. Cha, J. S. and Cooksey, D. A. (1991) Copper resistance in *Pseudomonas syringae* mediated by periplasmic and outer membrane proteins. *Proc. Natl. Acad. Sci. USA* **88,** 8915–8919.

16. Farrell, R. E., Germida, J. J., and Huang, P. M. (1993) Effects of chemical speciation in growth media on the toxicity of mercury (II). *Appl. Environ. Microbiol.* **59,** 1507–1514.

17. Lampinen, J., Koivisto, L., Wahlsten, M., Mäntsälä, P., and Karp, M. (1992) Expression of luciferase genes from different origins in *Bacillus subtilis. Mol. Gen. Genet.* **232,** 498–504.

18. Sambrook, J., Fritsch, E. F., and Maniatis, T. (1989) *Molecular Cloning: A Laboratory Manual.* Cold Spring Harbor Laboratory, Cold Spring Harbor, New York.

19. Ji, G. and Silver, S. (1992) Regulation and expression of the arsenic resistance operon from *Staphylococcus aureus* plasmid pI258. *J. Bacteriol.* **174,** 3684–3694.

20. Wood, K. V. and DeLuca, M. (1987) Photographic detection of luminescence in *Escherichia coli* containing the gene for firefly luciferase. *Anal. Biochem.* **161,** 501–507.

21. Dower, W. J., Miller, J. F., and Ragsdale, C. W. (1988) High efficiency transformation of *E. coli* by high voltage electroporation. *Nucleic Acid. Res.* **16,** 6126–6144.

22. Casabadan, M. J. and Cohen, S. N. (1978) Analysis of gene control signals by DNA fusion and cloning in *Escherichia coli. J. Mol. Biol.* **138,** 179–207.

23. Schenk, S., and Laddaga, R. A. (1992) Improved method for electroporation of *Staphylococcus aureus. FEMS Microbiol. Lett.* **94,** 133–138.

24. Kreiswirth, B. N., Lofdahl, M. J., O'Reilly, M., Schlievert, P. M., Bergdoll, M. S., and Novick, R. P. (1993) The toxic shock syndrome exotoxin structural gene is not detectably transmitted by a prophage. *Nature* **305,** 709.

20

Luminescence Facilitated Detection of Bioavailable Mercury in Natural Waters

Tamar Barkay, Ralph R. Turner,
Lasse D. Rasmussen, Carol A. Kelly, and John W. M. Rudd

1. Introduction

One of the major routes of human exposure to mercury is by the consumption of contaminated fish and shellfish. Mercury, in the form of methyl mercury (MM), accumulates in these biota by biomagnification through the aquatic food chain, to concentrations orders of magnitude higher than its levels in the water *(1,2)*. Dissolved MM is absorbed by unicellular organisms *(2,3)* at the base of the food chain, and since MM is only very slowly eliminated from the animal body, its concentration increases with the trophic level. The amount of dissolved MM available to the base of the food chain is critical, and this amount is determined by the rates of MM formation and degradation and by factors that directly and indirectly affect these rates. Thus, the concentration of bioavailable ionic mercury (Hg^{2+}) affects not only the methylation rate, but also the rate of the Hg^{2+} reduction and volatilization, reactions that compete with methylation for the same substrate *(4)*. Furthermore, Hg^{2+} is the inducer of a bacterial enzyme, organomercurial lyase, that degrades MM, as well as the reduction process *(5)*. Measuring bioavailable Hg^{2+} is essential for calculating methylation and reduction rates *in situ,* a measurement needed for evaluating the potential for MM accumulation and thus risk to public health. Total mercury levels presently serve as the basis for regulating mercury exposure. Because the majority of mercury in the environment is in a harmless inert form, accurate measurements of bioavailable Hg^{2+} may provide a basis for more realistic regulatory criteria.

State-of-the-art mercury analyses can measure subpicomolar concentrations of mercury and distinguish organic from inorganic forms of mercury *(6,7)*. They, cannot, however distinguish bioavailable from inert forms of mercury.

From: *Methods in Molecular Biology, Vol. 102: Bioluminescence Methods and Protocols*
Edited by: R. A. LaRossa © Humana Press Inc., Totowa, NJ

The only valid approach to measuring bioavailable mercury is by bioindicators, usually recombinant bacteria that contain a gene fusion between the regulatory region of the mercury resistance *(mer)* operon as a sensor for Hg^{2+} and a gene specifying an easily detected phenotype as a reporter. The reporters in most of the mercury bioindicators described to date are based on bacterial luminescence *(lux)* genes *(8–10)* or the firefly luciferase *luc* system *(11)*, but fusions to β-galactosidase are also in use. The bioindicator is induced when Hg^{2+} is present in the cytoplasm and interacts with the regulatory protein, MerR. MerR forms a complex with the *mer* operator/promoter and RNA polymerase in the absence or presence of Hg^{2+}. In the absence of Hg^{2+} the DNA in this complex is bent preventing the alignment of the −10 and −35 nucleotide regions, and thus initiation of transcription. When Hg^{2+} interacts with MerR, the complex changes configuration, the DNA rotates to align the promoter and transcription commences *(12)*. Because induction by Hg^{2+} takes place in the bacterial cytoplasm and this form of mercury is the substrate for methylation *(13)* and reduction *(5)* an active bioindicator suggests the presence of Hg^{2+} that is available as a substrate for these transformations. Furthermore, this response is quantitative; with more available Hg^{2+}, promoters are activated to a greater extent, yielding a higher titer of reporter molecules and an elevated response.

Applications of mercury biosensors in natural waters with p*M* concentrations of Hg^{2+} require precautions to prevent contaminations from exogenous sources in glassware and reagents. Furthermore, the distance between measuring bioavailable Hg^{2+} in a well-controlled experiment in the laboratory and a quantitative analysis in field samples is enormous. Mercury is highly reactive, forming complexes and ligands with dissolved organic and inorganic matter and with surfaces in the environment. The use of bioindicators to quantitate bioavailable Hg^{2+} in natural waters requires that calibration takes these interactions into account. The question, "does the measurement takes into account all the interactions that determine bioavailability in the analyzed sample?" must be positively answered if we are to accept results as absolute measurements of bioavailable Hg^{2+}. This problem is exasperated because calibration requires the addition of known concentrations of Hg^{2+} and we do not know that the bioavailable fraction of freshly added Hg^{2+} is similar to that of Hg^{2+} that has resided in the environment for extended periods of time. Here we describe an assay for the measurement of bioavailable Hg^{2+} using a *mer–lux* fusion in an *Escherichia coli* bioindicator (**Fig. 1**, **Subheadings 2.** and **3.**). Procedures to prevent contaminations and applications in natural waters are described as well (**Subheadings 2.**, **3.**, and **4.**). This bioindicator *(8)* has been employed to study the mode of Hg^{2+} transport through the bacterial cell wall *(15)*, how environmental factors modulate bioavailablity of Hg^{2+} *(16)*, and to measure bioavailable Hg^{2+} in a contaminated freshwater stream *(17)*.

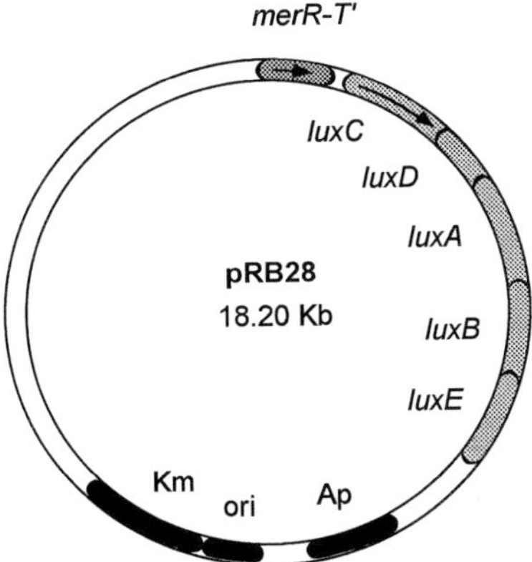

Fig. 1. Plasmid pRB28 encoding the *mer–lux* bioindicator. pRB28 was constructed by cloning a 0.7-kb *Eco*RI-*Bam*HI restriction fragment into the multiple cloning site of the *lux* vector plasmid pUCD615 *(14)*. Bacterial cells containing this plasmid produce light when Hg^{2+} is available in their cytoplasm *(see text)*.

2. Materials

2.1. Bioindicator Growth

1. Bacterial strains: *E. coli* HMS174 (F⁻ *recA1 rpoB331 hsdR19* λ-IN[*rrnD-rrnE]1*) *(18)*.
2. Plasmids: pRB28 (Hg^{2+} inducible *mer–lux* fusion) *(8)*, pRB27 (constitutive *mer–lux* mutant of pRB28) *(16)*.
3. Maintenance of strains: Frozen stocks are kept in 50% sterile glycerol at –70°C. Cultures are resuscitated by growth on LB medium with kanamycin (50 µg/mL) to assure plasmid maintenance. Plates with singly isolated colonies are used as a source of inoculum. Fresh plates are prepared from frozen stocks every 2 wk.
4. Growth medium is made in sterile Erlenmeyer flask from the stock solutions shown in **Table 1**.
5. All stock solutions are prepared and autoclaved individually, and stored at room temperature. Kanamycin is filter-sterilized and stored at –20°C.

2.2. mer–lux Assay

1. The assay medium (final volume of 2 mL) is made in 20-mL glass scintillation vials (for monitoring of light production by scintillation counting) or in disposable plastic tubes (for luminometer counting) and consists of: 5 m*M* pyruvate (20 µL of stock, **Table 1**), 67 m*M* Na,K-phosphate buffer, pH 6.8 (140 µL of stock, **Table 1**), and 0.091 m*M* (NH₄)₂SO₄ (20 mL of stock [10 g/100 mL distilled water] diluted 100-fold immediately prior to use).

**Table 1
Medium Constituents, Stock Solutions
and Preparation of Minimal Medium**

Constituent[a]	Stock solution	Vol, mL added to 40 mL growth medium[b]
Na, K-PO_4 (15X)	26.5 g K_2HPO_4, 11.7 g NaH_2PO_4, water[c] to 250 mL, pH 6.8	2.7 (67 mM PO_4)
$(NH_4)_2$ SO_4 (100X)	12 g, water to 100 mL	0.4 (0.12%)
R salts (200X)	8.0 g $MgSO_4 \cdot 7H_2O$; 0.2 g $FeSO_4 \cdot 7H_2O$, water to 100 mL	0.25
Pyruvate (100X)	5.5 g, water to 100 mL	0.4 (5 mM)
Trace elements (1000X)	100 mg $ZnSO_4 \cdot 7H_2O$, 30 mg $MnCl_2 \cdot 4H_2O$, 300 mg H_3BO_3, 10 mg $CoCl \cdot 6H_2O$, 20 mg $NiCl_2 \cdot 6H_2O$, 30 mg $Na_2MoO_4 \cdot 2H_2O$, water to 1 L	0.04
Kanamycin (200X)	10 mg/mL in water	0.2 (50 µg/mL)
Sterile water		36

[a]Numbers in parentheses indicate the concentration factors relative to the final concentrations in the media.

[b]Numbers in parentheses indicate final concentrations in the media.

[c]Distilled water.

2. Mercury stock solution is 5 mM Hg^{2+} as $Hg(NO_3)_2$ in 0.2 N HNO_3. Ten-fold dilutions are made in distilled water. To minimize loss, dilutions are made just before use.

3. Monitoring light emission: All light-monitoring devices are suited for this assay. We have been using the single photon count mode of a Tri-Carb 2500 TR (Packard Instruments, Meriden, CT) scintillation counter (counting conditions: count time per sample: 0.5 min; no. of cycles: 20–30; background correction: none; SPC %HV: 60) and a model BG-P luminometer (GEM Biomedical, Hampton, CT). The latter is particularly suited for field measurements, since it can be operated with a battery.

3. Methods

3.1. Bioindicator Growth

1. Day 1: Transfer 5 mL growth medium to sterile test tube and inoculate with a single colony of the bioindicator. All bioindicator incubations are performed at 37°C with shaking at 200 rpm.

2. Day 2: After 24 h of incubation, 0.5 mL of the culture is transferred to 4.5 mL of growth medium and reincubated over night.
3. Day 3: In the morning, the entire overnight culture is transferred to 20 mL growth medium in an Erlenmeyer flask and incubated for 3 h.
4. Cells are harvested by centrifugation at 12,000g for 10 min at 4°C.
5. Discard supernatant, resuspend pellet in 10 mL 67 mM phosphate buffer, pH 6.8, made from stock solutions in **Table 1**, and repeat centrifugation.
6. Resuspend pellet in 2 mL phosphate buffer.
7. Adjust the optical density of the final cell suspension to an OD_{660} corresponding to a cell density of approx. 2×10^8 cells/mL.
8. Make ten-fold dilutions in phosphate buffer.

3.2. mer–lux *Assay*

3.2.1. Standard Assay

1. Mix the assay medium (see **Subheading 2.2.**) directly in the scintillation vial or test tube.
2. Add sterile water to a final volume of 2 mL (subtract the volumes of cell suspension [0.1 mL] and mercury solution).
   Note: If assays are performed with natural waters, add the same volume of the water sample (*see* **Subheading 3.2.2.**).
3. Add 0.1 mL bioindicator suspension from the appropriate dilution.
   Note: Use the constitutive mutant as a control that light emission is not inhibited by assay conditions (especially important at very high mercury concentrations and in natural waters; *see* **Subheading 3.2.3.**).
4. Initiate assays by adding the appropriate volume of Hg^{2+} solution, and mix gently.
5. Immediately after the addition of mercury, transfer the samples to the scintillation counter and start measuring light emission.

3.2.2. Assays in Natural Waters (see **Note 2**)

1. Label a sufficient number of vials or tubes for the number of samples to be processed, including, as appropriate, replicates, reagent blanks, mercury calibration standards, mercury standard additions, and controls with the constitutive strain HMS174(pRB27) in distilled and natural water aliquots (*see* **Subheading 3.2.3.**). When designing an assay, remember that resolution of light emission kinetics requires that time intervals between recounts of each sample do not exceed 10 min; thus, the upper limit on the number of samples in a batch is determined by the light measurement system (about 10 for a scintillation counter and about 15 for a manually operated luminometer).
2. For a 2-mL assay, prepare a master mixture (3.5 mL Na, K-PO_4, 0.5 µL pyruvate [stocks as in **Table 1**], 5 µL[NH_4]$_2SO_4$, [*see* **Subheading 2.2.**] 0.495 mL distilled water). Pipet 180 µL of the mixture into each vial.
3. If distilled water mercury standards are to be run in the same batch, dispense the appropriate volume of distilled water into those tubes that are labeled as mercury

standards. Typically, a low-range standard series will consist of 0, 12.5, 25, 50, and 100 pM Hg^{2+} (prepared by adding 0, 5, 10, 20, and 40 µL of 5 nM Hg(NO$_3$)$_2$ stock, respectively). The corresponding volumes of distilled water are 1.72, 1.715, 1.71, 1.70, and 1.68 mL.

4. Collect natural water samples in appropriately clean containers (*see* **Note 4**), and quickly transport to the site where assay will be conducted. Use of large-volume containers (e.g., 1-L), even though only a few milliliters are needed for the assay, will reduce bottling effects on sample integrity. Rinse sample containers three times with site water before collection. If the assay is to be run with filtered aliquots, filter in the field as soon after collection as possible. Flush the filter apparatus with site water prior to collection to minimize contamination and to equilibrate the filter apparatus with the water. As discussed in **Note 4**, collection, processing, and handling of natural waters from pristine sites may require more Herculean measures to prevent contamination.

5. Dispense aliquots of each environmental sample into the appropriate vial. For samples that are not spiked with Hg^{2+}, the appropriate aliquot size is 1.72 mL if the cells are added in a volume of 100 µL of PO$_4$ buffer. Sample aliquots should, of course, be adjusted if a different volume of cells is used and if Hg^{2+} stock solution is added to the samples for quantitative analysis (standard additions, *see* **Subheading 3.3.4.**).

6. Add Hg^{2+} to the appropriate vials, and record the starting and ending times of this task. Protect the vials from exposure to bright light, especially sunlight, from this point onward. Incubate for at least 10 min from the starting time recorded above to let Hg^{2+} equilibrate in the water, before adding cells.

7. Cells are added in 30-s intervals (the time it takes to count each sample) and in the same sequence as they will be counted. Cap and thoroughly mix each vial as soon as cells are added.

8. Begin light emission measurements as soon as possible.

3.3. Data Analyses

3.3.1. Hg^{2+}-Dependent Light Production Curves

On the initiation of a *mer–lux* assay, a typical response is observed (**Fig. 2**). First, there is a lag period before light production begins. The length of this lag period depends on Hg^{2+} concentration and cell density. With 10^5 cells/mL, this lag can be as short as 20 min for >0.5 nM Hg^2, and longer than 70 min for ≤0.025 nM Hg^{2+}. In the example presented here (**Fig. 2**), a lag phase of 32 and 38 min was observed in assays containing 0.15 and 0.075 nM Hg^{2+}, respectively. At higher cell densities, a similar response is elicited by a higher range of Hg^{2+} concentrations (*see* **Note 1**). After this lag, light production rate increases logarithmically. With high Hg^{2+} concentrations, this increase is rapid, reaching optimum light production after 20–30 min. The rate of increase is slower with lower Hg^{2+} concentrations (**Figs. 2** and **3A**). Once optimal light yield is reached, the response stabilizes (**Fig. 3A**), and in some cases (mostly at

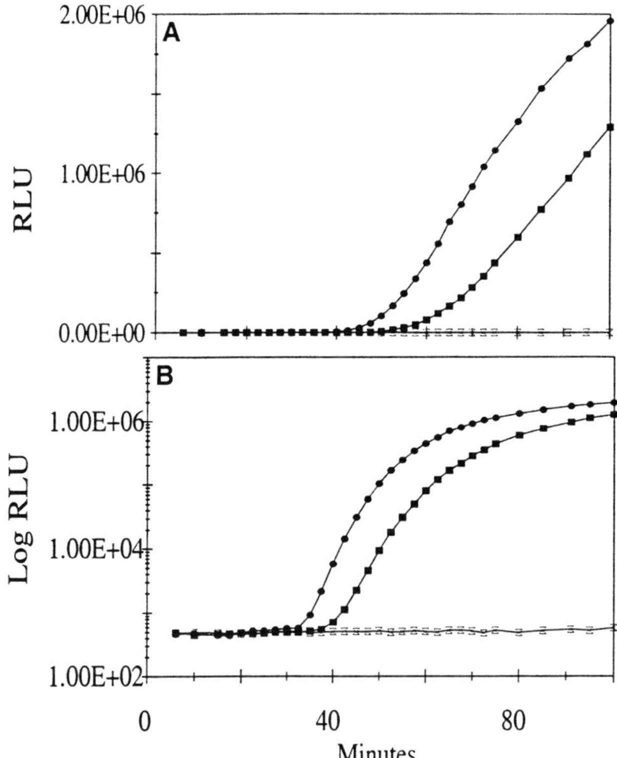

Fig. 2. Light emission by strain *E. coli* HMS174(pRB28) following induction at two different Hg^{2+} concentrations. Assays contained 10^5 cells/mL. Light output was recorded using a BG-P luminometer (*see* text) at two min intervals. Relative light units (RLU) are plotted on a linear **(A)** and logarithmic **(B)** scales. Note that by performing log transformation of light output data, the increase in light production is apparent earlier after induction relative to using untransformed data. , —✕— 0 Hg, —■— 0.075 m*M* Hg, —●— 0.15 m*M* Hg.

high Hg^{2+} concentrations), it even declines, most likely owing to depletion of essential precursors.

3.3.2. Quantitation of Hg²⁺ by Interpretation of Light Production Curves

Three characteristics of the light emission curve are influenced by the concentration of Hg^{2+}: the length of the lag period, the slopes of the increase in light production, and the optimal yield of light (i.e., value of the plateau) **(Fig. 3A–D)**. Of these three, the slope, or the maximal increase in rate of light production $(\delta L\ \delta T^{-1})_{max}$ **(Fig. 3B, C)**, was found most useful for quantitative

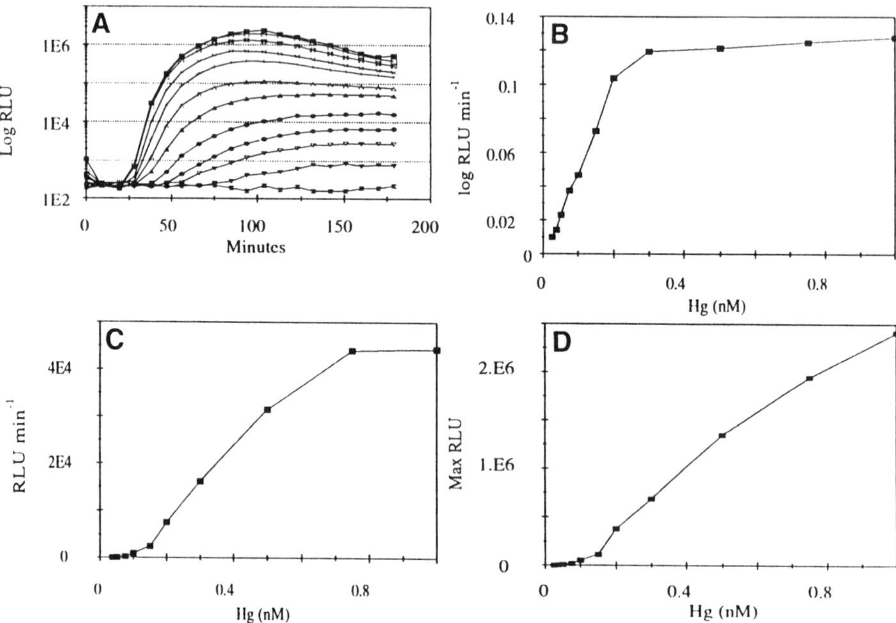

Fig. 3. Quantitation of Hg^{2+} by interpretation of light emission curves. Light emission induced by various concentrations of $Hg(NO_3)_2$ (0–1 nM) (**A**) was analyzed by obtaining expression factors (maximum rate of light increase) using log transformations of light output (**B**) and non transformed data (**C**). A plot of the maximal light output vs Hg^{2+} concentrations is depicted in (**D**).

analysis. This parameter, also termed the expression factor, is calculated from light production curves (**Fig. 3A**) as:

$$(\delta L\ \delta T^{-1})_{max} = (\text{light output at } t_2 - \text{light output at } t_1)/t_2 - t_1 \tag{1}$$

and has the units of quanta/min if photon counts are available (by multiplying counts by the Hastings-Weber constant *[19]*) or RLU/min if luminometer readings are used. t_1 and t_2 are selected in an area of the curves where light production logarithmically (or linearly) increases with time. When the log transformed data is used (**Fig. 3B**) $(\delta L\ \delta T^{-1})_{max}$ are expressed as (log quanta/min). When $(\delta L\ \delta T^{-1})_{max}$ values are obtained for various Hg^{2+} concentrations, a dose–response curve is generated (**Fig. 3B**). If light production data are used without log transformation (**Fig. 3C**), the relationships between Hg^{2+} concentration and expression factors is observed at a higher concentration range than with the log transformed data. These two modes of data presentation expand the range of useful Hg^{2+} concentrations, that can be obtained from each analysis. Overlapping curves are apparent at the highest Hg^{2+} concentrations (**Fig. 3A**), indicating that the reactions are no longer limited by the availability of the inducer.

The useful concentration range of the assay can be controlled by varying the number of cells that are added to the assay (*see* **Note 1**).

Although the optimal light output (**Fig. 3D**) is proportional to Hg^{2+} concentrations, these values are not reproducible and are therefore not useful, for quantitative measurements. Likewise, the length of the lag period is not useful because taking measurements at close intervals to allow accurate determinations is not practical with the described instrumentation.

For most aqueous environmental samples, the lower Hg^{2+} concentration range would be most often encountered. Even in highly polluted situations where the total mercury concentration is 0.5 n*M* or more, the bioavailable Hg^{2+} is likely to be only a fraction of this total. Thus, in most cases, the most useful factor expressing Hg^{2+} bioavailability is the logarithmic increase in rate of light production, $(\delta L\ \delta T^{-1})_{max}$.

3.3.3. Use of a Constitutive Luminescent Control

Changes in light production patterns during *mer–lux* assays might be the result of effects on the light-emitting reaction rather than alteration in Hg^{2+} availability, because the light-emitting reaction is highly sensitive to changing environmental conditions *(20,21)*. To rule out this possibility and to assure that the bioindicator performs well in natural waters that might contain inhibitors, an isogenic strain of HMS174(pRB28) containing a mutant plasmid that carries a constitutively expressed *lux,* designated pRB27, is used as a control. The nature of this mutation has not been investigated, but a small deletion upstream from the *mer* insert in pRB28 (**Fig. 1**) might be responsible for the altered phenotype *(16)*. Changes in patterns of light production by strain HMS174(pRB27) relative to a standard assay suggest that the experimental conditions affect light production, and consequently, patterns demonstrated by HMS174(pRB28) might not be exclusively owing to availability of Hg^{2+}. For example, dissolved organic carbon (DOC) at high, but not at low, concentrations inhibited light production by the constitutive control (**Fig. 4**), cautioning that this bioindicator could only be used to study how Hg^{2+} bioavailability is affected at low DOC levels. This experiment and its interpretation are described in further detail elsewhere *(16)*.

3.3.4. Applications of mer–lux Assays in Natural Water

An "apparent" bioavailable mercury concentration (*see* **Note 3**) using expression factors measured in natural water may be calculated using a standard curve constructed in distilled water. This approach assumes that all the mercury in the distilled water is bioavailable, and thus, the response of the indicator represents the maximum response for each mercury concentration. This assumption has not yet been experimentally verified, and there are reasons to believe it incorrect. At least one component (pyruvate) of the assay media may

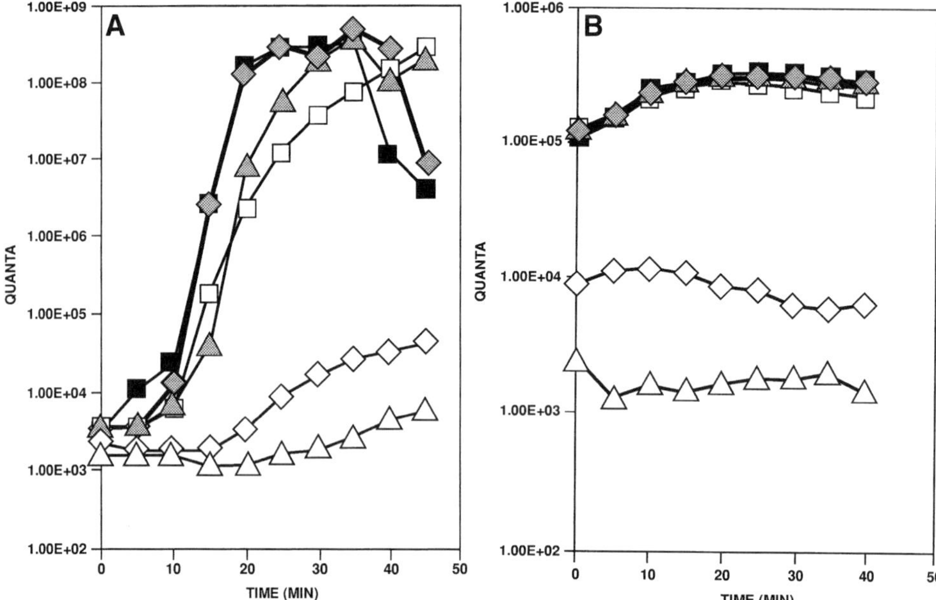

Fig. 4. The use of strain HMS174(pRB27) to verify that changes in light patterns are due to Hg^{2+}-dependent induction rather than to effect on the light-emitting reaction. *mer–lux* assays were performed in the presence of 0 (■), 2.0 (◇), 3.7 (△), 37.4 (□), 93.8 (◇), and 187.3 (△) µg/mL. Assays with the Hg^{2+} induced *mer–lux* in strain HMS174(pRB28) **(A)** and assays with the constitutive control strain HMS174(pRB27) **(B)** are presented.

attenuate bioavailability, and there is some evidence that additions of chloride affect bioavailability *(16)*. Thus, Hg^{2+} in the standard assay may not be entirely bioavailable.

An alternate approach to standardization of the assay in natural waters is taken from analytical chemistry where this method of quantitation is known as the "method of additions" (standard additions) and is practiced where complex sample matrices preclude quantitation using a standard curve developed in distilled water *(22,23)*. In the application to the *mer–lux* assay, increasing amounts of Hg^{2+} are added to water samples, and the resulting relationships (R^2 = 0.9965)between response and amount of Hg^{2+} added (**Fig. 5A**) are used to extrapolate to an equivalent amount of bioavailable Hg^{2+} in an unspiked sample (**Fig. 5B**). In the example presented in **Fig. 5**, the estimated concentration of bioavailable Hg^{2+} was 18 pM (3.6 ng/L). Total and total dissolved mercury concentrations measured for the same sample were 800 and 125 pM (160 and 25 ng/L). In contrast, use of a calibration curve obtained in distilled water as a basis of quantitating bioavailable Hg^{2+} resulted in an estimate of 4.5 pM (0.9 ng/L).

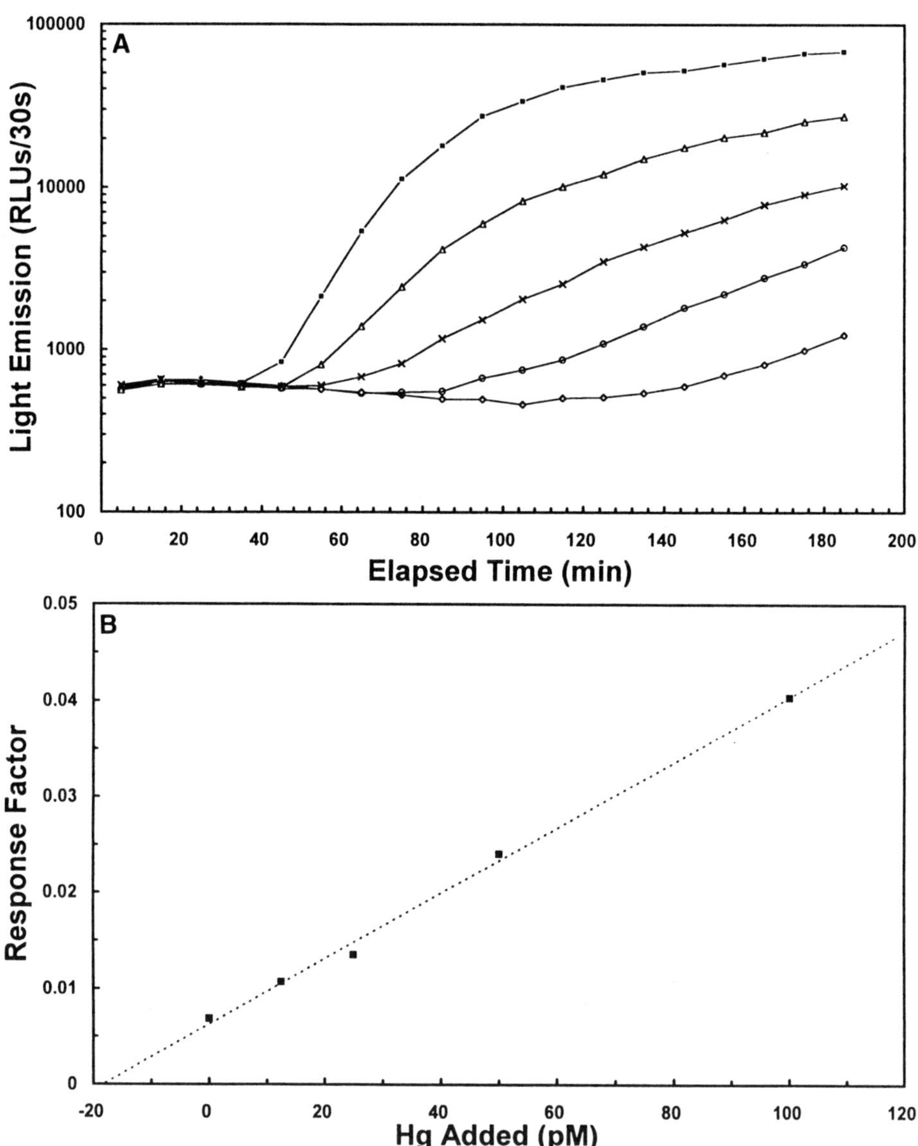

Fig. 5. Application of the standard addition approach to estimation of the concentration of bioavailable Hg^{2+} in a natural water sample. **(A)** The indicated concentrations of $Hg(NO_3)_2$ were added to aliquots of a water sample collected from a small stream contaminated with mercury inputs in the headwaters about 10 km upstream. Induction of light production with strain HMS174(pRB28) was followed using a luminometer. $\Diamond$–0; ■–100 p*M*; $\triangle$–50 p*M*; x–25 p*M*; $\bigcirc$–12.5 p*M* $Hg(NO_3)_2$. **(B)** Response factors ([log δL δT^{-1}]$_{max}$; expression factors) were calculated from the slope of each of the curves in **(A)** and plotted against the concentration of added Hg^{2+}. Extrapolation of the obtained curve estimated 18 p*M* of bioavailable Hg^{2+}.

The method of addition approach suits the *mer–lux* assay because Hg^{2+} added to environmental samples is rapidly sorbed to particles and complexed with ligands that are native to the sample. However, above some quantity of added mercury, the bioavailable fraction, as measured by this assay, will begin to increase in proportion to the amount added. That proportionality should capture the properties of the natural water sample that affect bioavailability of Hg^{2+} and should be useful in deriving an estimate of bioavailable mercury in the unamended sample.

4. Notes

1. The sensitivity range of the *mer–lux* assay is controlled by the density of biomass that is added to the assay. At each cell density, a plot of expression factors vs Hg^{2+} concentration yielded a sigmoidal response typical to a response that is controlled by activation of the *mer* promoter *(24)*. The range of Hg^{2+} concentrations at which this response occurred was higher when more cells where added to the assay (**Fig. 6**). Thus, the higher the cell number, the less sensitive the assay. At each cell density, there is a lower Hg^{2+} concentration below which no induction of light production occurs and an upper concentration above which the response is saturated with respect to Hg^{2+} (i.e., further addition of Hg^{2+} does not increase the expression factor). With 3×10^7 cells/mL, the linear range of increase in the values of expression factors spans 3–30 n*M*, with 10^7 cells/mL it spans 0.3–1 n*M*, and at cell density of $\leq 10^6$ cells/mL, linearity is observed between 0.03 and 0.3 n*M*. This inverse relationships of biomass to sensitivity of the assay is owing to competition for Hg^{2+}, between MerR and other cellular binding sites for Hg^{2+}, because the gradual addition of strain HMS174 to 10^6 cells of HMS174(pRB28) caused a decline in light production in assays containing 0.25 n*M* Hg^{2+} *(25)*.

   The practical application of cell density-dependent sensitivity of the *mer–lux* assay is in the ability to detect bioavailable Hg^{2+} in natural waters containing a wide range of mercury concentrations from samples collected at highly contaminated sites to those from some pristine locations. For example, Turner et al. *(17)* who employed the *mer–lux* bioindicator to measure bioavailable Hg^{2+} along a gradient in a contaminated stream, used high cell densities with samples collected near the contamination source and lower cell densities at downstream locations.

2. Performance of the assay in natural water entails additional considerations and some minor modifications of the basic assay procedure described in **Subheading 3.2.2.** It is paramount that natural water samples not be stored for very long prior to performance of the assay. The speciation of mercury can change very rapidly once a sample is bottled and moved from the environment. The inner surface of the bottle may adsorb bioavailable mercury and may also catalyze reduction of the bioavailable mercury to elemental mercury. In addition, mercury in natural water is vulnerable to photolysis *(26)* and thus, protection of samples from exposure to bright sunlight may be important. The best approach is to prepare all

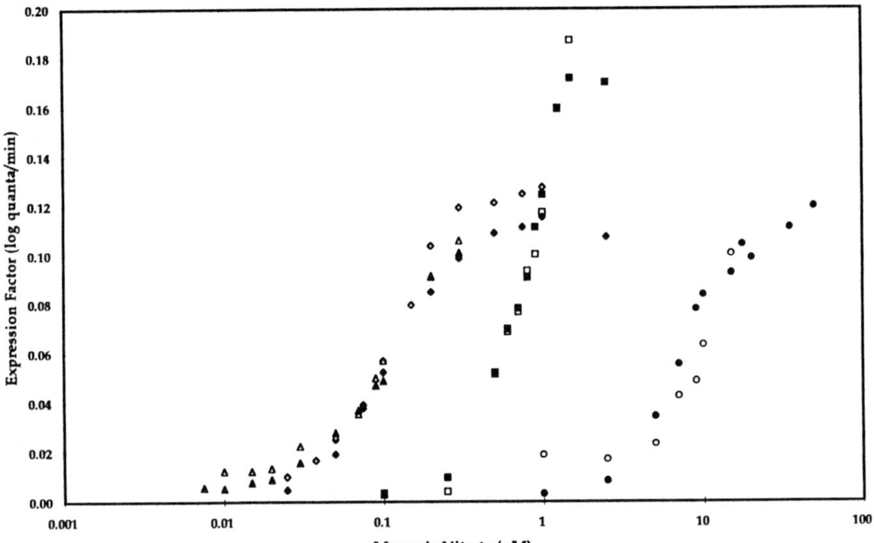

Fig. 6. Cell-density dependent sensitivity of the *mer–lux* assay. *mer–lux* assays were performed in distilled water supplemented with essential assay constituents employing 3×10^7 (○ and ●, duplicate experiments), 10^7 (■ and □), 10^6 (▲ and △), and 10^5 (◆ and ◇) cells/mL. With increased cell density, the response of strain HMS174(pRB28) was at a higher range of Hg^{2+} concentrations.

materials for the assay prior to collecting samples and then work quickly to minimize sample storage time. With environmental samples, it is highly likely that several sets of assays will have to be run to bracket the appropriate cell density and mercury calibration range (*see* **Note 1**). Thus, it is not possible to specify in advance the best assay conditions. The best approach is to initially select the most sensitive assay conditions (low cell density and low Hg calibration range) and then adjust subsequent assays as necessary to achieve optimal response.

3. The ultimate goal in the development of the *mer–lux* assay is a tested method for the determination of absolute values of bioavailable Hg^{2+}. Each of the two approaches described in **Subheading 3.3.4.** has its weaknesses. Calibration against a standard curve that is constructed in distilled water does not take into account effects of surfaces and ligands on availability of Hg^{2+}. Although the standard addition approach more clearly reflects interactions of Hg^{2+} in natural waters, the extrapolated values cannot be considered absolute concentrations of bioavailable Hg^{2+} because they rely on the concentrations of Hg^{2+} that were added, and not on those available. The fraction of the added Hg^{2+} that is available is unknown, and consequently, the bias that is introduced into the determinations by interactions cannot be taken into consideration. Further development and new approaches are needed if the *mer–lux* assay is to be used for absolute determinations.

However, expression factors calculated as described in section **Subheading 3.3.2.** and in conjunction with other analyses of mercury in the same water samples could provide important information on a comparative basis. For example, factors may decrease with increasing distance from a mercury source along a stream channel *(17)*. Similarly, expression factors may also show a high statistical correlation with total dissolved mercury and rates of biotic reduction of mercury. These patterns are useful even if results cannot be converted to absolute concentrations of bioavailable mercury.

4. Successful application of the assay depends significantly on maintaining very low mercury concentrations (sub part per trillion) in all assay solutions. Contamination can arise from numerous sources including:

   a. Reagents and the distilled water used to prepare nutrient solutions;
   b. Containers and transfer pipets; and
   c. The laboratory atmosphere where the assay solutions are prepared, stored, and used.

   All reagents used should be analytical reagent-grade chemicals or higher purity grades. Distilled water should be of the highest quality and should be immediately filtered before use. If deionization is employed as part of the process of producing high-quality water, beware of and avoid systems that are regularly sanitized with hypochlorite (major source of mercury contamination) and those that may use sodium-exchange resins charged with sodium produced by the mercury cell process. Assume that all labware used to prepare and store chemicals may be mercury-contaminated, and clean immediately before use. However, most disposable sterile pipets, pipet tips, and small test tubes are sufficiently clean when first removed from their sterile packaging, and do not cause problems with the assay. Glass scintillation vials may not be clean enough, and should be soaked for a few minutes in 2 N HNO$_3$ and then rinsed profusely with distilled water. Scintillation vial caps with aluminum liners should not be cleaned with acid, but simply rinsed with distilled water. Vials and caps can be air-dried preferably on a clean bench (specifically designed to remove mercury from air flow) or in a mercury-free drying oven. Other glass, plastic, and Teflon labware can be cleaned in the same manner. Muffling Pyrex glassware at 500°C overnight is also an effective means of decontamination. Laboratory air often contains much higher mercury vapor concentrations than outdoor air because of past mercury spillage in the laboratory. Drying ovens in which mercury thermometers may have been used in the past should be avoided or confirmed to be free of mercury contamination before use. The final measure of whether contamination has been prevented is no response in assays employing HMS174(pRB28) containing all assay components, but no added Hg^{2+}. Such "reagent blanks" should be run with each batch of assays.

Acknowledgments

Thanks are due to Paul Humeuchuk for technical assistance. Preparation of this chapter was supported by contract RP8021–10 between The Electric Power

Research Institute and Ramot of Tel Aviv University. The second author was supported by a Senior Research Associateship under the auspices of the National Research Council-USEPA/NHEERL at Gulf Breeze, Florida. Research by CAK and JWMR was supported by a grant from Alcoa (Pittsburgh, PA).

References

1. Hudson, R. J. M, Gherini, S. A. Watras, C. J., and Porcella, D. B. (1994) Modeling the biogeochemical cycle of mercury in lakes: The mercury cycling model (MCM) and its application to the MTL study lakes, in *Mercury Pollution: Integration and Synthesis* (Watras, C. J. and Huckabee, J. W., eds.), Lewis Publishers, Boca Raton. FL pp. 473–523.

2. Watras, C. J. and Bloom N. S. (1992) Mercury and methylmercury in individual zooplankton: implications for bioaccumulation. *Limnol. Oceanogr.* **37,** 1313–1318.

3. Mason, R. P., Reinfelder, J. R., and Morel, F. M. M. (1996) Uptake, toxicity, and trophic transfer of mercury in a coastal diatom. *Environ. Sci. Technol.* **30,** 1835–1845.

4. Barkay, T., Turner, R., Saouter, E., and Horn, J. (1992) mercury biotransformations and their potential for remediation of mercury contamination. *Biodegradation* **3,** 147–159

5. Silver, S. and Phung L. T. (1996) Bacterial heavy metal resistance: new surprises. *Annu. Rev. Microbiol.* **50,** 753–789.

6. Bloom, N. S. (1989). Determination of picogram levels of methylmercury by aqueous phase ethylation, followed by cryogenic gas chromatography with cold vapor atomic fluorescence detection. *Can. J. Fish. Aquat. Sci.* **46,** 1131–1140.

7. Fitzgerald, W. F. and Gill, G. A. (1979) Subnanogram determination of mercury by two-stage gold amalgamation and gas phase detection applied to atmospheric analysis. *Anal. Chem.* **15,** 1714–1720.

8. Selifonova, O., Burlage, R., and Barkay, T. (1993) Bioluminescent sensors for detection of bioavailable Hg(II) in the environment. *Appl. Environ. Microbiol.* **59,** 3083–3090.

9. Tescione, L. and Belfort, G. (1993) Construction and Evaluation of a metal ion biosensor. *Biotechnol. Bioeng.* **42,** 945–952.

10. Condee, C. W. and Summers, A. O. (1992) A *mer–lux* transcriptional fusion for real-time examination of in vivo gene expression kinetics and promoter response to altered superhelicity. *J. Bacteriol.* **174,** 8094–8101.

11. Virta, M., Lampinen, J., and Karp, M. (1995) A luminescence-based mercury biosensor. *Anal. Chem.* **67,** 667–669.

12. Summers, A. O. (1992) Untwist and Shout: a heavy metal-responsive transcriptional regulator. *J. Bacteriol.* **174,** 3097–3101.

13. Choi, S.-C., Chase, Jr. T., and Bartha, R. (1994) Metabolic pathways leading to mercury methylation in *Desulfovibrio desulfuricans* LS. *Appl. Environ. Microbiol.* **60,** 4072–4077.

14. Rogowsky, P. M., Close T. J., Chimera, J. A., Skaw, J. J., and Kado, C. I. (1987) Regulation of the vir genes of *Agrobacterium tumefaciens* plasmid pTiC58. *J. Bacteriol.* **169,** 5101–5112.

15. Selifonova, O. V. and Barkay, T. (1994) Role of Na^+ in transport of Hg^{2+} and induction of the Tn21 *mer* operon. *Appl. Environ. Microbiol.* **60,** 3503–3507.

16. Barkay, T., Gillman, M., and Turner, R. R. (1997) Effects of dissolved organic carbon and salinity on bioavailability of mercury. *Appl. Environ. Microbiol.,* in press.

17. Turner, R. R., Barkay, T., Bloom, N. S., and Rasmussen, L. D. Application of an indicator for the availability of mercury to microorganisms in natural water. *Chemosphere,* in preparation.

18. Campbell, J. L., Richardson, C. C., and Studier, F. W. (1978) Genetic recombination and complementation between bacteriophage T7 and cloned fragments of T7 DNA. *Proc. Natl. Acad. Sci. USA* **75,** 2276–2280.

19. Hastings, J. W. and Weber, G. (1963) Total quantum flux of isotropic sources. *J. Opt. Soc. Am.* **53,** 1410–1415.

20. Hastings, J. W, Potrikus, C. J., Gupta, S. C., Kurfürst, M., and Makemson, J. C. (1985) Biochemistry and physiology of bioluminescent bacteria. *Adv. Microb. Physiol.* **26,** 235–291.

21. Meighen, E. A. (1988) Enzymes and genes from the *lux* operons of bioluminescent bacteria. *Ann. Rev. Micorbiol.* **42,** 151–176.

22. Larsen, I. L., Hartmann, N. A., and Wagner, J. J. (1973) Estimating precision for the method of standard additions. *Anal. Chem.* **45,** 1511–1513.

23. Willard, H. H., Merritt, L. L., Jr., and Dean, J. A. (1958) *Instrumental Methods of Analysis.* D. Van Nostrand, Princeton, NJ.

24. O'Halloran, T. V. (1993) Transition metals in control of gene expression. *Science* **261,** 715–725

25. Rasmussen, L. D., Turner, R. R., and Barkay, T. (1997) cell-density dependent sensitivity of a *mer–lux* bioassay. *Appl. Environ. Microbiol.* **63,** 3291–3293.

26. Amyot, M., Mierle, G., Lean, D. R. S., and McQueen, D. J. (1994) Sunlight-induced formation of dissolved gaseous mercury in lake waters. *Environ. Sci. Technol.* **28,** 2366–2371.

21

A Panel of Stress-Responsive Luminous Bacteria for Monitoring Wastewater Toxicity

Shimshon Belkin

1. Introduction

In several chapters in this book (Chapters 7, 8, 12, 13, 17–22), as well as in other instances *(1–10)*, bacterial constructs were presented that respond by light emission to different environmental stresses. Such strains can serve as very useful tools for the study of bacterial responses to stress, but also as potentially efficient means for environmental monitoring. In the environmental category, one may envisage two general types of luminous constructs: those that respond to the presence of specific inducers, such as heavy metals *(4–7)* or certain organics *(8–10)*, and those in which luminescence is induced by broader classes of toxicants. In the latter group are included strains carrying fusions of a luminescent reporter to a promoter controlled by a global regulatory circuit *(1–3)*.

One of the potential applications of such strains is their use as test organisms for toxicity bioassays. Compared to the more developed organisms normally used for such purposes, such as fish or crustaceans, the advantages offered by microbial toxicity testing include high sensitivity, low costs, large homogenous test populations, and—most importantly—rapid responses *(11)*. Modern molecular biology techniques allow, in addition, the design of specific bacterial tester strains to detect selected classes of toxicants.

Several publications have recently promoted the use of a panel of genetically engineered stress-responsive luminous bacteria for ecotoxicity monitoring *(12–14)*. Members of such a panel can provide a spectrum of potential responses to a variety of toxicants; the data obtained should not only indicate the presence of toxic compounds, but also provide some information on their nature. In this chapter, the methodologies underlying this approach shall be presented, as related to the determination of toxicity in industrial wastewaters.

From: *Methods in Molecular Biology, Vol. 102: Bioluminescence Methods and Protocols*
Edited by: R. A. LaRossa © Humana Press Inc., Totowa, NJ

**Table 1
Four Stress-Responsive Luminous
Constructs Serving as the Toxicity Panel Members**

Strain	Regulatory circuit	Promoter used	Damage type indicated	Reference
DPD2511	*oxyR*	*katG*	Oxidative (peroxides)	*3*
DPD2515	*soxRS*	*micF*	Oxidative (superoxides)	*15*
DPD2794	"SOS" (*recA/lexA*)	*recA*	DNA damage	*15*
TV1061	"Heat shock" (*rpoH*)	*grpE*	General/protein damage	*1*

Selection of bacterial panel members can be dictated by the expected characteristics of the samples to be monitored on the one hand, and by available tester strains on the other hand. Limitations in both parameters may dictate the design and construction of additional strains, "tailored" to suit specific needs. For unknown samples, however, several "general responders" are recommended. In this chapter, the use of a group of strains, listed in **Table 1**, is presented.

All of these strains respond by increased luminescence to specific classes of toxicants, with varying degrees of specificities. Although the responses of strain DPD2794 *(recA'::lux)*, for instance, are mostly limited to compounds that endanger cellular DNA, strain TV1061 *(grpE'::lux)* responds to almost any compound that threatens the well-being of the cell. It is important to note that even without induction, there is a basal level of background luminescence, the intensity of which varies between the constructs. Thus, in addition to "lights on" effects by compounds acting on the specific *lux*-fused promoter, nonspecific "lights off" effects on basal luminescence can also be observed and quantified. Such decreases in the control luminescence can serve as a general toxicity indicator, similarly to the commercially available Microtox™ test *(16)*.

2. Materials

1. Luminometer: Numerous instruments are available for the sensitive quantification of photon fluxes (*see 17* for a recent review), and to a certain extent, details of the experimental procedure will be dictated by luminometer choice or availability. The methodology described here is structured to suit a microtiter-plate luminometer, which allows sensitive quantification of the light emitted from all 96 wells of a standard-size microtiter dish. Some simple modifications would be

needed to adapt the technique to other light measuring devices. Two microtiter plate instruments were used by the author: a Dynatech (Chantilly, VA) ML3000 and an Anthos Labtech (Salzburg, Austria) Lucy 1. Both allow incubation of the plate at a controlled temperature with shaking, and reading of the luminescence emitted from each of the wells at predetermined intervals. Similar instruments are available from other manufacturers, and the use of the specific models mentioned above does not imply their superiority to others.

2. Microtiter plates: Standard-size 96-well plates are used. To prevent light transfer between wells, it is essential that opaque (either white or black) plates are selected. Such plates are available from various manufacturers, and the author did not find one brand to be preferable. The methods outlined in this chapter refer to the standard A–H and 1–12 notation for rows and columns, respectively.

3. Bacterial strains: **Table 1** lists the four panel members serving as the example in this chapter. The last column refers to the publications in which the construction of the strains was reported in detail or in which their use was reported for the first time.

   Growth and maintenance of these strains are routinely carried out in LB medium *(18)*, in the presence of kanamycin (25 mg/L). Although routine strain maintenance can be carried out at 37°C, growth at 26°C is recommended in preparation for experiments (*see* **Note 4**).

4. Experimental media: Two simple sterile media are used in the procedures described here—single- and double-strength LB broth, without antibiotics.

5. Sample preparation: The procedure below allows for testing wastewater samples at a concentration of up to 25% of their original strength. Relatively simple modifications will be required to expand this range, as will be briefly described under **Subheading 4.**

   For wastewater samples of unknown composition, it is important that their total salinity is known prior to the test. This can be carried out by determining total dissolved solids *(19)*. A close approximation can be obtained by determining the samples' electrical conductivity. This information helps to differentiate between specific effects of wastewater components and nonspecific effects, which may be caused by parameters, such as the sample's salinity. In general, it is important that after dilution, the salinity of the highest sample concentration will have no significant effect on bacterial luminescence. This is normally achieved at a final dissolved solids concentrations of 0.5% or lower, thus limiting either the salinity of the samples to be tested or the highest concentration to be assayed. Even at total salts concentrations higher than 0.5%, however, simple sample-specific control tests may reveal no deleterious effects.

   To prevent undesired pH and turbidity effects, it is essential that the samples are neutralized to pH 7.0 and clarified, if necessary, of all suspended material by centrifugation or filtration.

6. Photometer or colorimeter: To preserve reproducibility, it is essential that cells from a constant growth phase are routinely used. It is therefore important to monitor growth for several hours until the desired cell density is reached, by any device allowing the determination of optical density. An old-fashioned Klett-Sumerson colorimeter, coupled with side-armed growth flasks, is very practical.

3. Methods

3.1. Experimental Design

1. It is important that the plate is designed well in advance, preferably with the aid of a blank 8 × 12 table.
2. When designing the screening of a large number of samples by a set of bacterial strains, one is often faced with the choice between two options: devoting a plate to as many samples as possible, thereby limiting the number of tester strains per plate, or increasing the number of strains simultaneously used, thus limiting the number of samples. Assuming a fixed number of panel members, the real magnitude of the problem can only be determined when the number of samples is known. Since it is likely that the number of samples will often greatly surpass the number of strains, the suggestions below reflect an attempt to maximize the number of samples using one strain per plate (*see* **Notes 1** and **2** for an additional discussion of this point).
3. Two sample procedures are described below, designed for testing either 7 double dilutions of each sample, with duplicates (yielding 6 samples/plate) or 4 double dilutions/sample, with no duplicates (allowing 24 samples/plate). With a proper design of sample preparation in the plate, almost any combination is possible.
4. If duplicates are planned for, they should be in adjacent columns.

3.2. Plate Preparation—Six Samples, Seven Dilutions Each (see Note 7)

1. Prepare the plate with the various dilutions of the tested samples in advance, so that when the cells are ready, they can be immediately introduced into the plate and luminescence monitoring initiated (*see* **Note 3**).
2. Use brand-new opaque white microtiter plates.
3. Clearly mark column pairs or well groups destined for each compound or strain.
4. Place 50 µL of a twofold concentrated LB medium in all wells in row A, and 50 µL of regular strength LB in all other wells (rows B–H).
5. Add 50 µL from the tested samples to wells in row A to yield a temporary concentration of 50%. Place duplicates in adjacent columns.
6. Generate a twofold dilution series "downward" along the columns by progressively transferring 50 µL from well to well, mixing lightly at each step (*see* **Note 5**). Avoid splatter. Discard 50 µL after mixing the contents of row G; **Do not touch row H**.
7. The plate should now contain 50 µL in all wells, each column containing a dilution series of the tested compound, from 50% in row A to 0.78% in row G. Row H, containing LB only, will serve as the zero control. Actual sample concentrations will be halved after addition of cells, to range between 25 and 0.39% (*see* **Note 6**).

3.3. Plate Preparation—24 Samples, 4 Dilutions Each (see Note 7)

1–3. As in **Subheading 3.2.**
4. Place 50 µL of a twofold concentrated LB medium in all wells in rows A and E, and 50 µL of regular strength LB in all other wells (rows B–D and F–H).

5. Add 50 µL from the tested samples to wells in rows A and E to yield a temporary concentration of 50% in a final volume of 100 µL. With no duplicates, 24 individual samples can be tested in this manner.

6. Generate twofold dilution series "downward" along the columns, by progressively transferring 50 µL from well to well, mixing lightly at each transfer (*see* **Note 5**). This should be carried out in two independent sets: rows A–D and rows E–H. Discard 50 µL after mixing the contents of rows D and H. Discard the tips, and load fresh ones after row D. Avoid splatter.

7. In this procedure, the dilution series does not contain a zero control. It is therefore essential that one of the samples tested will be an appropriate blank (LB, an appropriate buffer, or a mineral medium of a relevant ionic strength).

8. The plate should now contain 50 µL in all wells, each column containing two dilution series of two independent samples, at concentrations of 50, 25, 12.5, and 6.25% in rows A–D and E–H. Actual sample concentrations will be halved after addition of cells to range between 25 and 3.13% (*see* **Note 6**).

3.4. Cell Preparation and Experiment Initiation

1. Grow cells overnight in LB medium at 26°C (*see* **Note 4**), with shaking, in the presence of 25 mg/L kanamycin (or 50 mg/L ampicillin; resistance to both antibiotics is coded for by the plasmid containing the promoter::*lux* fusion).

2. Dilute the cells 100-fold into fresh LB without antibiotics, and reincubate at 26°C with shaking (*see* **Note 9**).

3. Follow growth of the culture for a few generations, until a predetermined cell concentration is reached. We have routinely used a cell density yielding 20–40 Klett units (filter 54), approximately corresponding to 2–4×10^8 cells/mL. Time from inoculation is ca. 3 to 4 hours (generation time of *Escherchia coli* at 26°C is close to 1 h).

4. Prepare the plate with the serially diluted samples in advance according to **Subheadings 3.2.** or **3.3.** above, or as dictated by your experimental needs. It is recommended that the plate is ready and at room temperature 30–60 min before the expected cell density is reached.

5. Remove the culture from the shaker, add 50 µL to each of the wells in the preprepared plate, insert into the luminometer, and immediately start monitoring luminescence. The luminometer should be set at 26°C, with shaking, and a reading taken approximately every 10 min for 3 h (*see* **Notes 8** and **10**).

3.5. Data Analysis

1. Different luminometers present data in different modes, not always immediately amenable to simple plotting of time-courses and dose–responses. It is therefore important to first transform the data generated by the luminometer into a table with the time-points in the first column and the luminescence reading for each well, column by column, in the following 96 data columns. In this seemingly trivial suggestion often lies the solution for a reasonable viewing of the enormous amount of data that can be generated in a single run (*see* **Note 11**). A simple

Excel (or equivalent) spreadsheet or macro can be designed for this purpose and routinely used. If samples were prepared in duplicate, as in **Subheading 3.2.**, the program should also calculate averages, yielding 48 data columns instead of 96.

2. Plot the kinetic response of the cells for each sample concentration. The data collected from a plate containing six samples, as in **Subheading 3.2.**, will now be presented in six figures, each containing eight curves. Each curve of a figure portrays the cellular response to a different concentration of a particular sample. A plate prepared as in **Subheading 3.3.** (24 samples, no duplicates) will potentially yield 24 figures with 4 curves each. Although it is not always essential, an observation of the kinetics of light development is often a good way to develop a feel for the effect of unknown samples on the induction or inhibition of luminescence in the tester strains.

   Time-course curves can take three general forms:
   a. If no induction takes place, luminescence should stay constant or not vary from the level obtained with the untreated control.
   b. If luminescence in the tester strain is induced, light emission should increase with time in a dose–responsive manner.
   c. If the sample is toxic to the cell in general or to the bioluminescence in particular, a decrease in bioluminescence will take place. This response should also be dose–dependent.

   It is likely that two or even all three effects may be expressed by a single sample, exerting no effect at the lowest concentration range, a toxic effect at the highest, and an inductive response in between. *Note:* This is the main reason why it is essential that a concentration gradinet of each sample is tested.

3. The time-course data plotted in **Step 2** above can be presented either as actual luminescence values (normally presented as the arbitrary light units of the specific instrument used) or as the as the ratio of the luminescence of the induced samples to that of the uninduced control, representing the degree to which luminescence was induced (for ratios higher than 1) or inhibited (ratios lower than 1).

4. Plot the responses as a function of sample concentration; this will reduce the data in each of the time-course figures into a single curve, and allow for the first time a quantitative estimate of the samples' effects. For this purpose, each of the time-course curves should be represented by a single data point. Several options exist for the selection of this representative value, including luminescence at a specific time-point, or maximal luminescence observed in the course of the run. Two other options are recommended here: maximal or minimal response ratios, for samples exhibiting inductive or toxic effects, respectively.

5. **Figure 1** presents, as an example, data collected from wastewater samples of two chemicals factories, before and after biological treatment, transformed as detailed in **Steps 1–4** above. The strain used in both cases was DPD2794, reporting on DNA damage hazards (*see* **Table 1**).

   For factory A, maximal ratios are presented (**Fig. 1A**), indicating that the raw influents were indeed DNA-threatening, thus being potentially mutagenic; this effect was strongest at 5%; at higher concentrations, a toxic effect was apparent

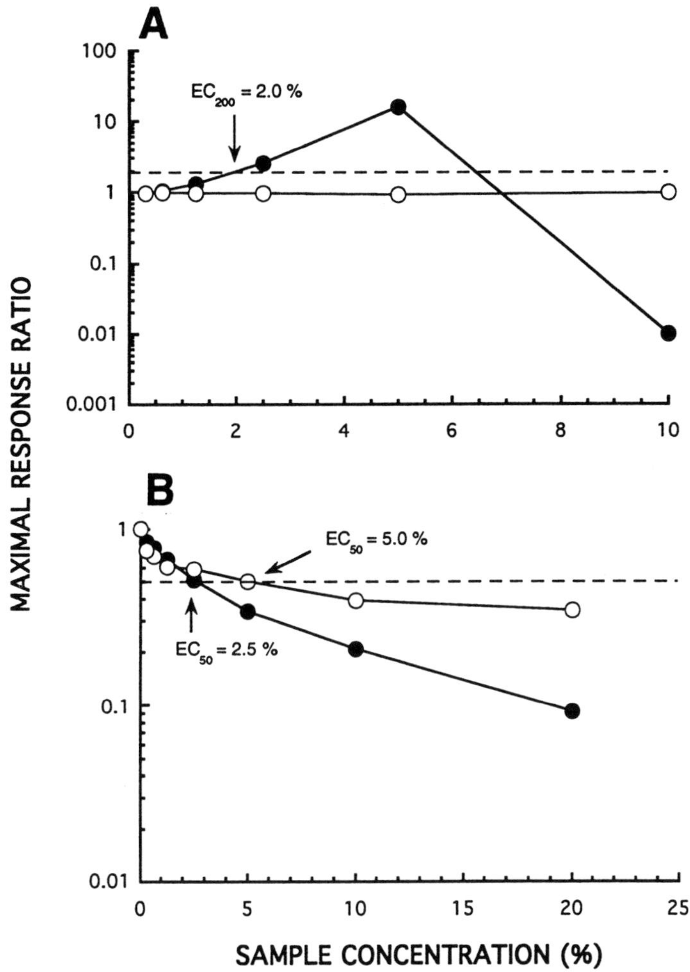

Fig. 1. Calculated response ratios of strain DPD2794 *(recA'::lux)* to raw and treated wastewaters of two chemical factories. **(A)** A clear inductive response, indicating potential genotoxicity, is abolished by treatment. **(B)** A nonspecific toxic response, partially removed by treatment. ——●—— Raw influent, ——○—— treated effluent.

that inhibited luminescence and masked the genotoxic potential. Both threats—toxicity and genotoxicity—were apparently removed during treatment, since the effluents exhibited a response ratio close to 1 at all concentrations. For factory B, no inductive effect was observed **(Fig. 1B)**, thus signifying no potential genotoxic hazards. Some of the components in the untreated wastewater, however, had a clear general toxic effect that could be quantified just as easily as the inductive one. This effect was only partially removed by treatment.

6. The response of each strain to the different samples can be further described by a single value, a constant for each sample/strain combination. This value describes the sample concentration causing a twofold increase in luminescence (in case of an inductive "lights on" effect, as in **Fig. 1A**) or the concentration causing a twofold decrease in luminescence (in the case of a toxic "lights off" effect, as in **Fig. 1B**). The latter parameter, analogous to an LD_{50}, is known as an EC_{50} *(16)* and has been well-characterized for other toxicity tests, especially Microtox. The former has recently been termed EC_{200} *(13,14)*. The lower this value is, the more toxic is the sample to the tester strain used. EC_{200} can be simply determined from the intersection of the dose–response curve, or its extrapolation, and the response ratio = 2 line. In the example presented in **Fig. 1**, the raw wastewater of factory A was characterized by an EC_{200} value of 2% for strain DPD2794. The wastewater of factory B, on the other hand, exhibited an EC_{50} value of 2.5 and 5.0% before and after treatment, respectively.

7. If sufficient data points are available, it is recommended that the EC_{200} is calculated to a higher degree of precision in the following manner *(13,14)*:

    a. for each sample concentration, calculate a gamma (G) value according to

    $$\Gamma = (I_s - I_o)/I_o \tag{1}$$

    where I_s is the maximal luminescence obtained for the given sample concentration s, and I_o is the luminescence of the control at the same time.

    b. plot Γ as a function of sample concentration. On a log-log scale, the dose-dependent segment of the response should generate a straight line.

    c. Since $s = EC_{200}$ when $\Gamma = 1$, as evident from the equation above, the intersection of this line with $\Gamma = 1$ should provide the EC_{200} value.

8. If the bacterial panel is exposed to a set of samples and the responses calculated as described above, an attempt to draw conclusions regarding the toxic nature of the samples can now be made. **Table 2** provides, as an example, a set of data obtained for the combined wastewater stream of a chemical factory, before and after biological treatment. The data are presented in a simplified manner to clarify the emerging pattern: in the raw influents, a general toxic effect (inhibition of luminescence) was observed across the panel. Only two of the strains also exhibited a specific inductive effect: DPD2515, responsive to superoxides, and DPD2794, the DNA damage sensor. Although it is tempting to hypothesize that the same wastewater constituent that caused the oxidative effect apparent by the DPD2515 response is the one responsible for the potential DNA damage indicated by DPD2794, this is by no means certain. What is clear, however, is that biological treatment succeeded in removing all hazards indicated by the panel bacteria.

4. Notes

1. Maximization of the amount of samples while limiting the number of strains per plate has two additional advantages:

    a. Since all plates will be identical, they can all be prepared in advance, and kept refrigerated and sealed until shortly before the assay (when no volatile organ-

Table 2
Responses of the Panel to an Industrial
Wastewater Sample Before and After Biological Treatment[a]

Strain	Raw influent		Treated effluent[d]
	Inductive effect[b]	Toxic effect[c]	
DPD2511	−	+	−
DPD2515	++	+	−
DPD2794	+	+	−
TV1061	−	+	−

[a]Highest concentration tested was 10%.
[b]−, No inductive effect; +, induction 2- to 10-fold; ++, induction > 10-fold.
[c]+, Inhibition of luminescence by over 50%.
[d]Neither an inductive nor a toxic effect was observed in the effluent samples.

ics are present). Special adhesive plate sealers can be obtained from various manufacturers; care should be taken when the seals are removed to avoid splatter. If the plates are indeed cooled, they should be warmed to room temperature before the bacteria are added.

  b. The use of one strain per plate normally solves the problem of "spilled over" photons from a highly to a weakly luminescent strain (*see* **Note 8**).

2. This, of course, does not preclude the possibility of designing a plate with all panel members, challenged with a limited number of samples. This arrangement provides the advantage that all tester strains are grown and prepared under identical conditions, avoiding variability emanating from such sources. When a limited number of samples is to be examined, this is certainly the option of choice.

3. Sterility: Since the actual assay procedure is only up to 3 h long and a relatively dense inoculum is used, the actual plate preparation procedure does not require sterile conditions; the microtiter dish, for instance, does not require presterilization. All media, however, should be sterile.

4. The assays described above are carried out at 26°C, a compromise between the optimal temperature for the host (37°C) and that for the luminescence apparatus (<20°C). Changes in the host strain or in the source of the luminescence genes can make this compromise unnecessary *(20)*.

5. In the preparation of the samples in the microtiter plates, as well as for the subsequent addition of the cells, the use of adjustable 12-channel multipipets and suitable reservoirs is highly recommended.

6. It is essential that the effect of each sample is tested in a dilution series rather then in a single concentration, for two main reasons:

  a. To characterize the dose–response; and

  b. To cover a broad concentration spectrum and identify the concentration above which a toxic effect may occur, masking or abolishing the induction.

7. LB concentration, cell density, and relative volumes added to the wells may be changed in order to allow for higher sample concentrations to be tested or for different dilution series to be generated.

8. For a significant increase in the number of samples screened simultaneously, the following approach may be adopted: prepare a large number of plates, and incubate them not in the luminometer, but rather in a suitable shaker at 26°C. In this case, all plates are in turn taken periodically out of the incubator, a single reading is taken in the luminometer, the data are immediately saved, and the plate is returned to the incubator. Since a single reading may last up to 2 min, at least six plates can be sequentially monitored while maintaining a 15-min interval between readings. Data collation is somewhat more laborious (unless a special computer program is generated), but for numerous samples the effort is certainly worthwhile.

9. The presence of some antibiotics drastically affects microbial bioluminescence; thus, although the presence of the drugs is essential for routine strain maintenance, it is avoided during the actual assay.

10. In all microtiter plate luminometers tested by the author, there is a potential danger of a "spillover" of photons to adjacent wells. This may amount to approx 0.1% of the original luminescence, and may therefore be considered a problem if luminescence is at least a thousand-fold higher then the controls. In such high-luminescence instances, it is recommended that empty columns of wells separate between samples or strains.

11. An enormous amount of data can be generated by a regular luminometer run: a plate read every 15 min for 3 h will yield over 1200 data points. To simplify handling of these data, it is recommended that a simple procedure is used to assimilate and reduce all the numbers to a conveniently handled format. Although many commercially available scientific data handling programs can carry this out, even a simple Excel (or equivalent) macro is sufficient. This macro should, ideally, carry out the following functions:
    a. Calculate averages of duplicates;
    b. Rearrange data, if necessary, in a plotable time-dependent matrice;
    c. Calculate response ratios for each time-point;
    d. Select maximal and minimal luminescence values obtained for each sample concentration;
    e. Calculate maximal and minimal response ratios for each sample concentration;
    f. Calculate Γ, EC_{50} and EC_{200} values; and
    g. Plot, on demand, the desired time-course and dose–response figures.

Acknowledgments

The procedures outlined in this chapter were developed in the laboratory of, and in cooperation with, R. A. LaRossa from DuPont Co. Central Research and Development, Wilmington, DE. His contribution, and that of his coworkers, T. K. Van Dyk, D. R. Smulski, and A. C. Vollmer (Swarthmore College, PA) were in many ways more significant than that of the author and are gratefully acknowledged.

References

1. Van Dyk, T. K., Majarian, W. R., Konstantinov, K. B., Young, R. M., Dhurjati, P. S., and LaRossa, R. A. (1994) Rapid and sensitive pollutant detection by induction of heat shock gene-bioluminescence gene fusions. *Appl. Environ. Microbiol.* **60,** 1414–1420.
2. Van Dyk, T. K., Smulski, D. R., Reed, T. R., Belkin, S., Vollmer, A. C., and LaRossa, R. A. (1995) Responses to toxicants of an Escherichia coli strain carrying a *uspA'::lux* genetic fusion and an *E. coli* strain carrying a *grpE'::lux* fusion are similar. *Appl. Environ. Microbiol.* **61,** 4124–4127.
3. Belkin, S., Smulski, D. R., Vollmer, A. C., Van Dyk, T. K., and LaRossa, R. A. (1996) Oxidative stress detection with *Escherichia coli* bearing a *katG'::lux* fusion. *Appl. Environ. Microbiol.* **62,** 2252–2256.
4. Corbisier, P., Ji G., Nuyts, G., Mergeay, M., and Silver, S. (1993) *LuxAB* gene fusions with the arsenic and cadmium resistance operons of *Staphylococcus aureus* plasmid pI258. *FEMS Microbiol. Lett.* **110,** 231–238.
5. Guzzo, A. and DuBow, M. S. (1994) A *luxAB* transcriptional fusion to the cryptic *celF* gene of *E. coli* displays increased luminescence in the presence of nickel. *Mol. Gen. Genet.* **242,** 455–460.
6. Guzzo, J., Guzzo, A., and DuBow, M. S. (1992) Characterization of the effects of aluminum on luciferase biosensors for the detection of ecotoxicity. *Toxicol. Lett.* **64, 65,** 687–693.
7. Selifonova, O., Burlage, R., and Barkay, T. (1993) Bioluminescent sensors for detection of bioavailable Hg (II) in the environment. *Appl. Environ. Microbiol.* **59,** 3083.
8. Heitzer, A., Webb, O. F., Thonnard J. E., and Sayler G. S. (1992) Specific and quantitative assessment of naphthalene and salicylate bioavailability by using a catabolic reporter bacterium. *Appl. Environ. Microbiol.* **58,** 1839–1846.
9. King, J. M. H., DiGrazia, P. M., Applegate, B., Burlage, R., Sanseverino, J., Dunbar, P., Larimer, F., and Sayler, G. S. (1990) Rapid, sensitive bioluminescent reporter technology for naphthalene exposure and biodegradation. *Science* **249,** 778–781.
10. Selifonova, O. V. and Eaton, R. W. (1996) Use of *ibp–lux* fusion to study regulation of the isopropylbenzene catabolism operon of *Pseudomonas putida* RE204 and to detect hydrophobic pollutants in the environment. *Appl. Environ. Microbiol.* **62,** 778–783.
11. Bitton, G. and Dutka, B. J. (1986) Introduction and review of microbial and biochemical toxicity screening procedures, in *Toxicity Testing Using Microorganisms* (Bitton, G. and Dutka, B. J., eds.), pp. 1–8. CRC, Boca Raton, FL.
12. Belkin, S., Van Dyk, T. K., Vollmer, A. C., Smulski, D. R., and LaRossa, R. A. (1996) Monitoring sub-toxic environmental hazards by stress-responsive luminous bacteria. *Environ. Toxicol. Water Quality* **11,** 179–185.
13. Belkin, S. (1998) Stress responsive luminous bacteria for toxicity and genotoxicity monitoring, in *Microscale Aquatic Toxicology—Advances, Techniques and Practice* (Wells, P. G., Lee, K., and Blaise, C., eds.), CRC Lewis, Boca Raton, Florida.
14. Belkin, S., Smulski, D. R., Dadon, S., Vollmer, A. C., Van Dyk, T. K., and LaRossa, R. A. (1997) A panel of stress-responsive luminous bacteria for the detection of specific classes of toxicants. *Wat. Res.,* in press.

15. Belkin, S., Vollmer, A. C., Van Dyk, T. K., Smulski, D. R., Reed, T. R., and LaRossa, R. A. (1994) Oxidative and DNA damaging agents induce luminescence in *E. coli* harboring *lux* fusions to stress promoters, in *Bioluminescence and Chemiluminescence: Fundamentals and Applied Aspects* (Campbell, A. K., Kricka, L. J., and Stanley, P. E., eds.), John Wiley, Chichester, pp. 509–512.
16. Ribo, J. M. and Kaiser, K. L. E. (1987) *Photobacterium phosphoreum* toxicity bioassay, I. Test procedures and applications. *Toxicity Assess.* **2,** 305–323.
17. Stanley, P. E. (1996) Commercially available luminometers and imaging devices for low-light level measurements and kits and reagents utilizing bioluminescence or chemiluminescence: survey update 4. *J. Bioluminescence Chemilminescence* **11,** 175–191.
18. Miller, J. H. (1972) *Experiments in Molecular Genetics.* Cold Spring Harbor Laboratory, Cold Spring Harbor, NY
19. Eaton, A. D., Clesceri, L. S., and Greenberg, A. E. (eds.) (1995) *Standard Methods for the Examination of Water and Wastewater,* 19th ed. American Public Health Association, Washington, DC.
20. Van Dyk, T. K. and Rosson, R. A. Chapter 7, this vol.

22

Organic Contaminant Detection
and Biodegradation Characteristics

Robert S. Burlage

1. Introduction

Catabolism of organic compounds has been extensively described in the literature *(1)*. Many different compounds can be utilized as carbon and energy sources by bacteria, including aliphatics and aromatics, both man-made and natural. Very often these compounds will be broken down to useable intermediates by a set of genes that are coordinately regulated by an operon. The advantage to the cell is that these specialized catabolic genes will only be expressed when the substrate is present, i.e., they are induced in the presence of the substrate. We have used this inducibility to form genetic fusions of the *lux* gene and catabolic genes of interest.

We have constructed these fusions for two general purposes: study of the expression of a gene of interest and for use as a monitoring tool (*see* **Note 1**). The latter use is dependent on the former, since the former describes the construction of a tool (the gene fusion). Use of the fusion for monitoring assumes that the tool is understood well enough to be of use under various conditions of interest. Monitoring can itself be divided into two categories: detection of contaminants and detection of biodegradation activity. These latter two procedures are clearly similar, although they have different objectives. Contaminant detection utilizes the bioreporter strain to find the presence of specific chemicals in soil or water samples, and this test requires reproducibility and effective use of controls. Detection of biodegradation activity is an on-line process, and specific conditions may not be exactly reproducible; confidence in the predictability of the bioreporter tool is essential in this case. Seen from a different angle, detection requires an assay where the majority of confounding factors is eliminated, whereas monitoring acknowledges the multiplicity of factors in an

From: *Methods in Molecular Biology, Vol. 102: Bioluminescence Methods and Protocols*
Edited by: R. A. LaRossa © Humana Press Inc., Totowa, NJ

environmental setting that may influence bioprocesses. Both techniques are useful, depending on the specific application.

Comparable data can be obtained without the use of bioluminescent reporter strains (e.g., using molecular techniques), although the use of these strains is a fast, reliable, and inexpensive means of obtaining data. We have constructed bioreporter strains for naphthalene and for toluene and the xylenes, and have extensively tested these strains. Methods for working with these strains are summarized below, and these methods should be generally applicable to other inducible genes. For instance, we have constructed a bioreporter of mercury cations and demonstrated that it is bioluminescent under inducing conditions *(2)*. Recently, we used a bioreporter of naphthalene *(3)* in an outdoor field release experiment. This was the first use of a genetically engineered microorganism in a bioremediation field test, and the success of this experiment suggests that future releases may become more frequent. These sorts of experiments must be sanctioned by the US Environmental Protection Agency through its Toxic Substances Control Act (TSCA). These bioreporter gene fusions may be attractive candidates for field use, particularly in waste sites that are being remediated, but it must be remembered that legal obstacles exist that must be addressed first. These regulations were put in place for the protection of the public and cannot be taken lightly.

2. Materials

2.1. Bacterial Strains

Strain construction is described elsewhere in this volume; however, some principles are worth restating. The gene fusion must incorporate the structural *lux* genes and the control region of the gene of interest. This means that the known promoter(s) and operator sites should be cloned intact and fused to the *lux* genes. There is always the possibility that undocumented regulatory regions exist, and every effort should be made to take a DNA fragment that is large enough to afford a comfort margin for any unknown sites. An example of a successful fusion is shown in **Fig. 1**.

2.2. Equipment

Light-detection equipment has been described in other sections of this book. The type of equipment needed is entirely dependent on the test system you construct (*see* **Note 2**). If sampling of the system is possible, then most any light detector will probably be useable, including ATP photometers (readily available and sensitive), photographic film, and scintillation counters. It may be that only a qualitative response is needed. That is, it may be sufficient to say that the gene is being expressed (light on) or is not being expressed (light off).

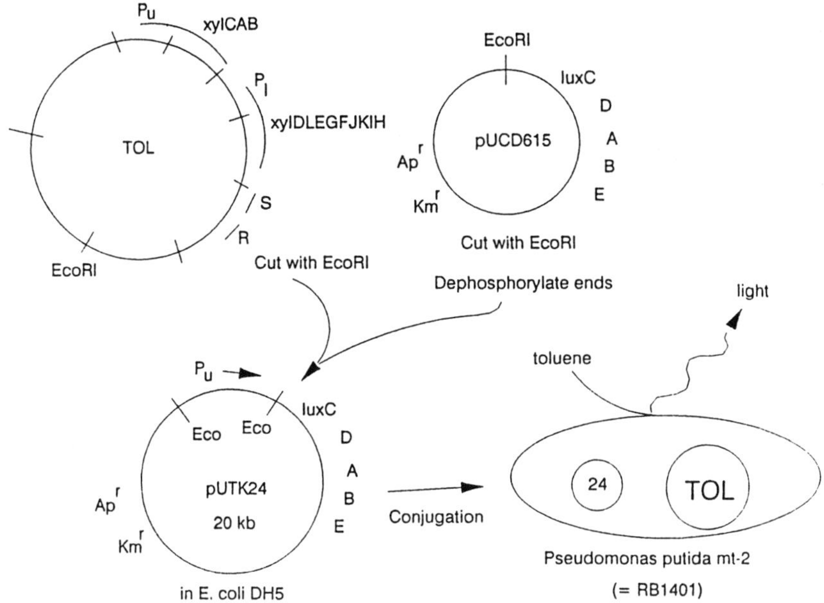

Fig. 1. The construction of a *xyl–lux* bioreporter plasmid is shown. The *xyl* gene fragment is derived from the TOL plasmid, which contains the genes needed for the degradation of toluene and the xylenes. The *lux* genes are located on the pUCD615 plasmid, which has a broad host range replicon. The constructed plasmid is introduced into the *Pseudomonas* host where the *xylR* gene is located on the intact TOL plasmid. This strain has proven effective in detecting low concentrations of the inducing substrates in environmental samples. The slashes represent multiple *Eco*RI restriction sites in the TOL plasmid and the single *Eco*RI site in plasmid pUCD615.

Although there is a possibility that the cells will show some background bioluminescence, after induction of the gene fusion the cells should show a marked increase in light production that is clearly distinguishable from background.

Photomultipliers and related electronic equipment (*see* **Note 3**) will produce a numerical indication of the light signal. This numerical value should be interpreted carefully, and should not be confused with a specific activity value. Specific activity, such as described for bioreporters of enzymatic activity, requires a rate that is normalized for the amount of material present. For instance, the rate might be micrograms of substrate consumed within a certain time, and then normalized for the number of cells in the reaction mix or (more commonly) the total protein in the mix. Photomultipliers report the amount of light at any given instant, and do not show the flux of light radiation over time. More sophisticated light detectors may be able to do this, provided they have the means to count photons during a set time. These systems are expensive, however, and it is easy to saturate the detector.

Photodetectors giving instantaneous readings are useful in certain situations, particularly those in which comparisons to identical test systems are contemporaneously made. Comparisons are more difficult when data from different days are compared, since different photodetectors may have been used, the efficiency of the detector may have changed, or slight variations in incubation conditions may drastically alter bioluminescence that only a reliable comparison of controls would detect. If contemporaneous measurements can be made, the normalization of results can be easily performed as described for enzymatic assays. We have incorporated a fluorescent gene product into our bioreporter strains that is easy to detect and quantify. A direct proportion between cell number and fluorescence has been constructed. By measuring both the fluorescence and bioluminescence, we are able to quickly normalize results.

3. Methods
3.1. Study of Genetic Expression

Gene expression can be difficult to predict. An inducible operon will have a regulatory mechanism associated with it to allow expression under conditions that are beneficial to the cell. Although some of these conditions can be found easily, other conditions may be surprising. For instance, the *nah* genes of *Pseudomonas putida* encode the enzymes for the degradation of naphthalene, a simple polyaromatic hydrocarbon. The actual inducer, however, is salicylate, an intermediate in naphthalene degradation. Surprisingly, the *nah* genes are expressed best when the culture is no longer growing exponentially, even if salicylate is present *(4)*. The reason for this growth-rate regulation is unclear.

The gene of interest may be fairly well-described, at least in terms of promoter location and available restriction sites. In this case, it may be relatively easy to clone the promoter fragment into a *lux* cloning vector and then reintroduce the plasmid back into the host strain. At other times, there may only be a single phenotype of interest, and one may wish to study its expression. Use of a *lux* transposon is preferable in this case, because a great number of mutants can be easily created, and these mutants can then be screened for bioluminescence in the presence of the inducing substrate. A bioluminescent response is no guarantee that there is a transposon insertion in the catabolic genes, however, since other genes may be triggered by the substrate. For example, if the chemical of interest is dissolved in an organic solvent (e.g., acetone) because of its low solubility in water, then it is possible that the gene is expressed because of the solvent, not the substrate. If the chemical(s) has a deleterious effect on the cells, they may express stress proteins. However, conventional molecular techniques can then be applied to the transposon insertion point to determine which gene has been interrupted. In addition, analysis of the chemical intermediates of the substrate may indicate whether the transposon has interrupted a catabolic gene.

As a general rule, the fusion will function best when the host strain is the strain from which the promoter was derived. This is not difficult to understand, since the host strain should provide the regulatory gene(s), σ factor, and polymerase that the promoter needs to function. There may be a problem encountered during the expression of the *lux* genes in this strain, since translation from *Vibrio* ribosome binding sites and folding of the nascent proteins must occur. Codon usage rates may also slightly vary from species to species, and affect efficiency of protein synthesis. In my experience, this has not been a practical limitation, although it is a possibility that cannot be overlooked. In addition, whether the host strain has an aldehyde that is an appropriate substrate for the luciferase reaction must be empirically determined, since means of determining this *a priori* are unavailable. This test is easily performed by random mutagenesis with a *lux* transposon or by introducing a fusion with a strong promoter. In the former case, some of the transposon insertions should have found constitutive promoters and thus produce noticeable bioluminescence. In the latter case, the transformants should be bioluminescent. If the colonies fail to bioluminesce, try adding a few drops of n-decyl aldehyde to the lid of the plate, and re-examine the colonies after 20 min.

General procedure:

1. Grow the bioreporter strain under the desired incubation conditions. This includes medium, temperature, aeration, and so forth. Keep in mind that some luciferases are very heat-sensitive and will be inactived at 37°C. However, cooling the cells to room temperature will allow the experiment to proceed.
2. Add the inducing substrate at an appropriate level. For many compounds, this means a concentration in the μM to low mM range. Some compounds are sparingly soluble in water, and a saturated solution may work well. Dissolving the chemical in another solvent before addition is another alternative.
3. At intervals, remove an aliquot, and determine both cell concentration and bioluminescence. Cell concentration is easily performed using optical density, whereas light readings can be taken using a variety of photodetectors. Intervals of 15 min are usually a good place to start. Use of a photomultiplier that constantly measures light output is a preferred means of recording light data. We have used a photomultiplier that has a RS232 port and that can transmit data to a computer for storage.
4. Always use a negative control in tandem with this experiment. If the cells have a small amount of light production, the amount of light will increase as the culture grows. The negative control will account for this increase.
5. Performing duplicate or triplicate sampling is always preferred, although it may not be practical for screening. If the light production increases, the trend will

probably be spotted over time. This will allow repetition of the experiment with greater attention to statistical analysis.

3.2. Contaminant Detection

As mentioned in the introduction, the success of this procedure is dependent on how well the bioreporter strain has been characterized. A bioreporter strain that appears to give inconsistent answers because the culture conditions are not understood is of little value for this work. The major disadvantage thus far is that relatively few bioreporter strains are available, and the appropriate strain for a specific contaminant might be difficult to construct (e.g., TCE).

It is also important to point out that the signal produced by these bacteria does not necessarily indicate the presence of the contaminant in the sample. It is more appropriate to say that it detects the bioavailable fraction of the contaminant in the sample (*see* **Note 1**). This is an important distinction, since chemicals can be tightly bound to soil (clay) and organic matter (humic acids), and thus be unavailable for uptake by the bacteria *(5)* . Although usually not a problem with water samples, it is a critical test for bioremediation of contaminated sites, since a high contaminant concentration with a low bioavailable concentration would strongly suggest that conventional biotreatments would be ineffective. The assay described below could then be used to determine whether additional treatments (e.g., surfactant addition) would be effective in releasing more of the contaminant from the soil.

These experiments are often run to screen soil and water samples for a specific contaminant, and are valuable as indices of site contamination *(6)*. A positive result can be taken as proof of site contamination. A census of sites can indicate the extent of contamination, as well as approximate quantities at each site. This work can be as accurate as conventional methods of chemical detection, and can be completed at a fraction of the cost.

General procedure *(7)*:

1. Cultivate the bioreporter strain in a liquid medium without the inducing substrate. A midexponential-phase culture is sometimes best, although it is often true that cells entering stationary phase are the most responsive to organic contaminants.
2. A small aliquot (1-mL) should be sufficient for each test vial. We have used scintillation vials in the past because of their convenient size. Make sure there is sufficient headspace for culture aeration.
3. Mix the strain and the test medium with a brief vortexing. Then let sit at the correct incubation temperature until the first time-point. Gentle shaking allows the culture to remain aerated. This is usually anywhere between 5 and 30 min. Since no volume is lost at time-point, there is no penalty for frequent data gathering.
4. Remember to include a positive and a negative control. The positive control should include the test material that has been spiked with the inducing chemical (probably around 1 m*M*). If all goes well, this sample should bioluminesce, unless

a toxic substance is present as well. The negative control should be a soil or water sample that has been demonstrated to be free of inducing substrates; it is likely that a certain baseline amount of light production will occur that is detected by this sample and subtracted from the test samples. Any appreciable light production, however, is suspect.

5. At each time-point, it is important to treat the samples as equally as possible. Small amounts of shaking will increase the aeration in the vials and give higher light readings. Therefore, it is probably best to run duplicate or (ideally) triplicate samples, and perform measurements on one full set of the test samples first before beginning again with the replicates. The data should tell you whether there is a significant change owing only to cell settling and oxygen deprivation during the sample reading. It may be preferable to take a limited number of samples from incubation and test them to avoid cell-settling problems.

6. Typically, light production will be induced, hit a high point, and then decrease (**Fig. 2**). The pattern resembles a normal curve. In some instances, all the cells are saturated, and a plateau of light production is seen for an extended period. In order to quantify the amount of inducer in this sample, it is best to make dilutions of the original material and repeat the test.

7. In our experiments, we have found that 3 h are usually sufficient for light detection, time-points at 15–20 min intervals *(6)*. If the light production has obviously peaked and is decreasing, it is relatively easy to determine how much longer the experiment should run. Bear in mind that our strains have intact catabolic pathways, so that the bacteria are actively degrading the substrate. In the absence of catabolism, the bioluminescence may continue far longer, giving the appearance of saturated conditions.

8. A set of samples spiked with known amounts of the contaminant can be used to produce a standard curve of light output vs concentration. The peak heights at each concentration can be used to produce the curve, or the area under the curve can be used (which may be more accurate if you think you missed the peak height by a substantial amount).

3.3. Bioremediation Monitoring

This goal of this task is essentially as described in **Subheading 3.2.** By monitoring the disappearance of a substrate, the extent of bioremediation can be evaluated. Since the number of different bioremediation scenarios is so vast, a generalized procedure cannot be listed. However, some principles to keep in mind are presented *(8)*.

The bioreporter bacteria must be delivered to the bioremediation venue in sufficient numbers to be detected. It must be assumed that a die-off of the introduced bacteria will take place soon after introduction, and that an order of magnitude may separate the initial and final bacterial concentrations. Preparation of the bacterial suspension must be considered carefully, so that appropriate numbers of bacteria are produced that are all at the right physiological state.

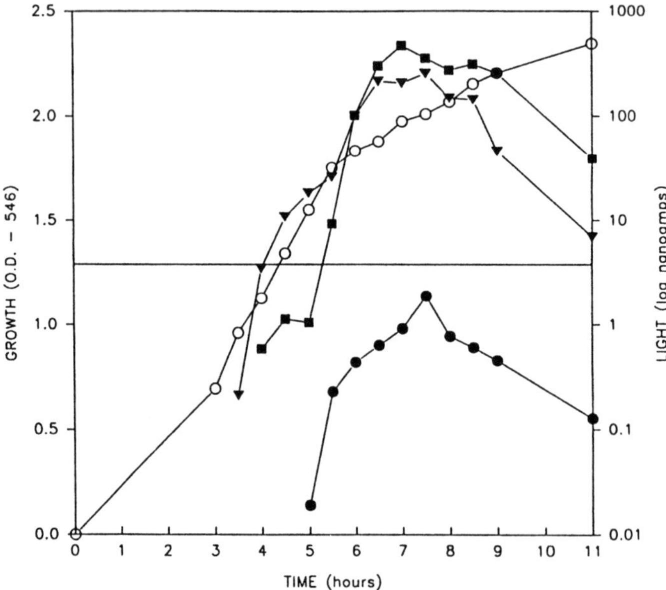

Fig. 2. The strain described in **Fig. 1** was tested with known inducers of the xyl genes. The growth curve is shown with open circles. Light production is shown with closed symbols: circles, uninduced control; triangles, toluene added; squares, methyl benzyl alcohol. Inducers were added at the 3-h time-point. Note that the bioluminescence from toluene induction occurs substantially before methyl benzyl alcohol.

The researcher must decide whether the cells should be washed before introduction into the bioremediation test, or whether the addition of more nutrients from the growth medium is acceptable.

Similarly, the photodetection apparatus must be placed to detect bioluminescence during the test, or a reliable sampling scheme must be developed to remove representative samples during the test. Fiber-optic cables can carry a bioluminescent signal for short distances without appreciable loss of signal. If the photodetector is located fairly close to the bacteria, then the electronic signal can be accurately transmitted over much longer distances. Miniaturized photodetectors are available that can perform this task. Immobilization of the bioreporter bacteria on the ends of these fiber-optic cables is also possible. For example, carrageenan gel may be used to immobilize the cells and provide a protected environment for them. Diffusion of the inducing substrate through the gel will trigger the cells, which will produce light. Since the number of bacteria in the gel is known, signal comparison can be accomplished relatively easily. The major drawback to this technique is that oxygen is also limiting in this environment, and the probe tip may need to be replaced more often than desired. This is still a relatively new area of research, and developments that will soon improve the technology are expected.

4. Notes

1. It is worth noting that in all of these procedures, the bioavailability of the inducing substrate is detected. That is, the fraction of the substrate that is able to be metabolized by the bacteria is found. This may be a small fraction, since organics can become complexed with soil constituents. Measurement of bioavailability has distinct advantages compared to wet chemistry detection, since only the bioavailable fraction can be bioremediated. Thus, the *lux* assay is a valuable index of bioremediation potential, as well as for a predictive tool for methods designed to increase the efficacy of bioremediation (such as surfactant addition).
2. In all cases where samples are examined, it is important that the light-detecting apparatus is located a standard distance from the sample. Light intensity rapidly decreases as distance increases. A vial holder that places the sample at a set distance from the detector is an enormous advantage.
3. Do not assume that your photodetecting equipment is free from light leaks, even if it comes straight from the manufacturer. Thoroughly check the equipment, and plug any leaks with electrical tape or a similar dense, dark material.
4. Many catabolic operons function anaerobically. Since the luciferase requires oxygen to produce bioluminescence, the *lux* genes are unsuitable for exploring gene expression in anaerobic systems.

Acknowledgments

Research was sponsored by the Office of Health and Environmental Research, US Department of Energy. Oak Ridge National Laboratory is managed by Lockheed Martin Energy Research Corp. for the US Department of Energy under contract number DE-AC05–96OR22464.

"The submitted manuscript has been authored by a contractor of the US government under contract no. DE-AC05–96OR22464. Accordingly, the US government retains a nonexclusive, royalty-free license to publish or reproduce the published form of this contribution, or allow others to do so, for US government purposes."

References

1. Atlas, R. M. and Cerniglia, C. E. (1995) Bioremediation of petroleum pollutants. *BioScience* **45,** 332–338.
2. King, J. M. H., DiGrazia, P. M., Applegate, B., Burlage, R., Sanseverino, J., Dunbar, P., Larimer, F., and Sayler, G. (1990) Bioluminescent reporter plasmid for naphthalene exposure and biodegradation. *Science* **249,** 778–781.
3. Selifonova, O., Burlage, R., and Barkay, T. (1993) Preparation of bioluminescent sensors for detection of Hg (II) in the environment. *Appl. Environ. Microbiol.* **59,** 3083–3090.

4. Burlage, R. S., Sayler, G. S., and Larimer, F. W. (1990) Monitoring of naphthalene catabolism by bioluminescence with *nah-lux* transcriptional fusions. *J. Bacteriol.* **172,** 4749–4757.
5. Burlage, R. S., Palumbo, A. V., Heitzer, A., and Sayler, G. S. (1993). Bioluminescent reporter bacteria detect contaminants in soil samples. *Appl. Biochem. Biotechnol.* **45/46,** 731–740.
6. Heitzer, A., Burlage, R. S., and Sayler, G. S. (1992) *lux* gene bioreporters, in *Bioremediation of Petroleum Contaminated Soil on Kwajalein Island: Microbiological Characterization and Biotreatability Studies,* (Adler, H. I., Jolley, R. L., and Donaldson, T. L., eds.), Oak Ridge National Lab, Oak Ridge, TN, pp. 14–28.
7. Heitzer, A., Webb, O. F., Thonnard, J. E., and Sayler, G. S. (1992) Specific and quantitative assessment of naphthalene and salicylate bioavailability by using a bioluminescent catabolic reporter bacterium. *Appl. Environ. Microbiol.* **58,** 1839–1846.
8. Sayler, G. S., King, J. M. H., Burlage, R., and Larimer, F. (1991) Molecular analysis of biodegradative bacterial populations: application of bioluminescence technology, in *Organic Substances and Sediments in Water,* vol. III, *Biological.* (Baker, R. A., ed.), Lewis Publishers, Chelsea, MI, pp. 299–314.

23

Detection of Firefly Luciferase-Tagged Bacteria in Environmental Samples

Annelie Möller and Janet K. Jansson

1. Introduction

In recent years, several molecular methods have been developed for tracking genetically engineered microorganisms (GEMs) in environmental samples *(1–3)*. The majority of the methods are based on monitoring of bacteria tagged with a marker gene, which provides the bacteria with a unique phenotype or DNA sequence for detection. The challenge has been to design molecular monitoring procedures that are quantitative in order to estimate the biomass of a specific microbial population in natural samples *(4)*. Quantitation of few cells of a given species becomes especially complicated when taking into account the vast natural diversity of microorganisms in nature; for example, it has been estimated that a single gram of soil contains thousands of distinct genotypes *(5)*. Therefore, it is important to have methods that are sensitive and specific for the tagged bacterium, and that can be used to quantitate the number of specific cells present in the sample.

Bioluminescent reporter genes, such as bacterial luciferase *(lux)* or eukaryotic luciferase *(luc)*, fullfill many of the criteria required for an optimal marker system. To begin with, bacteria tagged with bioluminescent markers can be easily identified on the basis of light production, and the light yield can be used to estimate the number of tagged cells in the sample. Also, since the *luc* gene is eukaryotic in origin, and therefore absent in the natural microbial population, it is a very specific marker. This specificity is one advantage over the bacterial luciferase marker, particularly in marine samples containing naturally luminescent bacterial cells.

The firefly luciferase gene has been used as a marker gene for detection of specific bacteria in different environmental samples, including freshwater *(6)*,

From: *Methods in Molecular Biology, Vol. 102: Bioluminescence Methods and Protocols*
Edited by: R. A. LaRossa © Humana Press Inc., Totowa, NJ

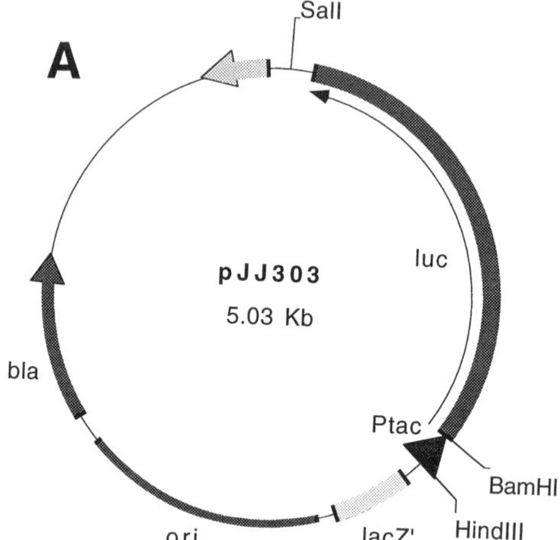

Fig. 1A. Plasmid pJJ303. *bla,* resistance to β-lactam antibiotics; *lacZ',* 'β-galactosi-dase; *ori,* origin of replication; *Ptac,* promoter driving transcription of the *luc* gene; some useful restriction enzyme sites are also indicated.

marine sediment *(6,7),* and soil *(8,9).* Bacteria tagged with the *luc* gene can be easily distinguished from the natural microbial population on the basis of light emission by the luciferase enzyme, after addition of the substrate luciferin, according to the following reaction:

$$\text{Luciferin} + \text{ATP} + \text{O}_2 \longrightarrow \text{oxyluciferin} + \text{AMP} + \text{PP}_i + \text{CO}_2 + hv \ (562 \text{ nm}) \quad (1)$$

The *luc*-tagged cells do not produce light unless luciferin is added to the sample. Therefore, expression of the luciferase protein does not provide the tagged cells with any selective advantages or conflict with their growth com-pared to the wild-type strain.

The amount of light produced by bacteria tagged with the *luc* gene varies according to the number of copies of *luc* introduced into the cell and with the strength of the promoter driving *luc* transcription. High light yields have been obtained when bacteria were tagged with *luc* cloned behind a strong promoter and introduced on a multicopy number plasmid. For example, fewer than 10 *Escherichia coli* cells could be detected when they were tagged with *luc* on the multicopy plasmid, pJJ303 (*6; see* **Fig. 1A**). Cebolla et al. observed a 10-fold increase in expression for *Rhizobium meliloti* cells when the *luc* gene was on a plasmid contained in 7.5 to 15 copies/cell, compared to cells with one single chromosomal insertion *(8).* However, to prevent the risk of plasmid transfer in nature, it is preferable to stably integrate the *luc* gene into the chro-mosome of bacteria intended for release into the environment, although lower

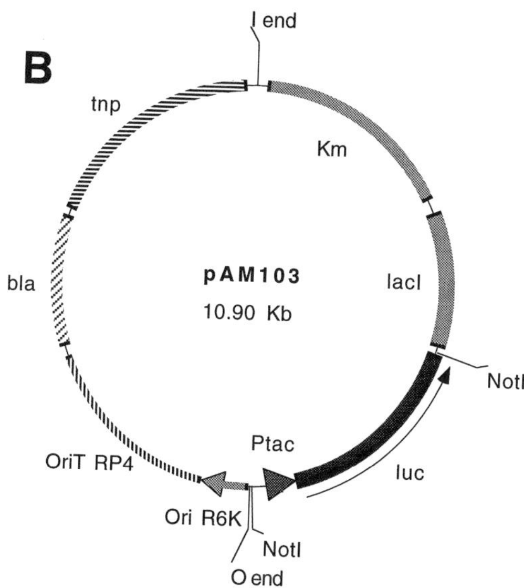

Fig. 1B. Plasmid pAM103. The *tnp* gene, encoding transposase, is located outside the insertion elements (I-end and O-end), and therefore not integrated into the chromosome; Km, kanamycin resistance; *bla,* resistance to β-lactam antibiotics; *lacI,* encodes the LacI repressor protein; *ori* R6K, plasmid R6K π protein-dependent origin of replication; *ori*T RP4, pRP4 transfer origin; P*tac,* promoter-driving transcription of the *luc* gene; *Not*I, restriction enzyme sites.

light yields are obtained. An ideal system for stable chromosomal integration of foreign DNA is the use of minitransposon delivery vectors *(10)*. An example of a minitransposon delivery vector for *luc* is plasmid pAM103, which has been used successfully to tag both Gram-negative and Gram-positive cells (*11; see* **Fig. 1B**).

Bacteria tagged with the *luc* gene can be detected and quantified using methods developed for quantitation of bioluminescence output (**Fig. 2**). All of the bioluminescence-based methods are rapid and simple techniques for quantitation of pure cultures of bacteria. However, it is more difficult to quantitate bioluminescence in environmental samples. Therefore, considerable effort has been spent on modification of available methods, or development of novel methods, for light quantitation in complex samples *(4,7)*. Additionally, it is possible to detect and quantitate the specific *luc*-DNA sequence by PCR amplification *(6,12)*, although DNA detection methods will not be described in this chapter. Therefore, *luc*-tagged cells can be monitored in environmental samples using several complementary approaches (*1; see* **Fig. 2**).

Cells tagged with the *luc* marker gene can be enumerated as bioluminescent colonies, after addition of luciferin at an appropriate pH *(6,8,13)*. Cells that

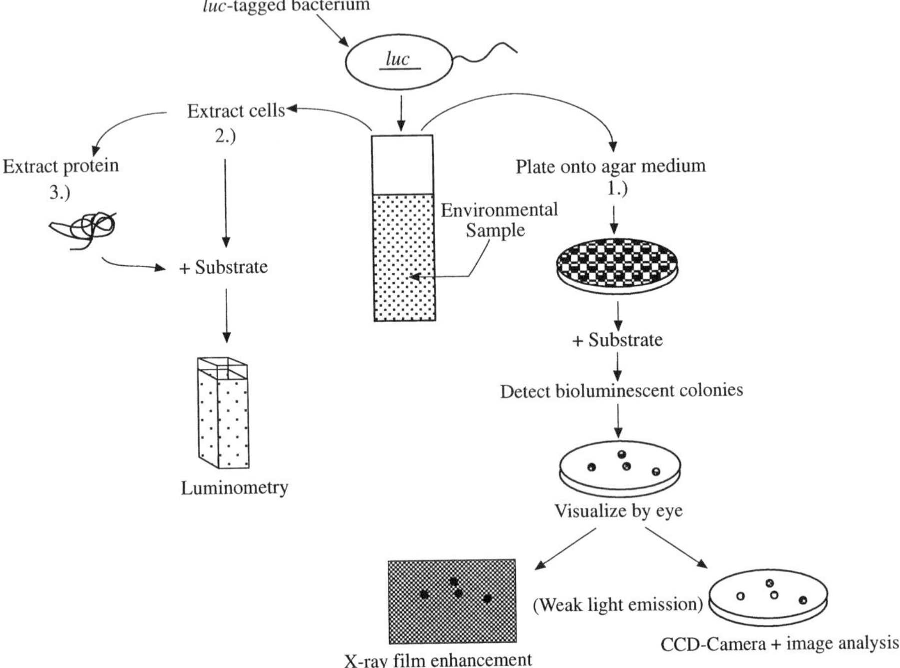

Fig. 2. Methods for detection and quantitation of bioluminescent bacterial cells in environmental samples. (1) Enumeration of bioluminescent colonies; (2) quantitation of luminescence in the bacterial cell fraction from an environmental sample; (3) quantitation of luminescence in protein extracted from an environmental sample.

have a high amount of luciferase (e.g., when *luc* is cloned on a multicopy plasmid and expressed from a strong promotor) usually have a high light output that enables luminescent colonies to be visualized by eye in a darkroom. For detection of bacteria emitting low levels of light, higher sensitivity is required in order to detect bioluminescent colonies. One option is to expose light emitting colonies to X-ray films *(8)*. Alternatively, the light signal can be enhanced using a chilled charge coupled devise (CCD) camera in a light-tight box (**Fig. 3**).

One drawback with cultivation-based detection methods is that not all cells remain culturable after prolonged incubation in the environment *(14)*. Therefore, enumeration of *luc*-tagged cells as bioluminescent colonies can be misleading, with lower values than the actual number of viable cells in a given sample. Another problem with enumeration of colonies from environmental samples is growth inhibition by the natural microbial flora, which can also lead to underestimation of the actual cell number *(7)*. Detection and enumeration of

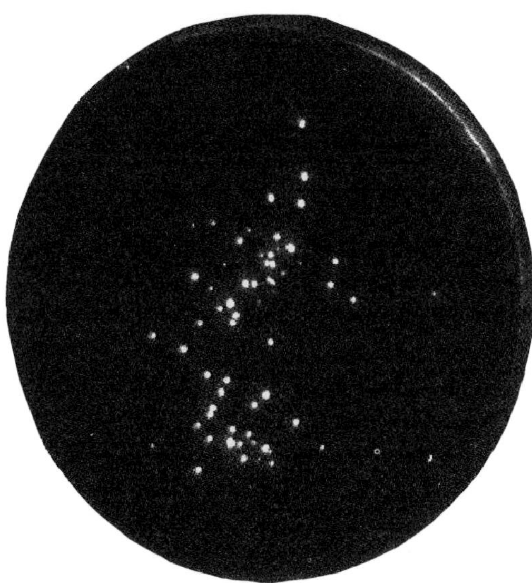

Fig. 3. CCD-enhanced image of colonies of *Arthrobacter* sp. A-6::*luc,* chromosomally tagged with a single copy of the *luc* gene. The *luc*-tagged cells were added to nonsterile soil, plated onto nonselective minimal medium, luciferin substrate was added, and bioluminescent colonies were visualized after 4 min of exposure with a peltier-cooled CCD camera in a light-tight box.

bioluminescent colonies remains, however, a convenient method for enumeration of *luc*-tagged cells in laboratory cultures and for screening of transformants.

An alternative method for detection and quantitation of bioluminescent cells in nature is to take a sample, add luciferin, and quantitate light production in a luminometer. Luminometers are known to be very sensitive and are useful for quantitation of low light levels (*see* **Fig. 4**). In addition, luminometry is very rapid and easy to perform. However, environmental samples, such as soil or sediment, contain large amounts of particulate material that can mask light production. Therefore, for environmental samples, higher light yields are obtained when the microbial cells are first partially separated from particles and other inhibitory material.

Depending on the sample type, different sorts of pretreatment are required prior to measurement of light output by luminometry. Bacteria can easily be concentrated from water samples by centrifugation *(6)* or by filtration *(15)*. However, for sediment and soil samples, more extensive purification is necessary to separate cells from particulate material and humic acids. Methods vary, and include differential centrifugation *(16)*, sedimentation of particulate matter *(7)*, or density gradient centrifugation *(17)*. It is possible to determine the cell

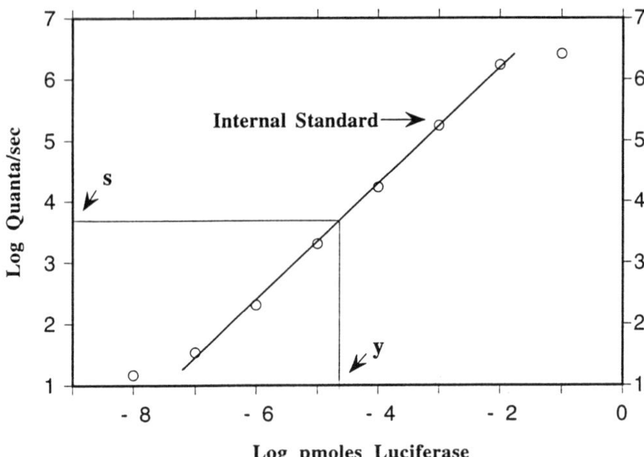

Fig. 4. Standard curve of light yield (log quanta/s) vs luciferase (pmol). The optimal light output for an internal standard is indicated by the arrow (*see* **Note 10**). For a given sample light output (s), one can interpolate to (y) a given amount of luciferase in picomoles (*see* **Note 18**).

extraction efficiency from environmental samples by comparison of the light output from the cells added to the sample to the light output recovered after cell extraction.

The bioluminescence reaction of firefly luciferase requires energy (ATP), and the light yield is higher in active cells than inactive or starved cells. This response is presumably owing to depletion of cellular ATP reserves *(6)*. Since the natural microbial population is usually in a state of starvation, the light yield tends to drop to low levels after prolonged incubation in environmental samples. Therefore, quantitation of bioluminescent cells in environmental samples by luminescence may underestimate the total population of *luc*-tagged cells. The light levels can be restored, to some extent, by addition of nutrients and a short incubation to activate the bacteria. The bioluminescence obtained after broth stimulation of bacteria, tagged with genes encoding bacterial luciferase, has been termed "potential luminesence" and can be used as a biomass indicator *(18)*.

An alternative to direct measurement of bioluminescent cells is to measure luciferase activity in cell lysates. The main advantage of this procedure is the independence of the assay on cellular ATP levels since ATP is directly added as a reagent. However, activity of the luciferase protein can also be inhibited by a range of substances found in environmental samples, such as chloride ions and salts *(19)*. Therefore, depending on the sample type, some protein purification from the sample might be necessary. A higher light output has been obtained from environmental samples by concentration of extracted protein onto membrane filters before light measurement *(7)*.

In order to avoid expensive and time-consuming protein purification steps, a rough purification is acceptable, if the amount of light quenching by sample impurities can be accurately determined. It is possible to calculate the amount of quenching by incorporation of an internal standard of pure luciferase directly to the sample *(7)* and remeasuring the luminescent output. The value of the light yield from the sample can then be adjusted to that for pure luciferase. This ability to rapidly determine the amount of quenching in environmental samples by luminometry is a particular advantage of luminescence determinations.

Finally, the light output from replicate samples is more consistent in cell lysates, compared to whole cells, resulting in improved accuracy and precision of the results *(7)*. However, it is more difficult to perform the additional steps required to break open the cells and to partially purify and concentrate protein compared to direct measurement of luminescence in whole cells.

This chapter will describe bioluminescence-based approaches we have developed for detection and quantitation of *luc*-tagged bacterial cells in environmental samples (**Fig. 2**). The three monitoring approaches described include:

1. Enumeration of bioluminescent colonies.
2. Quantitation of luciferase-tagged cells by luminometry.
3. Quantitation of luciferase protein in cell extracts by luminometry.

The methods focus on light measurement in complex samples; e.g. soil or sediment, which are notoriously difficult to analyze. In addition, methods are included for stable chromosomal tagging of bacteria with the *luc*-gene.

2. Materials

2.1. Stable luc *Tagging of Bacterial Cells*

1. pAM103 vector (**Fig. 1B**) in *E. coli* CC118(λ pir) host (*see* **Notes 1** and **2**).
2. Standard plasmid extraction protocol (*see* **Note 3**).
3. Appropriate liquid growth media (*see* **Note 4**).
4. Sterile tubes for centrifugation, for example, 50–100 mL plastic centrifuge tubes.
5. Ice-cold sterile distilled water.
6. TE buffer: 10 mM Tris-HCl, pH 8.0, 1 mM EDTA, pH 8.0.
7. Electroporation device, e.g., ECM-600 model Electroporation system (BTX, San Diego, CA).
8. Electroporation cuvets (2 mm).
9. LB (Luria, Bertani) broth: 10 g tryptone, 5 g yeast extract, 5 g NaCl, in 1 L water.
10. Agar growth medium (*see* **item 3**) contains the addition of 15 g agar/L. Agar plates should contain appropriate antibiotics as needed for selection.
11. Phosphate buffer: 0.1 M sodium phosphate buffer, pH 7.0.

2.2. Detection of Light-Emitting Colonies

1. Nitrocellulose membranes (e.g., 82-mm, circular, 0.45 μm pore size, Schleicher and Schuell, Dassel, Germany).
2. 10X Luciferin stock: resuspend click beetle (or firefly) luciferin (Sigma, St. Louis, MO) in sterile water to 10 mM and aliquot 50-μL portions into microcentrifuge tubes (*see* **Note 5**).
3. 1.0 M citrate buffer, pH 5.0: Mix 41 mL 1 M citric acid and 59 mL 1 M sodium citrate to pH 5.0 (*see* **Note 6**).
4. Whole-cell buffer (prepare fresh, *see* **Note 5**): Thaw a 50 μL aliquot of 10X luciferin stock (**step 2** above), add 50 μL citrate buffer and 400 μL dH$_2$O—enough for one Petri plate *(13)*.
5. Selective or nonselective agar media (*see* **Note 4**).
6. Photon-sensitive CCD camera with light-tight box (*see* **Note 7**) or X-ray film and developing solutions.

2.3. Extraction of Cells from Soil/Sediment

1. 0.5 g Soil/sediment sample containing *luc*-tagged bacteria.
2. Extraction buffer: 0.1 M sodium phosphate buffer, pH 6.5, (*see* **Note 8**).

2.4. Luminometric Detection of Whole-Cell Activity from Soil/Sediment

1. Citrate buffer and whole-cell buffer (*see* **Subheading 2.2**, **items 2–4** and **Note 5**).
2. 1-mL plastic cuvets.
3. Luminometer (*see* **Note 9**).

2.5. Luciferase Quantitation in Cell Extracts

1. Micro-tip sonicator (for example, model XL 2005, Mixonix, New York).
2. 70% Ethanol for washing of the sonicator microtip.
3. Lysis buffer: 0.1 M phosphate buffer, pH 7.8, 5% glycerol, 2 mM EDTA. Autoclave and add filter-sterilized BSA, 1 mg/mL.
4. Acid-washed polyvinyl polypyrrolidone (PVPP) *(16)*: Suspend 30 g insoluble PVPP (Sigma) in 0.4 L of 3 M HCl for 12–16 h at room temperature. Filter the suspension through MIRACLOTH (Chicopee Mills, Milltown, NJ). Suspend in 0.4 L of 20 mM potassium phosphate buffer (pH 7.4), and mix by stirring for 1–2 h. Repeat this process until the suspension reaches pH 7.0. Then filter through MIRACLOTH and dry overnight. Use a pestle to grind to a fine powder and store dry at room temperature until use.
5. Centricon 50 filter units with a mol-w cutoff at 50 kD (Amicon, Beverly, MA).
6. Tweezers and pliers.
7. Promega luciferase assay kit (Promega, Madison, WI), dilute according to the following protocol (*see* **Note 5** for handling and storage): 10X luciferase substrate stock: Reconstitute the luciferase assay substrate with 1 mL of the Promega luciferase assay buffer. Aliquot both the 10X luciferase substrate stock and the remaining 9 mL of luciferase assay buffer (50 and 450 μL/tube for 5 reactions, respectively) into separate microcentrifuge tubes.

8. Pure firefly luciferase (Sigma) reconstituted and diluted in lysis buffer (**step 3**, above) (for proper dilutions, *see* **Note 10** and **Fig. 4**). The enzyme is stable up to 2 years when stored at –80°C.
9. Luminometer (*see* **Note 9**).
10. Plastic 1-mL cuvets.

3. Methods

3.1. Stable luc Tagging of Bacterial Cells

1. Grow the bacteria of interest to the late exponential growth phase in 250 mL of the appropriate growth medium.
2. Collect cells by centrifugation at 5000g for 15 min, and resuspend the cells in ice-cold, sterile distilled water to one-half of the original volume.
3. Mix thoroughly, recentrifuge, and discard the supernatant.
4. Repeat the washing steps (**steps 2–3**) five times and sequentially reduce the volume each time to a final volume of 2–5 mL (approx 10^{10} cells/mL).
5. Aliquot the washed cells into microcentrifuge tubes (200 µL/tube), keep on ice or freeze in liquid nitrogen, and store at –80°C until use. The cells are now competent for DNA uptake by electroporation.
6. Add plasmid pAM103 DNA (5 µL at a concentration of 0.5 µg/µL dissolved in TE buffer) to 200 µL competent cells (*see* **Note 11**).
7. Incubate the mixture on ice 10 min.
8. Transfer the cell/DNA mixture to a 2-mm electroporation cuvet.
9. Electroporate the DNA into the cells using an electroporation device, adjusted to 2.5 kV, 25 µF, and 129 Ω, keeping the pulse length to 4–7 ms (*see* **Note 11**).
10. Immediately transfer the cells to 2 mL of LB broth.
11. Incubate the cells for 0.5–3 h (~1 generation time) at an appropriate growth temperature to allow the cells time to recover.
12. Pellet the cells by centrifugation at 5000g for 5 min.
13. Resuspend the cells in 0.5 mL of sterile sodium phosphate buffer, pH 7.0, and plate onto appropriate medium (*see* **Note 4**), for example, 1, 10, 100, and 400 µL/plate.

3.2. Detection of Light-Emitting Colonies

1. Grow colonies on suitable agar medium.
2. Place a circular nitrocellulose membrane filter onto the colonies.
3. Transfer the membrane (colony side up) onto the lid of a Petri plate containing 0.5 mL whole-cell buffer.
4. Cover the plate with plastic to keep the colonies moist, and incubate for 30–60 min at the appropriate growth temperature.
5. Visualize the luminescent colonies in a darkroom, by eye.

    Alternative 1: For weak luminescing colonies, place the membrane on top of an X-ray film, with a layer of plastic wrap between membrane and film, and expose for periods of 5 min to overnight. Develop film and enumerate blackened spots, corresponding to light-emitting colonies.

Alternative 2: Enhance the luminescent signal with a CCD camera. Place the filter with attached colonies (after **step 4**) in a light-tight box fitted with a chilled CCD camera. Usually exposure times of 5–30 min are optimal for detection of weak luminescence (*see* **Note 7**).

3.3. Extraction of Cells from Sediment/Soil

This section will only describe one protocol for extraction of bacteria from environmental samples. For water samples, *see* **Note 12**, and for other sediment/soil extraction protocols, *see* **Note 13**.

1. Suspend 0.5 g sediment or soil in 1 mL 0.1 M sodium phosphate buffer, pH 6.5.
2. Vortex vigorously for 1 min, and centrifuge for 6 min at 2200g in a microcentrifuge.
3. Transfer the supernatant to a new tube, and centrifuge at 10,000g for 10 min to pellet the bacteria.
4. Add a fresh 1-mL aliquot of 0.1 M sodium phosphate buffer, pH 6.5, to the soil/sediment pellet in the original tube, and repeat the low-speed centrifugation procedure (**steps 2** and **3**) two times.
5. Combine all supernatants containing bacterial fractions into one microcentrifuge tube, and centrifuge at 10,000g for 10 min until all the bacterial fractions from one sample are pelleted in one tube (*see* **Note 14**).

3.4. Whole Cell Luciferase Activity Determinations

1. Wash the bacterial pellet isolated from sediment or soil in 1 mL 0.1 M citrate buffer, and recentrifuge to collect the bacteria.
2. Add 100 µL whole-cell buffer to the purified cell pellet, mix by pipeting up and down, and transfer to a plastic cuvet.
3. Measure light output in a luminometer at a defined time after substrate addition (e.g., 5 min, *see* **Note 6**).

3.5. Luciferase Quantitation in Cell Extracts from Soil/Sediment

1. Resuspend the bacterial pellet isolated from sediment or soil in 0.5 mL lysis buffer.
2. Lyse the cells by sonication in an ice bath for 2 × 15 s at 70% power with a microtip sonicator (*see* **Note 15**). Wash the microtip with ethanol and water between each sample to avoid crosscontamination with luciferase.
3. To decrease the amount of humic acids in the sample *(16)*, add 25 mg acid-washed PVPP to the microcentrifuge tube.
4. Add 0.5 mL lysis buffer, and mix thoroughly on ice by pipeting the solution up and down. Centrifuge the suspension for 2 min at 1000g.
5. Transfer the supernant (containing the water-soluble protein fraction) to a new microcentrifuge tube on ice.
6. Repeat the washing step two times (**steps 4–5**).
7. After the last wash, centrifuge for 5 min at 1000g, and remove any remaining liquid with a micropipet tip.

8. Add the entire sample to a Centricon 50 filter unit.
9. Concentrate the larger material (including the 62-kDa luciferase protein) on the Centricon filter, and simultaneously desalt by centrifugation for 35–45 min at 5000*g*.
10. Transfer the concentrated protein fraction (the primary retenate, 25–45 μL) to a microcentrifuge tube on ice.
11. Wash the membrane + filter unit with 90 μL Promega luciferase assay buffer (without addition of the luciferin substrate).
12. Place the retenate cup over the Centricon filter unit, invert, centrifuge for 5 min at 5000*g* to collect the retained protein fraction (secondary retenate), and transfer it to the microcentrifuge tube containing the primary retenate.
13. Break the Centricon 50 unit at the ring junction with pliers, and carefully remove the membrane filter with a tweezers (*see* **Note 16**).
14. Place the filter into a plastic cuvet (with the attached material facing the light detector), along with the primary and secondary retenate fractions, and mix by vortexing.
15. Measure background light levels, usually some "electronic noise" from the luminometer.

Steps 16–18 should be done as quickly as possible.

16. Add 10 μL of 10X luciferase substrate stock underneath the filter, alongside the cuvet wall, and mix by vortexing (the final sample volume in the cuvet will be approx 125–140 μL). Make sure the filter is not blocking light from the retentate solution.
17. Immediately measure the light output as "quanta/sec" (or relative light units, RLU) in the luminometer.
18. Add 1 μL of pure firefly luciferase (internal standard) to the sample, and re-measure the light output (sample + internal standard).
19. In a separate cuvet, measure the light output from 1 μL of the internal standard after addition of 10 μL of luciferase substrate stock and 90 μL of luciferase assay buffer.
20. Subtract the background electronic noice (no substrate) from each measurement.
21. Calculate the amount of quenching in the sample according to the standard formula used for determination of scintillation counting efficiency:

$$Q = [(C - A)/B]\ 100 \tag{2}$$

 where Q = counting efficiency [%]; A = sample reading, B = known internal standard reading, C = sample + internal standard reading (*see* **Note 17**).
22. Correct the light values for quenching and report data as "quanta/sec" (or RLU)/g soil or sediment.
23. Use a standard curve (**Fig. 4**) after quenching correction to correlate the light output from the sample to the amount of luciferase protein in the sample (*see* **Note 18**).

4. Notes

1. The pAM103 vector was designed for stable integration of the *luc* gene plus a gene encoding kanamycin resistance for selection purposes into bacterial chromosomes. We have had success using this type of vector for both Gram-negative

and Gram-positive bacteria *(11)*, although the parent pUT minitransposon vector was designed primarily for Gram-negative bacteria *(10)*. In addition to electroporation, minitransposons, such as pAM103 may be mobilized into Gram-negative bacteria by conjugation *(20)*. A preincubation with IPTG can be used to increase the light output owing to the presence of the LacI repressor.

2. Minitransposon vectors having the R6K origin of replication, such as pAM103, require a λ*pir* lysogenic host cell, such as *E. coli* CC118(λ*pir*) for plasmid maintenance *(10)*.

3. Any plasmid extraction protocol can be used. For convenience, we have used Wizard Plus Minipreps (Promega) and the Qiagen Plasmid Maxi Kit (Qiagen, Chatsworth, CA).

4. The following considerations should be taken into account when designing a suitable medium for a particular microorganism:
   a. The wild-type strain must first be tested for sensitivity to kanamycin in order to determine the selective concentration for screening of transformants;
   b. If the microbial cells are to be selected on minimal medium, it is sometimes advisable to first plate onto a filter overlaid on top of nutrient rich agar and then, after a short incubation before visible colonies have formed, to transfer the filter to a selective minimal medium;
   c. Since the possibility of illegitimate recombination, although rare, does exist *(20)*, authentic transpositions should be confirmed by screening for ampicillin-sensitive colonies on plates containing ampicillin;
   d. No antibiotics should be added to plates from environmental samples if it is desired to maximize recovery of the indigenous microbial population; and
   e. In order to inhibit growth of fungi, cycloheximide (100 µg/mL) can be added to agar plates.

5. The luciferin substrate is light-sensitive and should be aliquoted in the dark, as quickly as possible, after reconstitution. Multiple freeze/thaw cycles should also be avoided in order to minimize breakdown of the substrate. Therefore, it is recommended to store small aliquots of all reagents. The reconstituted luciferin substrates are stable at least 1 yr when stored at –80°C and approx 1 mo at –20°C. The optimal reaction temperature of the luciferase enzyme is 20–25°C. Since the luminescent reaction is highly temperature-dependent, all reaction reagents should be equilibrated to room temperature before use (0.5 h in a water bath is enough). The use of cold reagents will reduce the enzyme activity by approx 5–10%.

6. The low pH of the whole cell buffer is thought to be necessary for protonation of the luciferin substrate, so that it penetrates the cell membrane *(13)*. However, some microorganisms take up luciferin at neutral pH (Möller, personal communication), so the pH optimum for uptake should be tested for novel species. In addition, the time required for uptake of substrate may vary between species, and therefore, preliminary trials should be undertaken to determine the time required for maximal light output. The optimal time for substrate uptake should be determined by measuring light output in 15-s intervals and plotting quanta/sec (or RLU) as a function of time.

7. There are several different types of CCD cameras to choose among for detection of luminescence. Photon-counting cameras and cooled CCD cameras are known to be

very sensitive detectors of luminescence. A digital CCD camera can be connected to a camera controlling unit for conversion of the image to a digital signal. Use of software (such as HiPic, Hamamatsu Photonic Deutschland GmbH, Hoveshing, Germany) is convenient for interphasing the control unit with the camera. A light-tight box (e.g., Image Box, Hamamatsu Photonics Systems, NJ) is necessary to prevent stray light background that would otherwise interfere with weak luminescent signals. To reduce background from random photons and electronic noise, during long exposures times, it is possible to substract a background (blank) image file from the sample image file.

8. The extraction buffer may need to be further optimized depending on the soil/sediment sample characteristics. Therefore, it is recommended to test a series of pH and phosphate concentrations to determine the optimal buffer for separation of cells into the supernatant.

9. The luminometer we use is a handmade unit equipped with a sensitive photomultiplier tube (Model R268, Hamamatsu TV, Tokyo, Japan). The photomultiplier detector unit is enclosed in a light-tight box. Luminometers are also commercially available from several sources.

10. Suitable dilutions of pure firefly luciferase for both standard curves and internal standards must be tested for each individual luminometer. A small aliqout (1-µL) of the optimal concentration of the internal standard should give a luminescent response in the upper range of the linear standard curve (**Fig. 4**). It is recommended to make a large batch of the proper dilution of the internal standard and to aliquot small amounts (e.g., 10 µL/tube) in order to avoid freeze/thaw cycles.

11. The electroporation system and criteria described are simply given as an example. The settings will most likely vary depending on the microorganism to be electroporated or on the electroporation unit used. The recommended pulse length (4–7 ms) for a given volume, cuvet distance, and ionic strength facilitates DNA uptake. The resistance should be adjusted in order to obtain the correct pulse length. More detailed information about transformation efficiencies can be obtained from the electroporation manual accompanying the unit.

12. There is usually no need to extract bacteria from water samples. Bacteria can simply be collected by centrifugation (for example: 1 mL, 10,000*g* for 10 min) or by filtration on a 0.2-µm membrane filter. Bacteria can then be recovered from the filter by addition of 1 mL citrate buffer to the filter in a microcentrifuge tube, vortexing for 1 min, and pelleting the cells by centrifugation as above.

13. There are several protocols available for extraction of bacteria from sediment or soil (e.g., *8,17,21*).

14. It is possible to do successive centrifugations in the same tube by adding more supernatant to the original bacterial pellet from the previous centrifugation, in order to collect all bacterial fractions in the same tube.

15. The exact setting required for sonication must be determined for each sonicator and optimized for a particular setup: soil/sediment type, bacterial species, and so on. For determination of the optimal sonication settings for lysis of cells, it is recommended to use the setting that yields the highest light output.

16. After the washing step, up to 30% of the initial amount of the luciferase protein is still bound to the membrane. Therefore, in order to account for the attached protein, the membrane must be included in the luminometer reading.
17. Quenching is a problem that is often encountered when measuring light output in an environmental sample. In particular, humic material and high salt concentrations are known to inhibit luciferase activity. Also, the particulate material in a sample can physically block the lightpath and add to quenching.
18. Depending on the extraction efficiency of luciferase protein from the sample, which can be experimentally determined, and the stability of the luciferase protein in the cells, it is theoretically possible to calculate the specific biomass of the *luc*-tagged cells in the sample using the following formula:

$$B = y(rq{\cdot}e{\cdot}c) \tag{3}$$

where B = biomass of the *luc*-tagged strain, q = counting efficiency $[(I + S) - S]/I$, where I = internal luciferase standard (quanta/sec) and S = sample (quanta/s), e = protein extraction efficiency (protein extracted/ total protein initially in the sample), c = pmol luciferase/cell, y = pmol luciferase for "s" sample quanta/s, (from a standard curve of quanta/s vs pmol of pure luciferase, *see* **Fig. 4**).

References

1. Jansson, J. K. (1995) Tracking genetically engineered microorganisms in nature. *Curr. Opinion Biotechnol.* **6,** 275–283.
2. Prosser, J. I. (1994) Molecular marker systems for detection of genetically engineered micro-organisms in the environment. *Microbiology* **140,** 5–17.
3. Lindow, S. E. (1995) The use of reporter genes in the study of microbial ecology. *Mol. Ecol.* **4,** 555, 566.
4. Jansson, J. K. and Prosser, J. (1997) Quantification of the presence and activity of specific microorganisms in nature. *Mol. Biotechnol.* **7,** 103–120.
5. Torsvik, V., Goksøyr, J., and Daae, F. L. (1990) High diversity in DNA of soil bacteria. *Appl. Environ. Microbiol.* **56,** 782–787.
6. Möller, A., Gustafsson, K., and Jansson, J. K. (1994) Specific monitoring by PCR amplification and bioluminescence of firefly luciferase gene-tagged bacteria added to environmental samples. *FEMS Microbiol. Ecol.* **15,** 193–206.
7. Möller, A., Norrby, A. M., Gustafsson, K., and Jansson, J. K. (1995) Luminometry and PCR-based monitoring of gene-tagged cyanobacteria in Baltic Sea microcosms. *FEMS Microbiol. Lett.* **129,** 43–50.
8. Cebolla, A., Ruiz-Berraquero, F., and Palomares, A. J. (1993) Stable tagging of *Rhizobium meliloti* with the firefly luciferase gene for environmental monitoring. *Appl. Environ. Microbiol.* **59,** 2511–2519.
9. Selbitschka, W., Hagen, M., Maier, S., and Pühler, A. (1992) The construction of bioluminescent and GUS-positive strains of *Rhizobium* for use in risk assessment studies, in *Proceedings of the 2nd International Symposium on the Biosafety Results of Field Tests of Genetically Modified Plants and Microorganisms* (Casper, R. and Landsmann, J., eds.), Goslar, Germany, pp. 267–272.

10. De Lorenzo, V., Eltis, L., Kessler, B., and Timmis, K. N. (1993) Analysis of *Pseudomonas* gene products using *lacI^q/Ptrp-lac* plasmids and transposons that confer conditional phenotypes. *Gene* **123,** 17–24.

11. Unge, A., Tombolini, R., Möller, A., and Jansson, J. K. (1996) Optimization of GFP as a marker for detection of bacteria in environmental samples, in *Biolumi- nescence and Chemiluminescence: Molecular Reporting with Photons* (Hastings, J. W., Kricka, L. J., and Stanley, P. E., eds.), John Wiley, Sussex, UK, in press.

12. Möller, A. and Jansson, J. K. (1997) Quantification of genetically-tagged cyano- bacteria in Baltic Sea sediment by competitive PCR. *BioTechniques* **22,** 512–518.

13. Wood, K. V. and DeLuca, M. (1987) Photographic detection of luminescence in *Escherichia coli* containing the gene for firefly luciferase. *Anal. Biochem.* **161,** 501–507.

14. Colwell, R. R., Brayton, P. R., Grimes, D. J., Roszak, D. B., Huq, S. A., and Palmer, L. M. (1985) Viable but non-culturable *Vibrio cholerae* and related patho- gens in the environment: implications for release of genetically engineered micro- organisms. *BioTechnology* **3,** 817–820.

15. Leser, T. D. (1995) Quantification of *Pseudomonas* sp. strain B13(FR1) in the marine environment by competitive polymerase chain reaction. *J. Microbiol. Methods* **22,** 249–262.

16. Holben, W. E., Jansson, J. K., Chelm, B. K., and Tiedje, J. M. (1988) DNA probe method for the detection of specific microorganisms in the soil bacterial commu- nity. *Appl. Environ. Microbiol.* **54,** 703–711.

17. Landahl, V. and Bakken, L. R. (1995) Evaluation of methods for extraction of bacteria from soil. *FEMS Microbiol. Ecol.* **16,** 135–142.

18. Meikle, A., Killham, K., Prosser, J. I., and Glover, L. A. (1992) Luminometric measurement of population activity of genetically modified *Pseudomonas fluorescens* in the soil. *FEMS Microbiol. Lett.* **99,** 217–220.

19. DeLuca, M. and McElroy, W. D. (1978) Purification and properties of firefly luci- ferase. *Methods Enzymol.* **57,** 3–15.

20. De Lorenzo, V., Herrero, M., Jacubizik, U., and Timmis, K. N. (1990) Mini-Tn5 transposon derivatives for insertion mutagenesis, promotor probing and chromosomal insertion of cloned DNA in Gram-negative eubacteria. *J. Bacteriol.* **172,** 6568–6572.

21. Torsvik, V., Daae, F. L., and Goksøyr, J. (1995) Extraction, purification and analy- sis of DNA from soil bacteria, in *Nucleic Acids in the Environment* (Trevors. J. T. and van Elsas, J. D., Eds.), Springer-Verlag, Heidelberg, pp. 29–48.

24

Monitoring of GFP-Tagged Bacterial Cells

Riccardo Tombolini and Janet K. Jansson

1. Introduction

In recent years, molecular tools have been developed to identify and quantify specific microbial cells, or specific microbial activities, in mixed populations *(1,2)*. These tools are especially valuable for analysis of specific bacteria in complex environmental samples. Increasingly, research has concentrated on the development of marker genes for tagging a particular bacterial species of interest, so that the cells can be specifically identified and monitored *(1–3)*. For example, the genes encoding bacterial luciferase *(luxAB)* or firefly luciferase *(luc)* have been found to be very useful markers, since tagged cells can be detected on the basis of their bioluminescent phenotype. Other examples of genes specifically used to mark bacteria include metabolic markers, such as *lacZY* (β-galactosidase and lactose permease), *gusA* (β-glucuronidase), and *xylE* (2,3–catechol dioxygenase), which are detected on the basis of unique colored products formed after growth of the cells on specific media *(1–3)*.

The search continues for other marker systems that can be used for *in situ* detection of specific bacteria, without the necessity for substrate addition or sample disturbance. Therefore, there was a tremendous burst of interest in the scientific community when the gene encoding green fluorescent protein (GFP) from the jellyfish *Aequorea victoria* was cloned and expressed in *Escherichia coli* and in a range of other cell types *(4)*. The beauty of GFP is that, unlike other marker gene phenotypes, GFP fluorescence is independent of any substrate or cellular energy reserves.

Wild-type GFP emits green fluorescence (Em_{max} = 509 nm) when excited at the appropriate wavelength (Ex_{max} = 395) without exogenously added substrates or cofactors. The GFP chromophore is generated via an oxygen-dependent, autocatalytic reaction within the protein amino acid backbone and

From: *Methods in Molecular Biology, Vol. 102: Bioluminescence Methods and Protocols*
Edited by: R. A. LaRossa © Humana Press Inc., Totowa, NJ

Table 1
A Selection of GFP Mutants Used as Markers or Reporters for Bacteria

GFP mutant designation	Mutation	Excitation maximum, nm	Emission maximum, nm	Reference
Wild type		395 (475)[a]	508	*10*
P11	Ile-167–Thr	471 (396)[a]	502 (507)	*8*
P4	Tyr-66–His	382	448	*8*
S65T	Ser-65–Thr	489	511	*11*
RSGFP4	Phe-64–Met	490	505	*12*
	Ser-65–Gly			
	Gln-69–Leu			
Mut 1	Phe-64–Leu	488	507	*9*
	Ser-65–Thr			
Mut 2	Ser-65–Thr	481	507	*9*
	Val-68–Leu			
Mut 3	Ser-65–Gly	501	511	*9*
	Ser-72–Ala			
GFP5	Val-163–Ala	395, 473[b]	509	*13*
	Iso-167–Thr			
	Ser-175–Gly			

[a]The value in parentheses is a minor peak.
[b]The excitation intensity is similar at both wavelengths.

involves formation of a cyclic tripeptide (Ser-Tyr-Gly). Apparently, once formed, GFP is extremely stable, which can partly be explained by examination of the three-dimensional protein structure that has recently been elucidated *(5,6)*. From this structure analysis, it was found that the protein motif represents a new protein fold with 11 β-sheets forming the wall of a cylinder and short segments of α helices capping the top and the bottom of the "β-can" *(5,6)*.

A series of GFP mutants have been constructed that have altered excitation and emission spectra compared to wild-type GFP *(7–9)* (**Table 1**). One of the mutants (P4) is a GFP variant with the same Ex_{max} as the wild-type, but with a blue shifted Em_{max} (blue GFP) (**Table 1**). In particular, the red-shifted mutants, having the excitation wavelength shifted toward a higher wavelength, have been found to be better adapted for visualization of cells by epifluorescence microscopy or flow cytometry, since the excitation wavelength more closely matches the FITC filter sets used in epifluorescence microscopy and the argon laser (488-nm wavelength) commonly used for flow cytometry and confocal scanning laser microscopy. In addition, having the major absorbance peak at 488 nm lowers the effect of photobleaching and increases stability of the GFP protein *(7)*.

There currently exist a variety of vectors (some also commercially available) bearing wild-type or mutant *gfp* genes. For environmental monitoring of specific bacteria, it is preferable to tag the cells with a marker gene that is stably integrated into the bacterial chromosome in order to reduce the risk of marker loss or marker transfer to other species. Delivery systems have been recently developed for single, or double *(14)*, *gfp* gene insertion into the chromosome of several bacterial species *(14–17)*. These "suicide" delivery systems are based on the use of minitransposon vectors that have the marker gene of choice flanked by insertion elements for chromosomal integration, whereas the transposase gene, responsible for transposition, is located outside the insertion sequences, thereby preventing further transposition *(18)*. Alternatively, to obtain the highest levels of GFP fluorescence, the *gfp* gene can be carried in bacteria on multicopy number plasmids *(15,17,19)* although at the risk of marker instability.

Recently, GFP has been used in a limited number of studies as a bacterial tag for environmental applications. Tombolini et al. demonstrated that GFP fluorescence is stable in *Pseudomonas fluorescens* bacterial cells, even after long-term starvation *(17)*. This finding implies that GFP is an ideal marker for use in the environment, where cells often are stressed or starving. Another study, with application to the environment, was the use of GFP-tagged cells to study bacterial transfer through sand columns as a model system for understanding the process of movement of bacteria through groundwater *(15)*. Christensen et al. studied horizontal transfer of plasmids between two populations of *Pseudomonas putida* at the level of single cells using GFP as tag on the conjugal TOL plasmid *(19)*. The potential for *in situ* monitoring of GFP-tagged bacteria on plant roots was demonstrated by a series of impressive confocal microscopic images of *Rhizobium meliloti* during early events of the nodulation process in alfalfa roots *(20)*. GFP-tagged *R. meliloti* cells were clearly visualized inside infection threads. These images provided the opportunity to measure the growth rate and to determine the pattern of growth of *R. meliloti* residing inside its host plant *(20)*. Other published studies, using GFP as a bacterial marker, include studies of host–pathogen interactions, with particular focus on mycobacterial interaction with macrophages *(21–23)*. Also, cell-specific gene expression and protein subcellular localization during sporulation in *Bacillus subtilis* have been achieved by employing *gfp* as a marker gene *(24)*.

There are different methods available for detection of specific GFP-tagged bacterial cells in environmental samples *(25)*. Visualization of GFP fluorescing colonies cultured on selective solid media is a straightforward approach for a rough evaluation of the presence and number of the specific tagged bacteria in a sample. GFP fluorescing colonies can be easily detected under light illumination that excites GFP fluorescence. Special lamps are commercially available that emit the appropriate wavelength for excitation of GFP fluorescence. Other light sources can be used for visualization of GFP-tagged colonies, such as UV transilluminators or slide projectors equipped with proper blue filters (bionet.molbio.proteins.fluorescent newsgroup; *see* **Note 1**). However, UV light illumination cannot be considered as an acceptable vital detection technique owing to the potential for DNA damage.

Individual GFP-tagged bacterial cells can be directly detected, in real time, by using techniques developed for visualization of fluorescent cells, such as epifluorescence microscopy, confocal scanning laser microscopy, and flow cytometry, each of which will be briefly described in the following sections.

Components of the epifluorescence microscope have been improved in recent years. For example, new high-gas-discharged light sources are now specifically produced for epifluorescence microscopy applications. In addition, a vast range of interference filters with very sharp barriers are available. Also, powerful software interfaces are commonly used for enhancement of digital images grabbed by sensitive charged coupled device (CCD) cameras. Therefore, epifluorescence microscopy has become a refined technique that is applicable to detection of GFP fluorescing cells. However, technical difficulties still exist. One difficulty is to distinguish weak GFP fluorescence apart from the background fluorescence always present in environmental samples containing organic matter. Background fluorescence from biological sources is excited from a wide range of wavelengths, producing light in the green to red range.

One way to partly overcome the problem with background fluorescence is to use filter sets that are optimized for GFP fluorescence, thereby improving the "signal-to-noise" ratio. Most of the optical devices for fluorescence visualization are equipped with a filter set optimized for FITC, with an excitation maximum at 490 nm and an emission maximum at 520 nm. This filter set also permits visualization of GFP. Nevertheless, it is possible to optimize detection of GFP fluorescence by using GFP specific filter sets. The choice of the optimal filter set is also dependent on the specific GFP mutant's excitation and emission maxima (*see* **Table 1**). If background autofluorescence is to be avoided, it is best to use excitation and emission band pass filters that exactly match the excitation and emission peaks of the specific mutant in use (*see* **Table 1**). If, on the other hand, some background autofluorescence can be maintained, to obtain an image of the sample (in addition to the GFP-tagged cells), a long-pass filter can be used for emitted light. Results from different filter sets have been published for eukaryotic cells, both free-living and fixed *(26,27)*. These results should be helpful in designing filters for bacterial applications as well.

Confocal laser scanning microscopy (CLSM) is increasingly becoming more common as a technique for visualization of objects with high resolution, which in

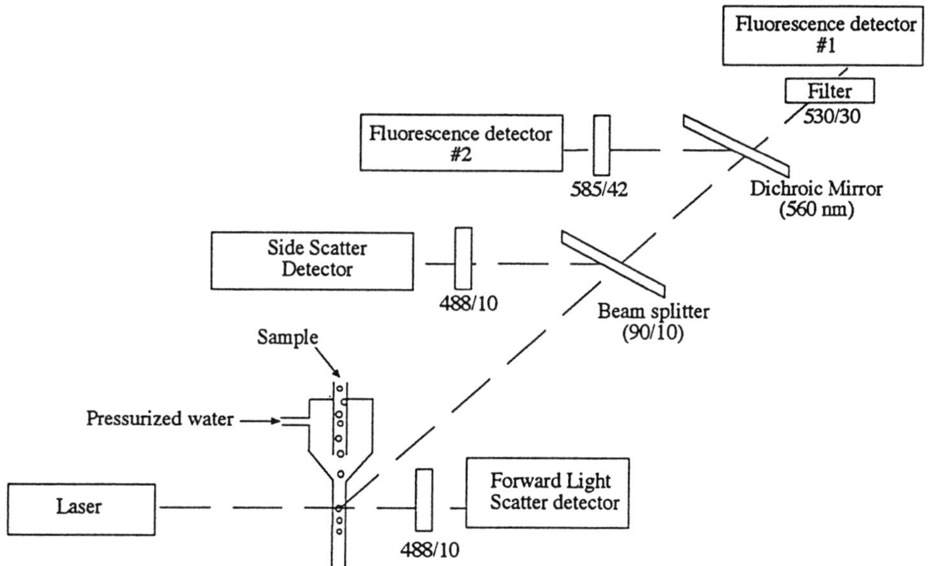

Fig. 1. Schematic representation of the optical components of a flow cytometer.

combination with 3D image analysis enables complex communities to be imaged in discrete sections and then reconstructed into a 3D visualization of the sample (*see* review in *28*). One big advantage of CLSM over epifluorescence microscopy is the possibility to reduce interference from background fluorescence *(29,30)*. CLSM has recently been demonstrated to be an excellent technique for visualization of root-associated GFP-tagged bacteria *(13,15)*. The cost of a confocal laser microscope is, however, prohibitive to routine use at this date.

A flow cytometer measures and analyzes optical properties of single cells passing through a focused laser beam (**Fig. 1**). Analysis of hundreds of cells per second provides a statistically significant picture of the samples physical and biochemical makeup. When cells pass through the laser beam, they disrupt and scatter the laser light, which is detected as forward scatter (FS) and side scattered (SS) light. FS light is related to cell size, whereas SS light is an indicator of internal cellular complexity. In addition to scatter, a cytometer measures fluorescence parameters (FL; **Fig. 2**). The reader is referred to more extensive recent reviews for additional information about flow cytometry *(31,32)*.

Although flow cytometers were originally designed for eukaryotic applications, modern flow cytometers are equipped with sensitive photomultipliers that permit prokaryotic cells to be detected *(32)*. Flow cytometry has been recently demonstrated to be an excellent technique for analysis and quantitation of GFP fluorescent bacterial cell populations *(17,33)* and

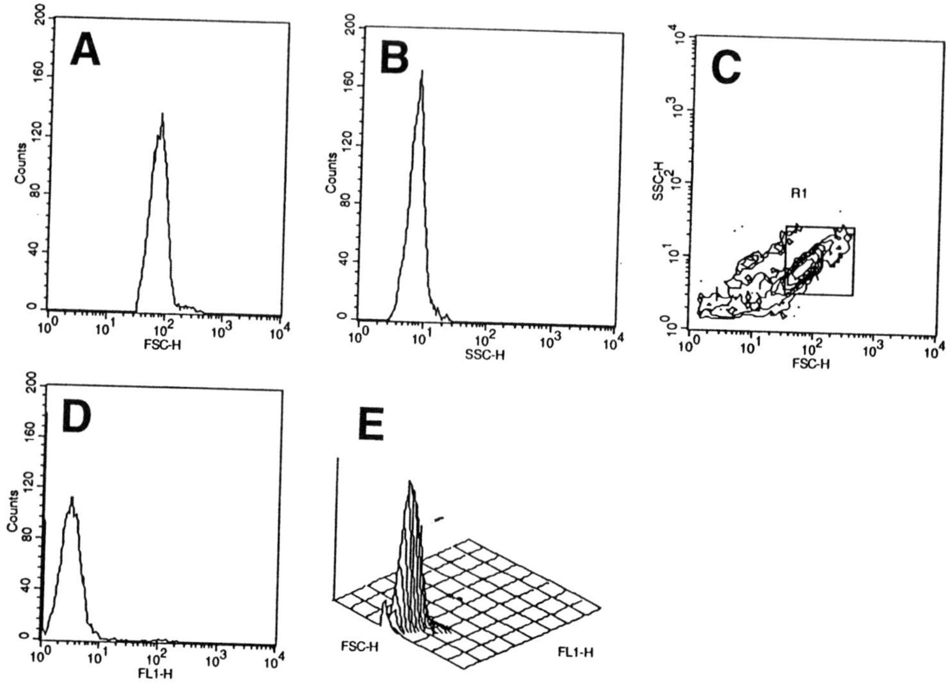

Fig. 2. Typical representation of flow cytometric data from a sample of GFP-tagged bacteria. **(A)** FS histogram; **(B)** SS histogram; **(C)** contour plot of SS and FS, R1 delineates the region used to gate events for histograms A, B, D, and E; **(D)** fluorescence (FL) histogram; **(E)** 3D plot of FS and FL.

for measuring the degree of association of GFP-tagged bacterial cells with mammalian cells *(22,23)*.

In this chapter, procedures are described for direct visualization of GFP-tagged bacterial colonies and for single-cell detection by epifluorescence microscopy and flow cytometry. The advantages of using GFP as a marker or reporter gene are reflected by the ease and simplicity of the described protocols.

2. Materials

2.1. Colony Visualization

1. Petri plates with appropriate agar growth medium.
2. "Black Light Blue" lamp (Philips, cat. no. 73411; Eindhoven, The Netherlands).

2.2. Epifluorescence Microscopy

1. Clean microscope slides.
2. Bunsen burner.

3. 1% Solution of polyetheleneimine (ICN Biomedical, Aurora, OH).
4. Mounting medium for epifluorescence microscopy (Vectashield®, Vector Laboratories, Burlingame, CA).
5. Immersion oil.
6. Epifluorescence microscope.
7. FITC, or GFP optimized, filter sets (e.g., Chroma filter set number 41017, Chroma Technology, Brattleboro, VT).
8. Camera attached to epifluorescence microscope with suitable films.
9. Peltier cooled CCD camera and image analysis software (optional, *see* **Note 2**).

2.3. Flow Cytometry

1. Flow cytometer.
2. Suitable software for acquiring and analyzing flow cytometer data (*see* **Note 3**).
3. 1.5X PBS (per liter): 12 g NaCl, 0.3 g KCl, 2.16 g Na_2HPO_4, 0.36 g KH_2PO_4. Adjust to pH 7.4, and sterilize by autoclaving. Before use, pass through a 0.22-μm filter to remove particles that would otherwise interfere with the flow cytometer readings.
4. Optimized band pass filter for GFP (optional) (*see* **Note 4**).
5. Internal standard beads: 2.2-μm polystyrene microspheres (e.g., Duke Scientific, Palo Alto, CA, USA) (*see* **Note 5**).

3. Methods

3.1. Colony Visualization on Plates

1. Streak out individual GFP-tagged bacterial strains or plate environmental samples containing GFP-tagged bacteria onto appropriate agar growth medium.
2. Incubate the inoculated plates at the appropriate growth temperature until distinct colonies are visible (*see* **Note 6**).
3. Illuminate the plates in a darkroom from above with a "Blue Light Black" lamp (*see* **Note 7**).
4. Visualize green fluorescing colonies by eye (*see* **Note 8**).

3.2. Epifluorescence Microscopy

1. Soak a clean glass microscope slide in 1% polyetheleneimine solution for 2 min.
2. Wash the slide for 30 s in sterile, double-distilled water and air-dry (*see* **Notes 9** and **10**).
3. Apply a drop of a sample containing GFP-tagged bacteria to the slide.
4. Air-dry for 10 min.
5. Fix the cells to the slide by passing rapidly through a Bunsen burner flame.
6. Wash for 30 s in sterile, double-distilled water to detach unfixed cells.
7. Air-dry for 15 min.
8. Add a drop of mounting medium for epifluorescence microscopy on top of the spot on the slide with attached cells, and apply a cover slip.
9. Use FITC, or GFP-optimized, filter sets to visualize the GFP-tagged cells in an epifluorescence microscope (for an example, *see* **Fig. 3**).

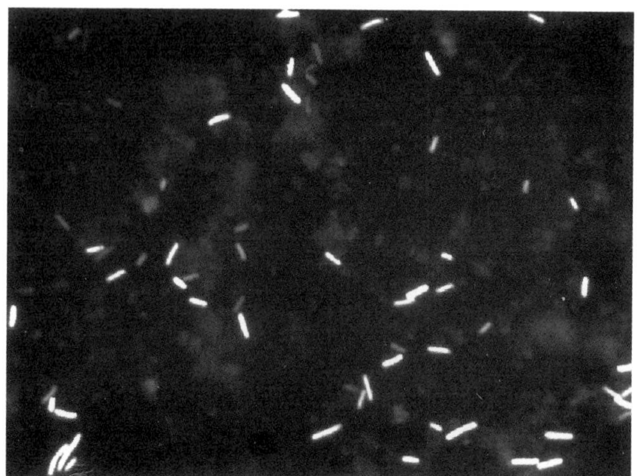

Fig. 3. A digital CCD-enhanced region of a fluorescent microscopic image of a soil suspension containing GFP-tagged *E. coli* DHα cells. The cells were tagged with GFP on the multicopy number plasmid, pRL765GFP, with *gfp* under control of the PpsbA promoter. The slide was prepared as described in **Subheading 3.2.** The sample was visualized using an Axiophot 2 Epifluorescence Microscope (Zeiss, Oberkochen, Germany). The objective used was an oil immersion 100 × objective with an optical aperture of 1.3. The microscopic image was digitalized with a peltier-cooled C4880 CCD camera (Hamamatsu Photonic, K.K., Hamamatsu City, Japan). The contrast of the digital image was enhanced using Adobe Photoshop 3.0 software (Adobe System Incorporated, Mountain View, CA).

10. If necessary, use a sensitive CCD camera and image analysis software to enhance the image (*see* **Note 11** and **Fig. 3**).

3.3. Flow Cytometry

The method below is for analysis of GFP-tagged cells by flow cytometry without using any stains or fixing agents. However, it is also possible to fix GFP-tagged cells prior to flow cytometry measurement (*see* **Note 12**). The ability to detect and quantitate GFP fluorescing cells by flow cytometry, without any sample prepreparation, is a particular advantage of the GFP-marker system.

3.3.1. Setup of the Flow Cytometer

1. Exchange the band pass filter on the flow cytometer fluorescence detector from an FITC filter to a GFP-optimized filter (*see* **Note 4**).

2. Set the amplification setting (gain) for the fluorescence photomultiplier of the flow cytometer in the presence of the wild-type bacteria (non GFP-tagged) in order to establish background fluorescence thresholds.
3. Increase the amplification factor until the background fluorescence is negligible (*see* **Note 13**).
4. Set the amplification setting for the FS and SS photomultipliers, according to the instrument manual accompanying the flow cytometer.

3.3.2. Sample Analysis by Flow Cytometry

1. Resuspend GFP-tagged bacterial cells in 1 vol of 1.5X PBS buffer (*see* **Notes 14** and **15**).
2. Centrifuge the cell suspension at 10,000*g* for 4 min.
3. Wash the cell pellet by resuspension in 1.5X PBS buffer, and centrifugation as above (repeat **steps 1** and **2**).
4. Dilute the cell suspension in 1.5X PBS buffer to a concentration equivalent to 1000–2000 events/s as determined by the flow cytometer (*see* **Note 16**).
5. For quantitation of the number of GFP-tagged cells in the sample, add a known concentration of fluorescent (or nonfluorescent) beads to the sample. The amount of beads added should be in the range of 10^5–10^6 beads/mL sample (*see* **Note 17**).
6. Introduce the sample into the flow cytometer within 1 h of sampling, since the cells have not been fixed.
7. Acquire data corresponding to 10,000 bacteria, using the software interfaced with the instrument (*see* **Note 3**).

3.3.3. Data Analysis (see **Note 18, Fig. 2**)

3.3.3.1. DETERMINATION OF BACTERIAL CONCENTRATION

1. Select the area on a FS/SS density (or contour) plot containing signals corresponding to both bacteria and beads, and gate, or define, the selected area as "R1" (**Fig. 4A**, R1).
2. Create an FS histogram of the events gated in the R1 region.
3. Separately define the peaks corresponding to bacteria (define as "M1") and beads (define as "M2") in the FS histogram (**Fig. 4B**).
4. Examine the statistics output from the flow cytometer software for the number of total events, and the number of events corresponding to bacteria (M1) and beads (M2) (**Fig. 4C**).
5. Calculate the bacterial concentration in the sample based on the known concentration of beads added to the sample.

3.3.3.2. DETERMINATION OF THE PERCENTAGE OF GREEN FLUORESCENT BACTERIA

1. Select the area on a FS/SS density (or contour) plot containing only bacterial signals and define the region as "R2" (**Fig. 4A**, R2).
2. Create an FL histogram of the events gated in the R2 region.
3. Select the bacterial peak on the FL histogram, and define this region as "M1" (**Fig. 4D**, M1).

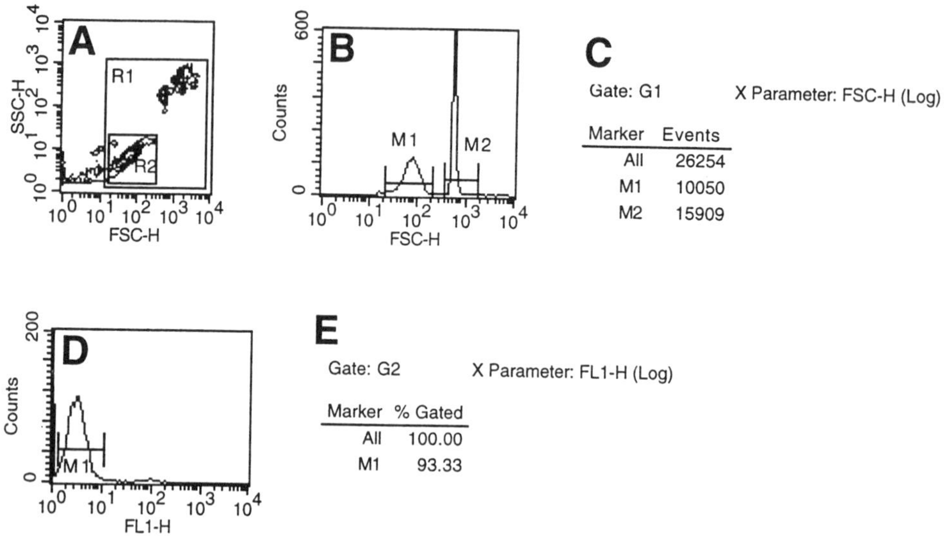

Fig. 4. Example of handling flow cytometric data. All data acquired from the experiment (~ 10,000 events) are plotted as FS vs SS in (**A**). The area containing signals corresponding to bacterial cells and beads are enclosed in region A-R1, whereas the area containing only bacterial signals is in A-R2. An FS histogram (**B**), gated in the R1 region is used for obtaining the number of events corresponding to bacteria (B-M1) and events corresponding to beads (B-M2). From the respective events number listed in the statistics panel (**C**), and by knowing the concentration of beads in the sample, the bacterial concentration can be derived. In (**D**) the fluorescence histogram gated with R2 is reported. The number of events falling in the D-M1 range correspond to the green fluorescing bacteria. The percentage of this number out of the total events gated with R2 is given in the statistics panel (**E**).

4. Examine the statistics from the flow cytometry software (**Fig. 4E**).
5. Calculate the percentage of green fluorescent bacteria over the total bacteria counted (*see* **Note 19**).

4. Notes

1. There is a newsgroup managed by the internet resource, BIOSCI, dealing primarily with matters related to GFP. The internet address is http://www.bio.net.
2. Many software options are available. The public domain NIH-image software developed at The US National Institute of Health is available on the internet at: http://rsb.info.nih.gov/nih-image/. A less technical, but still powerful program, is: Adobe Photoshop 3.0 software (Adobe System Incorporated, Mountain View, CA).
3. Appropriate software is usually provided by the flow cytometer vendor. There are several options commercially available. One good "freeware" option is the following: WinMDI version 2.3, which can be downloaded at: http://facs.scripps.edu/software/beta23.exe.

4. With optical devices that make use of a laser beam light source (typically an argon-ion laser), such as commonly used for confocal scanning laser microscopy and flow cytometry, it is possible to use emission band pass filters with a wavelength window very close to the emission wavelength of the light source *(21)*.

5. The beads may be nonfluorescent or fluorescent. The protocol for cell number determination we provide makes use of FS, which is a fluorescence-independent parameter and therefore does not require fluorescent beads as an internal standard. However, for control of the flow cytometer output, it is also possible to use fluorescent beads as an internal standard.

6. We have found that longer incubation times (a few days) may be necessary at 25°C for some strains to produce green fluorescent colonies, especially during screening of the exconjugants following genome tagging (Tombolini, unpublished data). This may be due to thermosensitivity of GFP folding, which has been found to be optimal at lower temperatures, although a new mutant, GFP5 (**Table 1**) has been recently constructed that is stable at higher temperatures *(13)*.

7. To avoid possible cell damage, it is advisable to always pretest the light source chosen. This can be done by plating 100–200 bacteria in two series of triplicate plates. One series should be subjected to illumination for different time intervals (e.g., 30 s, 1 min, 5 min, 10 min, 30 min). No statistical difference should be seen in the number of colonies from illuminated and nonilluminated plates.

8. In the case of faint or doubtful greenish fluorescence, a small portion of a colony can be smeared onto a slide and visualized by epifluorescence microscopy for more sensitive assessment of green fluorescent cells.

9. Although it is possible to quickly prepare a slide for microscopic visualization by putting a drop of a bacterial suspension onto the slide, it is usually necessary to have the cells stationary (fixed), and on the same plane of focus, in order to properly visualize individual cells. It is particularly important to have fixed samples for obtaining digital images with a CCD camera.

10. Other coating polymers can be used to pretreat the microscope slides, for example, polylysine (0.1%) or bovine serum albumin (5%).

11. A digital CCD camera can be used to obtain a digital image of the microscopic field, containing far more information, in terms of gray scale dynamic range, than the human eye can discriminate. Therefore, the use of a digital CCD camera can improve the sensitivity of the sample image and by computer image processing of the digitized image, the contrast can be enhanced.

12. Ethanol fixation abolishes GFP fluorescence in tagged cells. Fixation of cells in 1% formaldehyde or in 1:1 acetone/methanol is more compatible with GFP.

13. For example, in an FACScalibur flow cytometer (Becton Dickinson, Oxford, UK), the photomultiplier tube voltage for the fluorescence detector can be set at 600 V.

14. Running bacterial samples in the flow cytometer in 1.5X PBS buffer provides a more uniform FS and SS signal than using 1X PBS, thereby facilitating location of cells and gating of bacteria in histograms and plots.

15. For environmental samples, it is necessary to separate bacteria from particulate matter that could interfere with flow cytometry detection. Protocols have been developed for bacterial recovery from complex environmental samples (for example, *34*).
16. The number of events per second is determined by the flux of the sample introduction into the flow cytometer (μL/min) and the concentration of the bacterial suspension.
17. Since it is not reliable to determine the bacterial cell concentration based on the volume of the sample flux through the cytometer, it is recommended to use an internal standard (beads) for quantitative purposes.
18. Usually a population of particles is identified and selected (gated) by FS or SS in a histogram, or by both FS and SS in a dot, density, or contour plot (**Fig. 2C**). Then the fluorescence data are analyzed (**Fig. 2D,E**). Also, from the FS and SS data, it is possible to obtain complex morphological information about individuals of a particular population.
19. We hypothesize that owing to the long persistence of GFP, the fluorescent bacteria determined in this way account for the total number of GFP-tagged bacteria, irrespective of their growth status (e.g., living, viable but nonculturable, or dead), and we are currently performing experiments to confirm this hypothesis.

References

1. Prosser, J. I. (1994) Molecular marker systems for the detection of genetically modified microorganisms in the environment. *Microbiology* **140**, 5–17.
2. Jansson, J. K. (1995) Tracking genetically engineered microorganisms in nature. *Curr. Opinion Biotechnol.* **6**, 275–283.
3. Lindow, S. E. (1995) The use of reporter genes in the study of microbial ecology. Mol. Ecol. **4**, 555, 566.
4. Chalfie, M., Tu Y., Euskirchen, G., Ward, W. W., and Prasher, D. C. (1994) Green fluorescent protein as a marker for gene expression. *Science* **263**, 802–805.
5. Ormö, M., Cubitt, A. B., Kallio, K., Gross, L. A., Tsien, R. Y., and Remington, S. J. (1996) Crystal structure of the Aequorea victoria green fluorescent protein. *Science* **273**, 1392–1336.
6. Yang, F., Moss, L. G., and Phillips, G. N. (1996) The molecular structure of green fluorescent protein. *Nat. Biotechnol.* **14**, 1246–1251.
7. Cubitt, A. B., Heim, R., Adams, S. R., Boyd, A. E., Gross, L. A., and Tsien, R. Y. (1995) Understanding, improving and using green fluorescent proteins. *Trends Biochem. Sci.* **20**, 448–455.
8. Heim, R., Prasher, D. C., and Tsien, R. Y. (1994) Wavelength mutations and post-translational autoxidation of green fluorescent protein. *Proc. Natl. Acad. Sci. USA* **91**, 12,501–12,504.
9. Cormack, B. P., Valdivia, R. H., and Falkow, S. (1996) FACS-optimized mutants of green fluorescent protein (GFP). *Gene* **173**, 33–38.
10. Ward, W. W., Cody, C. W., Hart, R. C., and Cormier, M. J. (1980) Spectrophotometric identity of the energy transfer chromophores in *Renilla* and *Aequorea* green-fluorescent proteins. *Photochem. Photobiol.* **31**, 611–615.
11. Heim, R., Cubitt, A. B., and Tsien, R. Y. (1995) Improved green fluorescence. *Nature* **373**, 663, 664.

12. Delgrave, S., Hawtin, R. E., Silva, C. M., Yang, M. M., and Youvan, D. C. (1995) Red-shifted excitation mutants of the green fluorescent protein. *Bio-Technology* **13,** 151–154.

13. Kirby, R. S., Golbik, R., Sever, R., and Haseloff, J. (1996) Mutations that suppress the thermosensitivity of green fluorescent protein. *Curr. Biol.* **6,** 1653–1663.

14. Unge, A., Tombolini, R., Möller, A., and Jansson, J. K. (1997) Optimization of GFP as a marker for detection of bacteria in environmental samples, in *Bioluminescence and Chemiluminescence: Molecular Reporting with Photons* (Hastings, J. W., Kricka, L. J., and Stanley, P. E., eds.), John Wiley, Sussex, UK, pp. 391–394.

15. Burlage, R. S., Zamin, K. Y., and Mehlhorn, T. (1996) A transposon for green fluorescent protein transcriptional fusion: application for bacterial transport experiment. *Gene* **173,** 53–58.

16. Matthysse, A. G., Stretton, S., Dande, C., McClure, N. C., and Goodman A. E. (1996) Construction of GFP vectors for use in gram-negative bacteria other than *Escherichia coli. FEMS Microbiol. Lett.* **145,** 87–94.

17. Tombolini, R., Unge, A., Davey, M. E., de Bruijn, F. J., and Jansson, J. K. (1997) Flow cytometric and microscopic analysis of *gfp*-tagged *Pseudomonas fluorescens* bacteria. *FEMS Microbiol. Ecol.* **22,** 17–28.

18. Herrero, M., de Lorenzo, V., and Timmis, K. (1990) Transposon vectors containing non-antibiotic resistance selection markers for cloning and stable chromosomal insertion of foreign genes in gram-negative bacteria. *J. Bacteriol.* **172,** 6557–6567.

19. Christensen, B. B., Sternberg, C., and Molin, S. (1996) Bacterial plasmid conjugation on semi-solid surfaces monitored with the green fluorescent protein (GFP) from *Aequorea victoria* as a marker. *Gene* **173,** 59–65.

20. Gage, D. J., Bobo, T., and Long, S. R. (1996) Use of green fluorescent protein to visualize early events of symbiosis between *Rhizobium meliloti* and alfalfa *(Medicago sativa). J. Bacteriol.* **178,** 7159–7166.

21. Dhandayuthapani, S., Via, L. E., Thomas, C. A., Horowitz, P. M., Deretic, D., and Deretic, V. (1995) Green fluorescent protein as a marker for gene expression and cell biology of mycobacterial interactions with macrophages. *Mol. Microbiol.* **17,** 901–912.

22. Kremer, L., Baulard, A., Estaquier, J., Poulain-Godefroy, D., and Locht, C. (1995) Green fluorescent protein as a new expression marker in mycobacteria. *Mol. Microbiol.* **17,** 913–922.

23. Valdivia, R. H., Hromockyj, A. E., Monack, D., Ramakrishnan, L., and Falkow, S. (1996) Applications for green fluorescent protein (GFP) in the study of host-pathogen interactions. *Gene* **173,** 47–52.

24. Webb, C. D., Decatur, A., Teleman, A., and Losick, R. (1995) Use of green fluorescent protein for visualization of cell-specific gene expression and subcellular protein localization during sporulation in *Bacillus subtilis. J. Bacteriol.* **177,** 5906–5911.

25. Unge, A., Tombolini, R., Davey, M. E., de Bruijn, F. J., and Jansson, J. K. (1997) GFP as a marker gene, in *Molecular Microbial Ecology Manual* (Akkermans, A.

D. L., van Elsas, J. D., and de Bruijn, F. J., eds.), Kluwer Academic Publishers, Dordrecht, The Netherlands, in press.

26. Niswender, K. D., Blackman, S. M., Rohde, L., Magnuson, M. A., and Piston, D. W. (1995) Quantitative imaging of green fluorescent protein in cultured cells: comparison of microscopic techniques, use in fusion proteins and detection limits. *J. Microsc.* **180,** 109–116.

27. Zylka, M. J. and Schnapp, B. J. (1996) Optimized filter set and viewing conditions for the S65T mutant of GFP in living cells. *Biotechniques* **21,** 220–226.

28. Caldwell, D. E., Korber, D. R., and Lawrence, J. R. (1992) Confocal laser microscopy and digital image analysis in microbial ecology. *Adv. Microb. Ecol.* **12,** 1–67.

29. Assmus, B., Hutzler, P., Kirchhof, G., Amann, R., Lawrence, J. R., and Hartmann, A. (1995) *In situ* localization of *Azospirillum brasilense* in the rhizosphere of wheat with fluorescently labelled rRNA-targeted oligonucleotide probes and scanning confocal laser microscopy. *Appl. Environ. Microbiol.* **61,** 1013–1019.

30. Schloter, M., Borlinghaus, R., Bode, W., and Hartmann, A. (1993) Direct identification and localization of *Azospirillum* in the rhizosphere of wheat using fluorescence-labelled monoclonal antibodies and confocal scanning laser microscopy. *J. Microscopy* **171,** 173–177.

31. Troussellier, M., Courties, C., and Vaquer, A. (1993) Recent application of flow cytometry in aquatic microbial ecology. *Biol. Cell.* **78,** 111–121.

32. Davey, H. M. and Kell, D. B. (1996) Flow cytometry and cell sorting of heterogeneous microbial populations: the importance of single-cell analysis. *Microb. Rev.* **60,** 641–696.

33. Ropp, J. D., Donahue, C. J., Wolfgang-Kimball, D., Hooley, J., Chin, J. Y. W., Hoffman, R. A., Cuthbertson, R. A., and Bauer, K. D. (1995) *Aequorea* green fluorescent protein analysis by flow cytometry. *Mol. Microbiol.* **17,** 901–912.

34. Bakken, L. R. and Lindhal, V. (1995) Recovery of bacterial cells from soil, in *Nucleic Acids in the Environment* (Trevors, J. T. and van Elsas, J. D., eds.), Springer-Verlag, Berlin, pp. 9–27.

Index